Student Solutions Manual

to accompany

Differential Equations

An Introduction to Modern Methods and Applications

Third Edition

James R. Brannan
Clemson University

William E. Boyce
Rensselaer Polytechnic Institute

Prepared by Tamas Wiandt
Rochester Institute of Technology

WILEY

To order books or for customer service please, call 1-800-CALL WILEY (225-5945).

ISBN-13   978-1-118-98125-2

Printed in the United States of America

CONTENTS

Chapter 1

Introduction

1.1 Mathematical Models and Solutions

1.(a) Let $u(t)$ be the temperature of the coffee at time t (measured in minutes), and $T_0 = 70$ be the ambient temperature. The initial value problem is $u' = -k(u - T_0)$, $u(0) = 200$.

(b) According to the text, the solution of the differential equation is $u(t) = T_0 + ce^{-kt}$. $200 = u(0) = T_0 + c$, thus $c = 130$. Also, $190 = u(1) = 70 + 130e^{-k}$, thus $e^{-k} = 12/13$. The coffee reaches temperature 170 when $170 = 70 + 130e^{-kt} = 70 + 130(12/13)^t$, thus $t = \ln(10/13)/\ln(12/13) \approx 3.28$ minutes.

3. Let $t = 0$ be 11:09pm, and let us measure time in hours. The temperature of the body is then given by $u(t) = 68 + 12e^{-kt}$. Also, $78.5 = u(1) = 68 + 12e^{-k}$. This gives that $e^{-k} = 7/8$. The time of death is given by the equation $98.6 = 68 + 12e^{-kt}$; we obtain that $t = \ln(51/20)/\ln(7/8) \approx -7.01$ hours, i.e. at around 4:08pm.

5.(a) The general solution is $p(t) = 900 + ce^{t/2}$. Plugging in for the initial condition, we have $p(t) = 900 + (p_0 - 900)e^{t/2}$. With $p_0 = 850$, the solution is $p(t) = 900 - 50e^{t/2}$. To find the time when the population becomes extinct, we need to find the time T when $p(T) = 0$. Therefore, $900 = 50e^{T/2}$, which implies $e^{T/2} = 18$, and, therefore, $T = 2\ln 18 \approx 5.78$ months.

(b) Using the general solution, $p(t) = 900 + (p_0 - 900)e^{t/2}$, we see that the population will become extinct at the time T when $900 = (900 - p_0)e^{T/2}$. That is, $T = 2\ln[900/(900 - p_0)]$ months.

(c) Using the general solution, $p(t) = 900 + (p_0 - 900)e^{t/2}$, we see that the population after 1 year (12 months) will be $p(6) = 900 + (p_0 - 900)e^6$. If we want to know the initial population which will lead to extinction after 1 year, we set $p(6) = 0$ and solve for p_0. Doing so, we have $(900 - p_0)e^6 = 900$ which implies $p_0 = 900(1 - e^{-6}) \approx 897.8$.

7.(a) The general solution of the equation is $Q(t) = ce^{-rt}$. Given that $Q(0) = 100$, we have $c = 100$. Assuming that $Q(1) = 82.04$, we have $82.04 = 100e^{-r}$. Solving this equation for r, we have $r = -\ln(82.04/100) = 0.19796$ per week or $r = 0.02828$ per day.

(b) Using the form of the general solution and r found above, we have $Q(t) = 100e^{-0.02828t}$.

(c) Let T be the time it takes the isotope to decay to half of its original amount. From part

1

(b), we conclude that $0.5 = e^{-0.2828T}$ which implies that $T = -\ln(0.5)/0.2828 \approx 24.5$ days.

9.(a) Following the discussion in the text, the equation is given by $mv' = mg - kv^2$.

(b) After a long time, $v' \to 0$. Therefore, $mg - kv^2 \to 0$, or $v \to \sqrt{mg/k}$.

(c) We need to solve the equation $\sqrt{0.025(9.8)/k} = 35$. Solving this equation, we see that $k = 0.0002$ kg/m.

11. Using the model from the text, we obtain the equation $Q' = (1 + (\sin t)/2)/2 - 2Q/100$, $Q(0) = 50$.

13.(a) We obtain that $C(t) = C_0 e^{-kt}$; this function satisfies $dC/dt = -kC$ and $C(0) = C_0$.

(b) $C_2 = C_0 + C(T) = C_0 + C_0 e^{-kT} = C_0(1 + e^{-kT})$

(c) Using induction, $C_n = C_0 + C_{n-1}e^{-kT} = C_0(1 + e^{-kT} + e^{-2kT} + \ldots + e^{-(n-1)kT}) = C_0(1 - e^{-nkT})/(1 - e^{-kT})$. Thus $\lim_{n \to \infty} C_n = C_0/(1 - e^{-kT})$.

15. We compute: $P'(t) = (P_0 + k/r)e^{rt}r = r[(P_0 + k/r)e^{rt} - (k/r)] + k = rP + k$, and $P(0) = P_0 + k/r - k/r = P_0$.

17. Using the model from problem 15, $k = -200(12)$, $P_0 = 20,000$ and $r = 0.05$. Then solving the equation $P(t) = (P_0 + k/r)e^{rt} - k/r = 0$ we obtain $t \approx 10.78$ years; the total amount paid is $(10.78)(12)(200) = \$25,872$.

19. The surface area of a spherical raindrop of radius r is given by $S = 4\pi r^2$. The volume of a spherical raindrop is given by $V = 4\pi r^3/3$. Therefore, we see that the surface area $S = cV^{2/3}$ for some constant c. If the raindrop evaporates at a rate proportional to its surface area, then $dV/dt = -kV^{2/3}$ for some $k > 0$.

1.2 Qualitative Methods: Phase Lines and Direction Fields

1. $y = 0$, $y = 2$ unstable, $y = 1$ asymptotically stable.

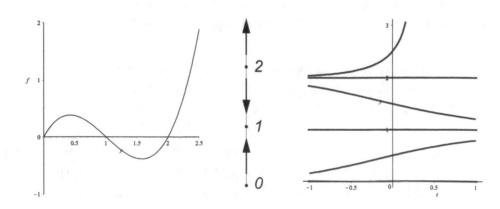

3. $y = 0$ asymptotically stable.

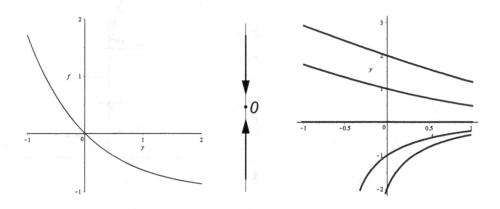

5. $y = -1$ asymptotically stable, $y = 0$ semistable, $y = 3$ unstable.

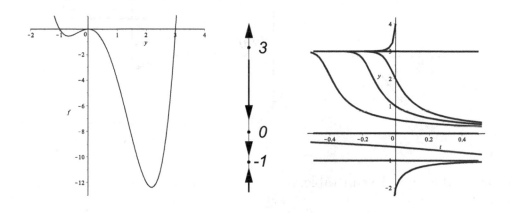

7. $y = 0$ unstable, $y = -a/b$ asymptotically stable.

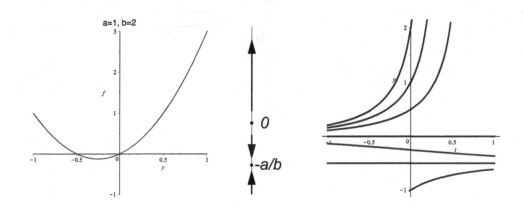

9. $y = -1$ asymptotically stable, $y = 0$ semistable, $y = 1$ unstable.

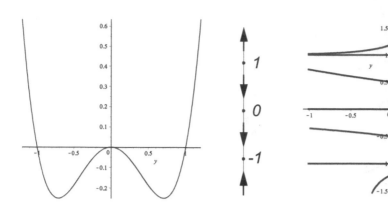

11. $y = 0$ asymptotically stable, $y = (b/a)^2$ unstable.

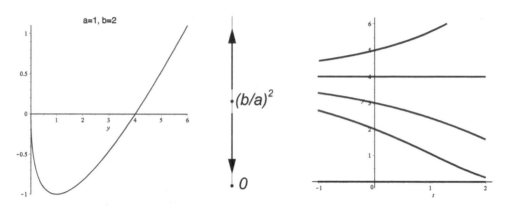

13. $y = 0$ semistable, $y = 1$ semistable.

15.

For $y > 3/2$, the slopes are positive, therefore the solutions increase. For $y < 3/2$, the slopes are negative, therefore the solutions decrease. As a result, y diverges from $3/2$ as $t \to \infty$ if $y(0) \neq 3/2$.

17.

For $y > -1/2$, the slopes are negative, therefore the solutions decrease. For $y < -1/2$, the slopes are positive, therefore the solutions increase. As a result, $y \to -1/2$ as $t \to \infty$ for all initial conditions y_0.

19.

For $y > -2$, the slopes are positive, therefore the solutions increase. For $y < -2$, the slopes are negative, therefore the solutions decrease. As a result, y diverges from -2 as $t \to \infty$ if $y(0) \neq -2$.

21.

$y = 0$ and $y = 5$ are equilibrium solutions; y diverges from 5 if the initial value is greater than 5; $y \to 0$ if the initial value is less than 5.

23.

$y = 0$ and $y = 2$ are equilibrium solutions; y diverges from 0 if the initial value is negative; $y \to 2$ if the initial value is between 0 and 2; y diverges from 2 if the initial value is greater than 2.

25. (c) - only equilibrium is $y = 2$; $y' < 0$ when $y < 2$.

27. (b) - only equilibrium is $y = -2$; $y' < 0$ when $y < -2$.

29. (e) - equilibria at $y = 0$, $y = 3$; $y' < 0$ when $0 < y < 3$.

31. Using the fact that

$$R \sin(\omega t - \delta) = R \cos \delta \sin \omega t - R \sin \delta \cos \omega t,$$

where $R^2 \cos^2 \delta + R^2 \sin^2 \delta = R^2 = A^2 + B^2$, the desired result follows.

33.(a) $y = 0$ unstable, $y = K$ asymptotically stable. The figure shows the case $r = 2$, $K = 4$.

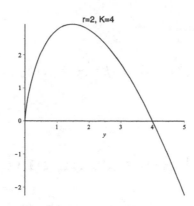

(b) We have to show that for $0 < y \leq K$, $ry(1 - y/K) \leq ry \ln(K/y)$. This is the same as $1 - y/K \leq \ln(K/y)$; thus we have to show that $e^{1-y/K} \leq K/y$, which is equivalent to $e^{y/K-1} \geq y/K$. The function e^x is concave up, thus the tangent line at $x = 0$ is below the function, i.e. $e^x \geq 1 + x$ for every x value. Plugging in $x = y/K - 1$ we obtain the inequality we need.

35.(a) We compute:

$$\frac{dC}{d\tau} = \frac{dC}{dt}\frac{dt}{d\tau} = \frac{1}{c_i}\frac{dc}{dt}\frac{dt}{d\tau} = \frac{1}{c_i}\left(\frac{r_i}{V}c_i - r_i\frac{c}{V} - kc\right)\frac{V}{r_i} = 1 - \frac{c}{c_i} - \frac{kV}{r_i}\frac{c}{c_i} = 1 - C - \alpha C.$$

(b) The equilibrium is at $C = 1/(1 + \alpha)$; it is asymptotically stable.

37.(a)

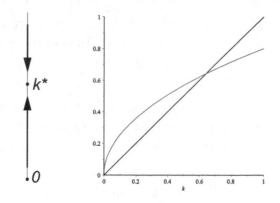

(b) We compute, using the product rule:

$$\frac{dY}{dt} = \frac{dA}{dt}Lf(k) + A\frac{dL}{dt}f(k) + ALf'(k)\frac{dk}{dt} = gALf(k) + nALf(k) = (n+g)Y,$$

because $dk/dt = 0$ at $k = k^*$.

1.3   Definitions, Classification, and Terminology

1. The differential equation is second order, since the highest derivative in the equation is of order two. The equation is linear since the left hand side is a linear function of y and its derivatives and the right hand side is just a function of t.

3. The differential equation is fourth order since the highest derivative in the equation is of order four. The equation is linear since the left hand side is a linear function of y and its derivatives and the right hand side does not depend on y.

5. The differential equation is second order since the highest derivative in the equation is of order two. The equation is nonlinear because of the term $\sin(t+y)$ which is not a linear function of y.

7. $a_0 = 1$, $a_1 = 1/(1+t)$, $g = 2\sin t$; nonhomogeneous.

9. $a_0 = x^2$, $a_1 = -3x$, $a_2 = 4$, $g = \ln x$; nonhomogeneous.

11. $a_0 = 1$, $a_1 = 0$, $a_2 = \cos t$, $a_3 = 0$, $a_4 = 1$, $g = e^{-t}\sin t$; nonhomogenous.

13. If $y_1 = e^t$, then $y_1' = e^t$ and $y_1'' = e^t$. Therefore, $y_1'' - y_1 = 0$. Also, $y_2 = \cosh t$ implies $y_2' = \sinh t$ and $y_2' = \cosh t$. Therefore, $y_2'' - y_2 = 0$.

15. If $y = 3t + t^2$, then $y' = 3 + 2t$. Therefore, $ty' - y = t(3 + 2t) - (3t + t^2) = t^2$.

17. If $y_1 = t^{1/2}$, then $y_1' = t^{-1/2}/2$ and $y_1'' = -t^{-3/2}/4$. Therefore, $2t^2 y_1'' + 3ty_1' - y_1 = 2t^2(-t^{-3/2}/4) + 3t(t^{-1/2}/2) - t^{1/2} = (-1/2 + 3/2 - 1)t^{1/2} = 0$. Also, $y_2 = t^{-1}$ implies $y_2' = -t^{-2}$ and $y_2'' = 2t^{-3}$. Therefore, $2t^2 y_2'' + 3ty_2' - y_2 = 2t^2(2t^{-3}) + 3t(-t^{-2}) - t^{-1} = (4 - 3 - 1)t^{-1} = 0$.

19. If $y = (\cos t)\ln\cos t + t\sin t$, then $y' = -(\sin t)\ln\cos t + t\cos t$ and $y'' = -(\cos t)\ln\cos t - t\sin t + \sec t$. Therefore, $y'' + y = -(\cos t)\ln\cos t - t\sin t + \sec t + (\cos t)\ln\cos t + t\sin t = \sec t$.

21. Let $y = e^{rt}$. Then $y' = re^{rt}$. Substituting these terms into the differential equation, we have $y' + 2y = re^{rt} + 2e^{rt} = (r+2)e^{rt} = 0$. This equation implies $r = -2$.

23. Let $y = e^{rt}$. Then $y' = re^{rt}$ and $y'' = r^2 e^{rt}$. Substituting these terms into the differential equation, we have $y'' + y' - 6y = (r^2 + r - 6)e^{rt} = 0$. In order for r to satisfy this equation, we need $r^2 + r - 6 = 0$. That is, we need $r = 2, -3$.

25. Let $y = t^r$. Then $y' = rt^{r-1}$ and $y'' = r(r-1)t^{r-2}$. Substituting these terms into the differential equation, we have $t^2 y'' + 4ty' + 2y = t^2(r(r-1)t^{r-2}) + 4t(rt^{r-1}) + 2t^r = (r(r-1) + 4r + 2)t^r = 0$. In order for r to satisfy this equation, we need $r(r-1) + 4r + 2 = 0$. Simplifying this expression, we need $r^2 + 3r + 2 = 0$. The solutions of this equation are $r = -1, -2$.

27. If $y = Ce^{-2t}$, then $y' = -2Ce^{-2t}$. Thus $y' + 2y = 0$. Also, $1 = y(0) = Ce^0 = C$.

29. If $y = \sin t/t^2 + C/t^2$, then $y' = \cos t/t^2 - 2\sin t/t^3 - 2C/t^3$. Thus $y' + (2/t)y = \cos t/t^2$. Also, $1/2 = y(1) = \sin 1 + C$ and then $C = 1/2 - \sin 1$.

31. If $y = e^{-t^2/4} \int_0^t e^{s^2/4} \, ds + Ce^{-t^2/4}$, then $y' = -(t/2)e^{-t^2/4} \int_0^t e^{s^2/4} \, ds + e^{-t^2/4}e^{t^2/4} - (t/2)Ce^{-t^2/4}$. Thus $2y' + ty = 2$. Also, $1 = y(0) = Ce^0 = C$.

33. If $\phi(t) = c_1 e^t + c_2 t e^t$, then $\phi'(t) = c_1 e^t + c_2 e^t + c_2 t e^t$ and $\phi''(t) = c_1 e^t + 2c_2 e^t + c_2 t e^t$. Thus $\phi'' - 2\phi' + \phi = 0$.

(a) $3 = y(0) = c_1$ and $1 = y'(0) = c_1 + c_2$, thus $c_2 = -2$.

(b) $1 = y(0) = c_1$ and $-4 = y'(0) = c_1 + c_2$, thus $c_2 = -5$.

Chapter 2

First Order Differential Equations

2.1 Separable Equations

1. Rewriting as $y\,dy = x^4 dx$, then integrating both sides, we have $y^2/2 = x^5/5 + c$, or $5y^2 - 2x^5 = c$; $y \neq 0$.

3. Rewriting as $y^{-3}dy = -\sin x\,dx$, then integrating both sides, we have $-y^{-2}/2 = \cos x + c$, or $y^{-2} + 2\cos x = c$ if $y \neq 0$. Also, $y = 0$ is a solution.

5. Rewriting as $\sec^2 y\,dy = \sin^2 2x\,dx$, then integrating both sides, we have $\tan y = x/2 - (\sin 4x)/8 + c$, or $8\tan y - 4x + \sin 4x = c$ as long as $\cos y \neq 0$. Also, $y = \pm(2n+1)\pi/2$ for any integer n are solutions.

7. Rewriting as $(y/(1+y^2))dy = xe^{x^2}dx$, then integrating both sides, we obtain $\ln(1+y^2) = e^{x^2} + c$. Therefore, $y^2 = ce^{e^{x^2}} - 1$.

9. Rewriting as $(1+y^2)dy = x^2 dx$, then integrating both sides, we have $y + y^3/3 = x^3/3 + c$, or $3y + y^3 - x^3 = c$.

11. Rewriting as $y^{-1/2}dy = 4\sqrt{x}\,dx$, then integrating both sides, we have $y^{1/2} = 4x^{3/2}/3 + c$, or $y = (4x^{3/2}/3 + c)^2$. Also, $y = 0$ is a solution.

13.(a) Rewriting as $y^{-2}dy = (1 - 12x)dx$, then integrating both sides, we have $-y^{-1} = x - 6x^2 + c$. The initial condition $y(0) = -1/8$ implies $c = 8$. Therefore, $y = 1/(6x^2 - x - 8)$.

(b)

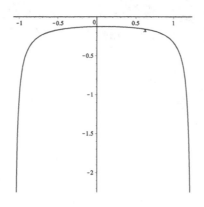

(c) $(1 - \sqrt{193})/12 < x < (1 + \sqrt{193})/12$

15.(a) Rewriting as $xe^x dx = -y dy$, then integrating both sides, we have $xe^x - e^x = -y^2/2 + c$. The initial condition $y(0) = 1$ implies $c = -1/2$. Therefore, $y = \sqrt{2(1-x)e^x - 1}$.

(b)

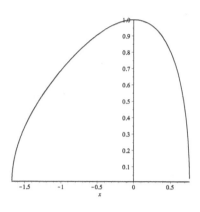

(c) $-1.68 < x < 0.77$, approximately

17.(a) Rewriting as $y dy = 3x/(1 + x^2)dx$, then integrating both sides, we have $y^2/2 = 3\ln(1 + x^2)/2 + c$. The initial condition $y(0) = -7$ implies $c = 49/2$. Therefore, $y = -\sqrt{3\ln(1 + x^2) + 49}$.

(b)

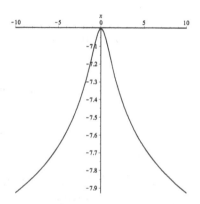

(c) $-\infty < x < \infty$

19.(a) Rewriting as $y^{-2}dy = (2x + 4x^3)dx$, then integrating both sides, we have $-y^{-1} = x^2 + x^4 + c$. The initial condition $y(1) = -2$ implies $c = -3/2$. Therefore, $y = 2/(3 - 2x^4 - 2x^2)$.

(b)

(c) $\sqrt{(-1+\sqrt{7})/2} < x < \infty$

21.(a) Rewriting as $dy/(1+y^2) = \tan 2x\,dx$, then integrating both sides, we have $\arctan y = -\ln(\cos 2x)/2 + c$. The initial condition $y(0) = -\sqrt{3}$ implies $c = -\pi/3$. Therefore, $y = -\tan(\ln(\cos 2x)/2 + \pi/3)$.

(b)

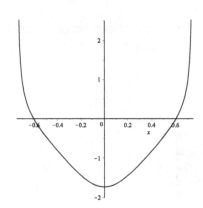

(c) $-\pi/4 < x < \pi/4$

23.(a) Rewriting as $(2y-11)dy = (3x^2-e^x)dx$, then integrating both sides, we have $y^2-11y = x^3 - e^x + c$. The initial condition $y(0) = 11$ implies $c = 1$. Completing the square, we have $(y - 11/2)^2 = x^3 - e^x + 125/4$. Therefore, $y = 11/2 + \sqrt{x^3 - e^x + 125/4}$.

(b)

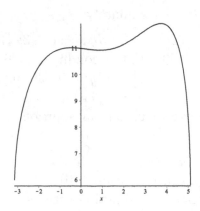

(c) $-3.14 < x < 5.10$, approximately

25.(a) Rewriting as $(3+4y)dy = (e^{-x}-e^x)dx$, then integrating both sides, we have $3y+2y^2 = -(e^x + e^{-x}) + c$. The initial condition $y(0) = 1$ implies $c = 7$. Completing the square, we have $(y + 3/4)^2 = -(e^x + e^{-x})/2 + 65/16$. Therefore, $y = -3/4 + (1/4)\sqrt{65 - 8e^x - 8e^{-x}}$.

(b)

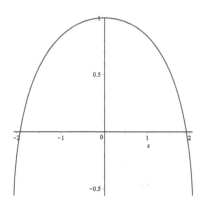

(c) $-\ln 8 < x < \ln 8$

27.(a) Rewriting as $\cos 3y dy = -\sin 2x dx$, then integrating both sides, we have $(\sin 3y)/3 = (\cos 2x)/2 + c$. The initial condition $y(\pi/2) = \pi/3$ implies $c = 1/2$. Thus we obtain that $y = (\pi - \arcsin(3\cos^2 x))/3$.

(b)

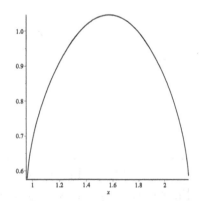

(c) $\pi/2 - 0.62 < x < \pi/2 + 0.62$, approximately

29. Rewriting the equation as $(12y^2 - 12y)dy = (1 + 3x^2)dx$ and integrating both sides, we have $4y^3 - 6y^2 = x + x^3 + c$. The initial condition $y(0) = 2$ implies $c = 8$. Therefore, $4y^3 - 6y^2 - x - x^3 - 8 = 0$. When $12y^2 - 12y = 0$, the integral curve will have a vertical tangent. This happens when $y = 0$ or $y = 1$. From our solution, we see that $y = 1$ implies $x = -2$; this is the first y value we reach on our solution, therefore, the solution is defined for $-2 < x < \infty$.

31. Rewriting the equation as $y^{-2}dy = (2 + x)dx$ and integrating both sides, we have $-y^{-1} = 2x + x^2/2 + c$. The initial condition $y(0) = 1$ implies $c = -1$. Therefore, $y = -1/(x^2/2 + 2x - 1)$. To find where the function attains it minimum value, we look where

$y' = 0$. We see that $y' = 0$ implies $y = 0$ or $x = -2$. But, as seen by the solution formula, y is never zero. Further, it can be verified that $y''(-2) > 0$, and, therefore, the function attains a minimum at $x = -2$.

33. Rewriting the equation as $(10 + 2y)dy = 2\cos 2x dx$ and integrating both sides, we have $10y + y^2 = \sin 2x + c$. By the initial condition $y(0) = -1$, we have $c = -9$. Completing the square, it follows that $y = -5 + \sqrt{\sin 2x + 16}$. To find where the solution attains its maximum value, we need to check where $y' = 0$. We see that $y' = 0$ when $2\cos 2x = 0$. This occurs when $2x = \pi/2 + 2k\pi$, or $x = \pi/4 + k\pi$, $k = 0, \pm 1, \pm 2, \ldots$.

35.(a) First, we rewrite the equation as $dy/(y(4-y)) = tdt/3$. Then, using partial fractions, after integration we obtain

$$\left| \frac{y}{y-4} \right| = Ce^{2t^2/3}.$$

From the equation, we see that $y_0 = 0$ implies that $C = 0$, so $y(t) = 0$ for all t. Otherwise, $y(t) > 0$ for all t or $y(t) < 0$ for all t. Therefore, if $y_0 > 0$ and $|y/(y-4)| = Ce^{2t^2/3} \to \infty$, we must have $y \to 4$. On the other hand, if $y_0 < 0$, then $y \to -\infty$ as $t \to \infty$. (In particular, $y \to -\infty$ in finite time.)

(b) For $y_0 = 0.5$, we want to find the time T when the solution first reaches the value 3.98. Using the fact that $|y/(y-4)| = Ce^{2t^2/3}$ combined with the initial condition, we have $C = 1/7$. From this equation, we now need to find T such that $|3.98/.02| = e^{2T^2/3}/7$. Solving this equation, we obtain $T \approx 3.29527$.

37. We can write the equation as

$$\left(\frac{cy + d}{ay + b} \right) dy = dx,$$

which gives

$$\left(\frac{cy}{ay + b} + \frac{d}{ay + b} \right) dy = dx.$$

Now we want to rewrite these so in the first component we can simplify by $ay + b$:

$$\frac{cy}{ay + b} = \frac{\frac{1}{a}cay}{ay + b} = \frac{\frac{1}{a}(cay + bc) - bc/a}{ay + b} = \frac{1}{a}c - \frac{\frac{bc}{a}}{ay + b},$$

so we obtain

$$\left(\frac{c}{a} - \frac{bc}{a^2 y + ab} + \frac{d}{ay + b} \right) dy = dx.$$

Then integrating both sides, we have

$$\frac{c}{a}y - \frac{bc}{a^2} \ln |a^2 y + ab| + \frac{d}{a} \ln |ay + b| = x + C.$$

Simplifying, we have

$$\frac{c}{a}y - \frac{bc}{a^2} \ln |a| - \frac{bc}{a^2} \ln |ay + b| + \frac{d}{a} \ln |ay + b| = x + C,$$

which implies that

$$\frac{c}{a}y + \left(\frac{ad - bc}{a^2}\right)\ln|ay + b| = x + C.$$

Note, in this calculation, since $\frac{bc}{a^2}\ln|a|$ is just a constant, we included it with the arbitrary constant C. This solution will exist as long as $a \neq 0$ and $ay + b \neq 0$.

2.2   Linear Equations: Method of Integrating Factors

1.(a)

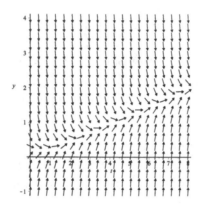

(b) All solutions seem to converge to an increasing function as $t \to \infty$.

(c) The integrating factor is $\mu(t) = e^{4t}$. Then

$$e^{4t}y' + 4e^{4t}y = e^{4t}(t + e^{-2t})$$

implies that

$$(e^{4t}y)' = te^{4t} + e^{2t},$$

thus

$$e^{4t}y = \int (te^{4t} + e^{2t})\, dt = \frac{1}{4}te^{4t} - \frac{1}{16}e^{4t} + \frac{1}{2}e^{2t} + c,$$

and then

$$y = ce^{-4t} + \frac{1}{2}e^{-2t} + \frac{t}{4} - \frac{1}{16}.$$

We conclude that y is asymptotic to the linear function $g(t) = t/4 - 1/16$ as $t \to \infty$.

3.(a)

(b) All solutions appear to converge to the function $g(t) = 1$.

(c) The integrating factor is $\mu(t) = e^t$. Therefore, $e^t y' + e^t y = t + e^t$, thus $(e^t y)' = t + e^t$, so

$$e^t y = \int (t + e^t) \, dt = \frac{t^2}{2} + e^t + c,$$

and then

$$y = \frac{t^2}{2} e^{-t} + 1 + ce^{-t}.$$

Therefore, we conclude that $y \to 1$ as $t \to \infty$.

5.(a)

(b) Some of the solutions increase without bound, some decrease without bound.

(c) The integrating factor is $\mu(t) = e^{-2t}$. Therefore, $e^{-2t} y' - 2e^{-2t} y = 3e^{-t}$, which implies $(e^{-2t} y)' = 3e^{-t}$, thus

$$e^{-2t} y = \int 3e^{-t} \, dt = -3e^{-t} + c,$$

and then $y = -3e^t + ce^{2t}$. We conclude that y increases or decreases exponentially as $t \to \infty$.

7.(a)

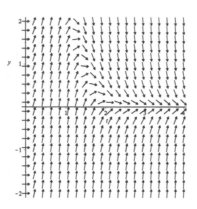

(b) For $t > 0$, all solutions seem to eventually converge to the function $g(t) = 0$.

(c) The integrating factor is $\mu(t) = e^{t^2}$. Therefore,

$$(e^{t^2}y)' = e^{t^2}y' + 2tye^{t^2} = 16t,$$

thus

$$e^{t^2}y = \int 16t\, dt = 8t^2 + c,$$

and then $y(t) = 8t^2 e^{-t^2} + ce^{-t^2}$. We conclude that $y \to 0$ as $t \to \infty$.

9.(a)

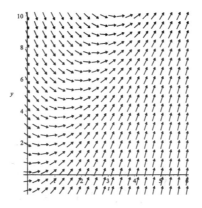

(b) All solutions increase without bound.

(c) The integrating factor is $\mu(t) = e^{t/2}$. Therefore, $2e^{t/2}y' + e^{t/2}y = 3te^{t/2}$, thus

$$2e^{t/2}y = \int 3te^{t/2}\, dt = 6te^{t/2} - 12e^{t/2} + c,$$

and then $y = 3t - 6 + ce^{-t/2}$. We conclude that y is asymptotic to $g(t) = 3t - 6$ as $t \to \infty$.

11.(a)

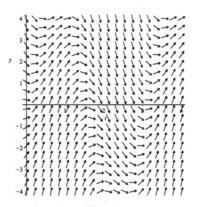

(b) All solutions appear to converge to an oscillatory function.

(c) The integrating factor is $\mu(t) = e^t$. Therefore, $e^t y' + e^t y = 5e^t \sin 2t$, thus $(e^t y)' = 5e^t \sin 2t$, which gives

$$e^t y = \int 5e^t \sin 2t \, dt = -2e^t \cos 2t + e^t \sin 2t + c,$$

and then $y = -2\cos 2t + \sin 2t + ce^{-t}$. We conclude that y is asymptotic to $g(t) = \sin 2t - 2\cos 2t$ as $t \to \infty$.

13. The integrating factor is $\mu(t) = e^{-t}$. Therefore, $(e^{-t} y)' = 2te^t$, thus

$$y = e^t \int 2te^t \, dt = 2te^{2t} - 2e^{2t} + ce^t.$$

The initial condition $y(0) = 1$ implies $-2 + c = 1$. Therefore, $c = 3$ and $y = 3e^t + 2(t-1)e^{2t}$.

15. Dividing the equation by t, we see that the integrating factor is $\mu(t) = t^4$. Therefore, $(t^4 y)' = t^5 - t^4 + t^3$, thus

$$y = t^{-4} \int (t^5 - t^4 + t^3) \, dt = \frac{t^2}{6} - \frac{t}{5} + \frac{1}{4} + \frac{c}{t^4}.$$

The initial condition $y(1) = 1/4$ implies $c = 1/30$, and $y = (10t^6 - 12t^5 + 15t^4 + 2)/60t^4$.

17. The integrating factor is $\mu(t) = e^{-2t}$. Therefore, $(e^{-2t} y)' = 1$, thus

$$y = e^{2t} \int 1 \, dt = e^{2t}(t + c).$$

The initial condition $y(0) = 2$ implies $c = 2$ and $y = (t + 2)e^{2t}$.

19. After dividing by t^3, we see that the integrating factor is $\mu(t) = t^4$. Therefore, $(t^4 y)' = te^{-t}$, thus

$$y = t^{-4} \int te^{-t} \, dt = t^{-4}(-te^{-t} - e^{-t} + c).$$

The initial condition $y(-1) = 0$ implies $c = 0$ and $y = -(1 + t)e^{-t}/t^4$.

21.(a)

The solutions appear to diverge from an oscillatory solution. It appears that $a_0 \approx -1$. For $a > -1$, the solutions increase without bound. For $a < -1$, the solutions decrease without bound.

(b) The integrating factor is $\mu(t) = e^{-t/3}$. From this, we get the equation $y'e^{-t/3} - ye^{-t/3}/3 = (ye^{-t/3})' = 3e^{-t/3}\cos t$. After integration, $y(t) = (27\sin t - 9\cos t)/10 + ce^{t/3}$, where (using the initial condition) $c = a + 9/10$. The solution will be sinusoidal as long as $c = 0$. Therefore, $a_0 = -9/10$.

(c) y oscillates for $a = a_0$.

23.(a)

Solutions eventually increase or decrease without bound, depending on the initial value a_0. It appears that $a_0 \approx -1/8$.

(b) Dividing the equation by 3, we see that the integrating factor is $\mu(t) = e^{-2t/3}$. From this, we get the equation $y'e^{-2t/3} - 2ye^{-2t/3}/3 = (ye^{-2t/3})' = 2e^{-\pi t/2 - 2t/3}/3$. After integration, the general solution is $y(t) = e^{2t/3}(-(2/3)e^{-\pi t/2 - 2t/3}(1/(\pi/2 + 2/3)) + c)$. Using the initial condition, we get $y = ((2 + a(3\pi + 4))e^{2t/3} - 2e^{-\pi t/2})/(3\pi + 4)$. The solution will eventually behave like $(2 + a(3\pi + 4))e^{2t/3}/(3\pi + 4)$. Therefore, $a_0 = -2/(3\pi + 4)$.

(c) $y \to 0$ for $a = a_0$.

25.(a)

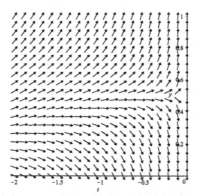

It appears that $a_0 \approx .4$. That is, as $t \to 0$, for $y(-\pi/2) > a_0$, solutions will increase without bound, while solutions will decrease without bound for $y(-\pi/2) < a_0$.

(b) After dividing by t, we see that the integrating factor is $\mu(t) = t^2$. After multiplication by μ, we obtain the equation $t^2 y' + 2ty = (t^2 y)' = \sin t$, so after integration, we get that the general solution is $y = -\cos t/t^2 + c/t^2$. Using the initial condition, we get the solution $y = -\cos t/t^2 + \pi^2 a/4t^2$. Since $\lim_{t \to 0} \cos t = 1$, solutions will increase without bound if $a > 4/\pi^2$ and decrease without bound if $a < 4/\pi^2$. Therefore, $a_0 = 4/\pi^2$.

(c) For $a_0 = 4/\pi^2$, $y = (1 - \cos t)/t^2 \to 1/2$ as $t \to 0$.

27. The integrating factor is $\mu(t) = e^{t/2}$. Therefore, the general solution is $y(t) = (4\cos t + 8\sin t)/5 + ce^{-t/2}$. Using our initial condition, we have $y(t) = (4\cos t + 8\sin t - 9e^{t/2})/5$. Differentiating, we obtain

$$y' = (-4\sin t + 8\cos t + 4.5e^{-t/2})/5$$
$$y'' = (-4\cos t - 8\sin t - 2.25e^{t/2})/5.$$

Setting $y' = 0$, the first solution is $t_1 \approx 1.3643$, which gives the location of the first stationary point. Since $y''(t_1) < 0$, the first stationary point is a local maximum. The coordinates of the point are approximately $(1.3643, 0.8201)$.

29.(a) The integrating factor is $\mu(t) = e^{t/4}$. The general solution is $y(t) = 12 + (8\cos 2t + 64\sin 2t)/65 + ce^{-t/4}$. Applying the initial condition $y(0) = 0$, we arrive at the specific solution $y(t) = 12 + (8\cos 2t + 64\sin 2t - 788e^{-t/4})/65$. As $t \to \infty$, the solution oscillates about the line $y = 12$.

(b) To find the value of t for which the solution first intersects the line $y = 12$, we need to solve the equation $8\cos 2t + 64\sin 2t - 788e^{-t/4} = 0$. The value of t is approximately 10.0658.

31. The integrating factor is $\mu(t) = e^{-3t/2}$ and the general solution of the equation is $y(t) = -2t - 4/3 - 4e^t + ce^{3t/2}$. The initial condition implies $y(t) = -2t - 4/3 - 4e^t + (y_0 + 16/3)e^{3t/2}$. The solution will behave like $(y_0 + 16/3)e^{3t/2}$ (for $y_0 \neq -16/3$). For $y_0 > -16/3$, the solutions will increase without bound, while for $y_0 < -16/3$, the solutions will decrease without bound. If $y_0 = -16/3$, the solution will decrease without bound as the solution will be $-2t - 4/3 - 4e^t$.

33. The integrating factor is $\mu(t) = e^{at}$. First consider the case $a \neq \lambda$. Multiplying the equation by e^{at}, we have $(e^{at}y)' = be^{(a-\lambda)t}$, which implies

$$y = e^{-at}\int be^{(a-\lambda)t} = e^{-at}\left(\frac{b}{a-\lambda}e^{(a-\lambda)t} + c\right) = \frac{b}{a-\lambda}e^{-\lambda t} + ce^{-at}.$$

Since a, λ are assumed to be positive, we see that $y \to 0$ as $t \to \infty$. Now if $a = \lambda$ above, then we have $(e^{at}y)' = b$, which implies $y = e^{-at}(bt + c)$ and similarly $y \to 0$ as $t \to \infty$.

35. We notice that $y(t) = ce^{-t} + 4 - t$ approaches $4 - t$ as $t \to \infty$. We just need to find a first order linear differential equation having that solution. We notice that if $y(t) = f + g$, then $y' + y = f' + f + g' + g$. Here, let $f = ce^{-t}$ and $g(t) = 4 - t$. Then $f' + f = 0$ and $g' + g = -1 + 4 - t = 3 - t$. Therefore, $y(t) = ce^{-t} + 4 - t$ satisfies the equation $y' + y = 3 - t$. That is, the equation $y' + y = 3 - t$ has the desired properties.

37. We notice that $y(t) = ce^{-t} + 2 - t^2$ approaches $2 - t^2$ as $t \to \infty$. We just need to find a first-order linear differential equation having that solution. We notice that if $y(t) = f + g$, then $y' + y = f' + f + g' + g$. Here, let $f = ce^{-t}$ and $g(t) = 2 - t^2$. Then $f' + f = 0$ and $g' + g = -2t + 2 - t^2 = 2 - 2t - t^2$. Therefore, $y(t) = ce^{-t} + 2 - t^2$ satisfies the equation $y' + y = 2 - 2t - t^2$. That is, the equation $y' + y = 2 - 2t - t^2$ has the desired properties.

39.(a) The integrating factor is $e^{\int p(t)\,dt}$. Multiplying by the integrating factor, we have

$$e^{\int p(t)\,dt}y' + e^{\int p(t)\,dt}p(t)y = 0.$$

Therefore,

$$\left(e^{\int p(t)\,dt}y\right)' = 0,$$

which implies

$$y(t) = Ae^{-\int p(t)\,dt}$$

is the general solution.

(b) Let $y = A(t)e^{-\int p(t)\,dt}$. Then in order for y to satisfy the desired equation, we need

$$A'(t)e^{-\int p(t)\,dt} - A(t)p(t)e^{-\int p(t)\,dt} + A(t)p(t)e^{-\int p(t)\,dt} = g(t).$$

That is, we need

$$A'(t) = g(t)e^{\int p(t)\,dt}.$$

(c) From equation (iv), we see that

$$A(t) = \int_0^t g(\tau)e^{\int p(\tau)\,d\tau}\,d\tau + C.$$

Therefore,

$$y(t) = e^{-\int p(t)\,dt}\left(\int_0^t g(\tau)e^{\int p(\tau)\,d\tau}\,d\tau + C\right).$$

41. Here, $p(t) = 1/t$ and $g(t) = 3\cos 2t$. The general solution is given by

$$
\begin{aligned}
y(t) &= e^{-\int p(t)\,dt}\left(\int_0^t g(\tau)e^{\int p(\tau)\,d\tau}\,d\tau + C\right) = e^{-\int \frac{1}{t}\,dt}\left(\int_0^t 3\cos 2\tau\, e^{\int \frac{1}{\tau}\,d\tau}\,d\tau + C\right) \\
&= \frac{1}{t}\left(\int_0^t 3\tau\cos 2\tau\,d\tau + C\right) = \frac{1}{t}\left(\frac{3}{4}\cos 2t + \frac{3}{2}t\sin 2t + C\right).
\end{aligned}
$$

43. Here, $p(t) = 1/2$ and $g(t) = 3t^2/2$. The general solution is given by

$$
\begin{aligned}
y(t) &= e^{-\int p(t)\,dt}\left(\int_0^t g(\tau)e^{\int p(\tau)\,d\tau}\,d\tau + C\right) = e^{-\int \frac{1}{2}\,dt}\left(\int_0^t \frac{3\tau^2}{2}e^{\int \frac{1}{2}\,d\tau}\,d\tau + C\right) \\
&= e^{-t/2}\left(\int_0^t \frac{3\tau^2}{2}e^{\tau/2}\,d\tau + C\right) = e^{-t/2}\left(3t^2 e^{t/2} - 12te^{t/2} + 24e^{t/2} + C\right) \\
&= 3t^2 - 12t + 24 + Ce^{-t/2}.
\end{aligned}
$$

2.3 Modeling with First Order Equations

1. Let $Q(t)$ be the quantity of dye in the tank. We know that

$$
\frac{dQ}{dt} = \text{rate in} - \text{rate out}.
$$

Here, fresh water is flowing in. Therefore, no dye is coming in. The dye is flowing out at the rate of $(Q/150)$ grams/liters $\cdot$ 3 liters/minute $= (Q/50)$ grams/minute. Therefore,

$$
\frac{dQ}{dt} = -\frac{Q}{50}.
$$

The solution of this equation is $Q(t) = Ce^{-t/50}$. Since $Q(0) = 450$ grams, $C = 450$. We need to find the time T when the amount of dye present is 2% of what it is initially. That is, we need to find the time T when $Q(T) = 9$ grams. Solving the equation $9 = 450e^{-T/50}$, we conclude that $T = 50\ln(50) \approx 195.6$ minutes.

3. Let $Q(t)$ be the quantity of salt in the tank. We know that

$$
\frac{dQ}{dt} = \text{rate in} - \text{rate out}.
$$

Here, water containing 1/4 lb/gallon of salt is flowing in at a rate of 4 gallons/minute. The salt is flowing out at the rate of $(Q/160)$ lb/gallon $\cdot$ 4 gallons/minute $= (Q/40)$ lb/minute. Therefore,

$$
\frac{dQ}{dt} = 1 - \frac{Q}{40}.
$$

The solution of this equation is $Q(t) = 40 + Ce^{-t/40}$. Since $Q(0) = 0$ grams, $C = -40$. Therefore, $Q(t) = 40(1 - e^{-t/40})$ for $0 \leq t \leq 8$ minutes. After 8 minutes, the amount of salt in the tank is $Q(8) = 40(1 - e^{-1/5}) \approx 7.25$ lbs. Starting at that time (and resetting the time variable), the new equation for dQ/dt is given by

$$\frac{dQ}{dt} = -\frac{3Q}{80},$$

since fresh water is being added. The solution of this equation is $Q(t) = Ce^{-3t/80}$. Since we are now starting with 7.25 lbs of salt, $Q(0) = 7.25 = C$. Therefore, $Q(t) = 7.25e^{-3t/80}$. After 8 minutes, $Q(8) = 7.25e^{-3/10} \approx 5.37$ lbs.

5.(a) Let $Q(t)$ be the quantity of salt in the tank. We know that

$$\frac{dQ}{dt} = \text{rate in} - \text{rate out}.$$

Here, water containing $\dfrac{1}{4}\left(1 + \dfrac{1}{2}\sin t\right)$ oz/gallon of salt is flowing in at a rate of 2 gal/minute. The salt is flowing out at the rate of $(Q/100)$ oz/gallon$\cdot$2 gallons/minute $= (Q/50)$ oz/minute. Therefore,

$$\frac{dQ}{dt} = \frac{1}{2} + \frac{1}{4}\sin t - \frac{Q}{50}.$$

This is a linear equation with integrating factor $\mu(t) = e^{t/50}$. The solution of this equation is $Q(t) = (12.5\sin t - 625\cos t + 63150e^{-t/50})/2501 + c$. The initial condition, $Q(0) = 50$ oz implies $C = 25$. Therefore, $Q(t) = 25 + (12.5\sin t - 625\cos t + 63150e^{-t/50})/2501$ oz.

(b)

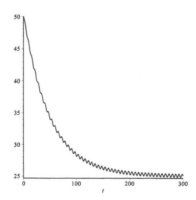

(c) The amount of salt approaches a steady state, which is an oscillation of amplitude $25\sqrt{2501}/5002 \approx 0.24995$ about a level of 25 oz.

7.(a) The equation describing the water volume is given by $V' = G - 0.0005V$. Thus the equilibrium volume is $V_e = 2000G$. The figure shows some possible sketches for $V(t)$ when $G = 5$.

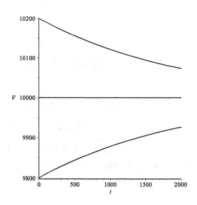

(b) The differential equation $V' = G - 0.0005V$ is linear with integrating factor $\mu = e^{t/2000}$. The general solution we obtain is $V(t) = 2000G + ce^{-t/2000}$. If $V(0) = 1.01V_e = 2020G$, then $c = 20G$, and the solution is $V = 2000G + 20Ge^{-t/2000}$.

(c) From part (a), $12000 = V_e = 2000G$, thus $G = 6$ gallons per day.

9.(a) The differential equation for the amount of poison in the keg is given by $Q' = 5 \cdot 0.5 - 0.5 \cdot Q/500 = 5/2 - Q/1000$. Then using the initial condition $Q(0) = 0$ and the integrating factor $\mu = e^{t/1000}$ we obtain $Q(t) = 2500 - 2500e^{-t/1000}$.

(b) To reach the concentration 0.005 g/L, the amount $Q(T) = 2500(1 - e^{T/1000}) = 2.5$ g. Thus $T = 1000 \ln(1000/999) \approx 1$ minute.

(c) The estimate is 1 minute, because to pour in 2.5 grams of poison without removing the mixture, we have to pour in a half liter of the liquid containing the poison. This takes 1 minute.

11.(a) The equation for S is given by

$$\frac{dS}{dt} = rS + k.$$

This is a linear equation with solution $S(t) = \dfrac{k}{r}(e^{rt} - 1)$.

(b) Using the function in part (a), we need to find k so that $S(42) = 1,000,000$ assuming $r = 0.055$. That is, we need to solve

$$1,000,000 = \frac{k}{0.055}(e^{0.055(42)} - 1).$$

The solution of this equation is $k \approx \$6061$.

(c) Now we assume that $k = 4000$ and want to find r. Our equation becomes

$$1,000,000 = \frac{4000}{r}(e^{42r} - 1).$$

The solution of this equation is approximately 6.92%.

13.(a) Let $S(t)$ be the balance due on the loan at time t. Taking into account that t is measured in years, we rewrite the monthly payment as $800(1 + t/10)$ where t is now in years.

The equation for S is given by

$$\frac{dS}{dt} = 0.09S - 12(800)(1 + t/10).$$

This is a linear equation. Its solution is $S(t) = 225185.23 + 10666.67t + ce^{0.09t}$. The initial condition $S(0) = 100,000$ implies $c = -125185.23$. Therefore, the particular solution is $S(t) = 225185.23 + 10666.67t - 125185.23e^{0.09t}$. To find when the loan will be paid, we just need to solve $S(t) = 0$. Solving this equation, we conclude that the loan will be paid off in 11.28 years (135.36 months).

(b) From part (a), we know the general solution is given by $S(t) = 225185.23 + 10666.67t + ce^{0.09t}$. Now we want to find c such that $S(20) = 0$. The solution of this equation is $c = -72486.67$. Therefore, the solution of the equation will be $S(t) = 225185.23 + 10666.67 - 72846.67e^{0.09t}$. Therefore, $S(0) = 225185.23 - 72486.67 = 152,698.56$.

15.(a) The general solution is $Q(t) = Q_0 e^{-rt}$. If the half-life is 5730, then $Q_0/2 = Q_0 e^{-5730r}$ implies $-5730r = \ln(1/2)$. Therefore, $r = 1.2097 \times 10^{-4}$ per year.

(b) Therefore, $Q(t) = Q_0 e^{-1.2097 \times 10^{-4} t}$.

(c) Given that $Q(T) = Q_0/2$, we have the equation $1/2 = e^{-1.2097 \times 10^{-4} T}$. Solving for T, we have $T = 5,729.91$ years.

17.(a) The solution of this separable equation is given by $y(t) = \exp(2/10 + t/10 - 2\cos t/10)$. The doubling-time is found by solving the equation $2 = \exp(2/10 + t/10 - 2\cos t/10)$. The solution of this equation is given by $\tau \approx 2.9632$.

(b) The differential equation will be $dy/dt = y/10$ with solution $y(t) = y(0)e^{t/10}$. The doubling time is found by setting $y(t) = 2y(0)$. In this case, the doubling time is $\tau \approx 6.9315$.

(c) Consider the differential equation $dy/dt = (0.5 + \sin(2\pi t))y/5$. This equation is separable with solution $y(t) = \exp((1 + \pi t - \cos 2\pi t)/10\pi)$. The doubling time is found by setting $y(t) = 2$. The solution is given by $\tau \approx 6.3804$.

(d)

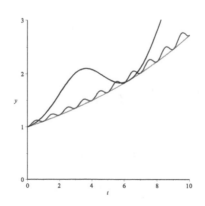

19. Let $T(t)$ be the temperature of the coffee at time t. The governing equation is given by

$$\frac{dT}{dt} = -k(T - 70).$$

This is a linear equation with solution $T(t) = 70 + ce^{-kt}$. The initial condition $T(0) = 200$ implies $c = 130$. Therefore, $T(t) = 70 + 130e^{-kt}$. Using the fact that $T(1) = 190$, we see that $190 = 70 + 130e^{-k}$ which implies $k = -\ln(12/13) \approx 0.08$ per minute. To find when the temperature reaches 150 degrees, we just need to solve $T(t) = 70 + 130e^{\ln(12/13)t} = 150$. The solution of this equation is $t = \ln(13/8)/\ln(13/12) \approx 6.07$ minutes.

21.(a) The differential equation for Q is

$$\frac{dQ}{dt} = kr + P - \frac{Q(t)}{V}r.$$

Therefore,

$$V\frac{dc}{dt} = kr + P - c(t)r.$$

The solution of this equation is $c(t) = k + P/r + (c_0 - k - P/r)e^{-rt/V}$. Therefore $\lim_{t \to \infty} c(t) = k + P/r$.

(b) In this case, we will have $c(t) = c_0 e^{-rt/V}$. The reduction times are $T_{50} = \ln(2)V/r$ and $T_{10} = \ln(10)V/r$.

(c) Using the results from part (b), we have: Superior, $T = 430.85$ years; Michigan, $T = 71.4$ years; Erie, $T = 6.05$ years; Ontario, $T = 17.6$ years.

23.(a) We have $mdv/dt = -v/30 - mg$. Given the conditions from problem 22, we see that the solution is given by $v(t) = -73.5 + 97.5e^{-t/7.5}$. The ball will reach its maximum height when $v(t) = 0$. This occurs at $t = 2.12$ seconds. The height of the ball is given by $s(t) = 757.25 - 73.5t - 731.25e^{-t/7.5}$. When $t = 2.12$ seconds, we have $s(2.12) = 50.24$ meters, the maximum height.

(b) The ball will hit the ground when $s(t) = 0$. This occurs when $t = 5.57$ seconds.

(c)

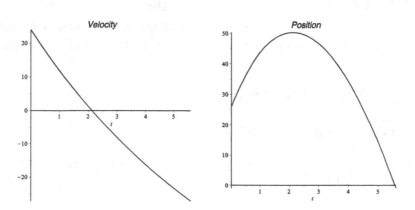

25.(a) Measure the positive direction of motion downward. Then the equation of motion is given by

$$m\frac{dv}{dt} = \begin{cases} -0.75v + mg & 0 < t < 10 \\ -12v + mg & t > 10. \end{cases}$$

For the first 10 seconds, the equation becomes $dv/dt = -v/7.5 + 32$ which has solution $v(t) = 240(1 - e^{-t/7.5})$. Therefore, $v(10) = 176.7$ ft/s.

(b) Integrating the velocity function from part (a), we see that the height of the skydiver at time t $(0 < t < 10)$ is given by $s(t) = 240t + 1800e^{-t/7.5} - 1800$. Therefore, $s(10) = 1074.5$ feet.

(c) After the parachute opens, the equation for v is given by $dv/dt = -32v/15 + 32$ (as discussed in part (a)). We will reset t to zero. The solution of this differential equation is given by $v(t) = 15 + 161.7e^{-32t/15}$. As $t \to \infty$, $v(t) \to 15$. Therefore, the limiting velocity is $v_l = 15$ feet/second.

(d) Integrating the velocity function from part (c), we see that the height of the sky diver after falling t seconds with his parachute open is given by $s(t) = 15t - 75.8e^{-32t/15} + 1150.3$. To find how long the skydiver is in the air after the parachute opens, we find T such that $s(T) = 0$. Solving this equation, we have $T = 256.6$ seconds.

(e)

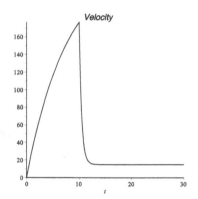

27.(a) Measure the positive direction of motion upward. The equation of motion is given by $mdv/dt = -kv - mg$. The solution of this equation is given by $v(t) = -mg/k + (v_0 + mg/k)e^{-kt/m}$. Solving $v(t) = 0$, we see that the mass will reach its maximum height $t_m = (m/k)\ln[(mg+kv_0)/mg]$ seconds after being projected upward. Integrating the velocity equation, we see that the position of the mass at this time will be given by the position equation

$$s(t) = -mgt/k + \left[\left(\frac{m}{k}\right)^2 g + \frac{mv_0}{k}\right](1 - e^{-kt/m}).$$

Therefore, the maximum height reached is

$$x_m = s(t_m) = \frac{mv_0}{k} - g\left(\frac{m}{k}\right)^2 \ln\left[\frac{mg + kv_0}{mg}\right].$$

(b) These formulas for t_m and x_m come from the fact that for $\delta << 1$, $\ln(1+\delta) = \delta - \frac{1}{2}\delta^2 + \frac{1}{3}\delta^3 - \frac{1}{4}\delta^4 + \ldots$, which is just Taylor's formula.

(c) Consider the result for t_m in part (b). Multiplying the equation by $\frac{g}{v_0}$, we have

$$\frac{t_m g}{v_0} = \left[1 - \frac{1}{2}\frac{kv_0}{mg} + \frac{1}{3}\left(\frac{kv_0}{mg}\right)^2 - \ldots\right].$$

The units on the left must match the units on the right. Since the units for $t_m g/v_0 = (\text{s·m/s}^2)/(\text{m/s})$, the units cancel. As a result, we can conclude that kv_0/mg is dimensionless.

29.(a) The equation of motion is given by

$$m\frac{dv}{dt} = -6\pi\mu a v + \rho'\frac{4}{3}\pi a^3 g - \rho\frac{4}{3}\pi a^3 g.$$

We can rewrite this equation as

$$v' + \frac{6\pi\mu a}{m}v = \frac{4}{3}\frac{\pi a^3 g}{m}(\rho' - \rho).$$

Multiplying by the integrating factor $e^{6\pi\mu a t/m}$, we have

$$(e^{6\pi\mu a t/m}v)' = \frac{4}{3}\frac{\pi a^3 g}{m}(\rho' - \rho)e^{6\pi\mu a t/m}.$$

Integrating this equation, we have

$$v = e^{-6\pi\mu a t/m}\left[\frac{2a^2 g(\rho' - \rho)}{9\mu}e^{6\pi\mu a t/m} + C\right] = \frac{2a^2 g(\rho' - \rho)}{9\mu} + Ce^{-6\pi\mu a t/m}.$$

Therefore, we conclude that the limiting velocity is $v_L = (2a^2 g(\rho' - \rho))/9\mu$.

(b) By the equation above, we see that the force exerted on the droplet of oil is given by

$$Ee = -6\pi\mu a v + \rho'\frac{4}{3}\pi a^3 g - \rho\frac{4}{3}\pi a^3 g.$$

If $v = 0$, then solving the above equation for e, we have

$$e = \frac{4\pi a^3 g(\rho' - \rho)}{3E}.$$

31.(a) The equation of motion is given by $mdv/dt = -GMm/(R + x)^2$. By the chain rule,

$$m\frac{dv}{dx} \cdot \frac{dx}{dt} = -G\frac{Mm}{(R + x)^2}.$$

Therefore,

$$mv\frac{dv}{dx} = -G\frac{Mm}{(R + x)^2}.$$

This equation is separable with solution $v^2 = 2GM(R + x)^{-1} + 2gR - 2GM/R$. Here we have used the initial condition $v_0 = \sqrt{2gR}$. From physics, we know that $g = GM/R^2$. Using this substitution, we conclude that $v(x) = \sqrt{2g}\,R/\sqrt{R + x}$.

(b) By part (a), we know that $dx/dt = v(x) = \sqrt{2g}\,R/\sqrt{R + x}$. We want to solve this differential equation with the initial condition $x(0) = 0$. This equation is separable with solution $x(t) = [\frac{3}{2}(\sqrt{2g}Rt + \frac{2}{3}R^{3/2})]^{2/3} - R$. We want to find the time T such that $x(T) = 240,000$. Solving this equation, we conclude that $T \approx 50.6$ hours.

33.(a) The initial conditions are $v(0) = u \cos A$ and $w(0) = u \sin A$. Therefore, the solutions of the two equations are $v(t) = (u \cos A)e^{-rt}$ and $w(t) = -g/r + (u \sin A + g/r)e^{-rt}$.

(b) Now $x(t) = \int v(t)\, dt = \frac{u}{r}(\cos A)(1 - e^{-rt})$, and

$$y(t) = \int w(t)\, dt = -\frac{gt}{r} + \left(\frac{u}{r}\sin A + \frac{g}{r^2}\right)(1 - e^{-rt}) + h.$$

(c) Below we have plotted the trajectory of the ball in the cases $\pi/6$, $\pi/5$, $\pi/4$, and $\pi/3$, respectively.

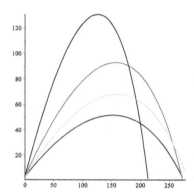

(d) Let T be the time it takes the ball to go 350 feet horizontally. Then from above, we see that $e^{-T/5} = 1 - 70/u \cos A$. At the same time, the height of the ball is given by $y(T) = -160T + (800 + 5u \sin A)70/u \cos A + 3$. Therefore, u and A must satisfy the inequality

$$800 \ln\left(1 - \frac{70}{u \cos A}\right) + 350 \tan A + \frac{56000}{u \cos A} + 3 \geq 10.$$

Using graphical techniques, we identify the minimum velocity necessary is 145.3 ft/s and the optimal angle necessary is 0.644 radians.

2.4   Differences between Linear and Nonlinear Equations

1. Rewriting the equation as

$$y' + \frac{\ln t}{t - 3}y = \frac{2t}{t - 3}$$

and using Theorem 2.4.1, we conclude that a solution is guaranteed to exist in the interval $0 < t < 3$.

3. By Theorem 2.4.1, we conclude that a solution is guaranteed to exist in the interval $\pi/2 < t < 3\pi/2$.

5. Rewriting the equation as

$$y' + \frac{2t}{4 - t^2}y = \frac{3t^2}{4 - t^2}$$

and using Theorem 2.4.1, we conclude that a solution is guaranteed to exist in the interval $-2 < t < 2$.

7. Using the fact that

$$f = \frac{t - y}{2t + 5y} \quad \text{and} \quad f_y = -\frac{7t}{(2t + 5y)^2},$$

we see that the hypotheses of Theorem 2.4.2 are satisfied as long as $2t + 5y \neq 0$.

9. Using the fact that

$$f = \frac{\ln |ty|}{1 - t^2 + y^2} \quad \text{and} \quad f_y = \frac{1 - t^2 + y^2 - 2y^2 \ln |ty|}{y(1 - t^2 + y^2)^2},$$

we see that the hypotheses of Theorem 2.4.2 are satisfied as long as $y, t \neq 0$ and $1 - t^2 + y^2 \neq 0$.

11. Using the fact that

$$f = \frac{1 + t^2}{3y - y^2} \quad \text{and} \quad f_y = -\frac{(1 + t^2)(3 - 2y)}{(3y - y^2)^2},$$

we see that the hypotheses of Theorem 2.4.2 are satisfied as long as $y \neq 0, 3$.

13.(a) We know that the family of solutions given by equation (19) are solutions of this initial-value problem. We want to determine if one of these passes through the point $(1, 1)$. That is, we want to find $t_0 > 0$ such that if $y = [\frac{2}{3}(t - t_0)]^{3/2}$, then $(t, y) = (1, 1)$. That is, we need to find $t_0 > 0$ such that $1 = \frac{2}{3}(1 - t_0)$. But, the solution of this equation is $t_0 = -1/2$. Therefore the solution does not pass through $(1, 1)$.

(b) From the analysis in part (a), we find a solution passing through $(2, 1)$ by solving $1 = \frac{2}{3}(2 - t_0)$. We obtain $t_0 = 1/2$, and the solution is $y = [\frac{2}{3}(t - 1/2)]^{3/2}$.

(c) Since we need $y_0 = \pm[\frac{2}{3}(2 - t_0)]^{3/2}$, we must have $|y_0| \leq (\frac{4}{3})^{3/2}$.

15. The equation is separable, $y\,dy = -4t\,dt$. Integrating both sides, we conclude that $y^2/2 = -2t^2 + y_0^2/2$ for $y_0 \neq 0$. The solution is defined for $y_0^2 - 4t^2 \geq 0$.

17. The equation is separable and can be written as $dy/y^3 = -dt$. Integrating both sides, we arrive at the solution $y = y_0/(\sqrt{2ty_0^2 + 1})$. Solutions exist as long as $2y_0t + 1 > 0$. If $y_0 \neq 0$, the solution exists for $t > -\frac{1}{2y_0^2}$ and if $y_0 = 0$, $y(t) = 0$ for all t.

19.

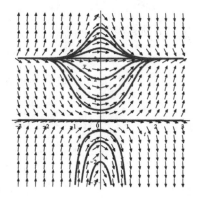

If $y_0 > 0$, then $y \to 3$. If $y_0 = 0$, then $y = 0$. If $y_0 < 0$, then $y \to -\infty$.

21.

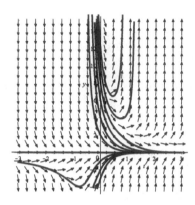

If $y_0 > 9$, then $y \to \infty$. If $y_0 \leq 9$, then $y \to 0$.

23.(a) $\phi(t) = e^{2t}$, thus $\phi' = 2e^{2t}$. Therefore, $\phi' - 2\phi = 0$. Since $(c\phi)' = c\phi'$, we see that $(c\phi)' - 2c\phi = 0$. Therefore, $c\phi$ is also a solution.

(b) $\phi(t) = 1/t$, thus $\phi' = -1/t^2$. Therefore, $\phi' + \phi^2 = 0$. If $y = c/t$, then $y' = -c/t^2$. Therefore, $y' + y^2 = -c/t^2 + c^2/t^2 = 0$ if and only if $c^2 - c = 0$; that is, if $c = 0$ or $c = 1$.

25. Let $y = y_1 + y_2$, then $y' + p(t)y = y_1' + y_2' + p(t)(y_1 + y_2) = y_1' + p(t)y_1 + y_2' + p(t)y_2 = 0$.

27. The solution of the initial value problem $y' + 2y = 1$ is $y = 1/2 + ce^{-2t}$. For $y(0) = 0$, we see that $c = -1/2$. Therefore, $y(t) = \frac{1}{2}(1 - e^{-2t})$ for $0 \leq t \leq 1$. Then $y(1) = \frac{1}{2}(1 - e^{-2})$. Next, the solution of $y' + 2y = 0$ is given by $y = ce^{-2t}$. The initial condition $y(1) = \frac{1}{2}(1 - e^{-2})$ implies $ce^{-2} = \frac{1}{2}(1 - e^{-2})$. Therefore, $c = \frac{1}{2}(e^2 - 1)$ and we conclude that $y(t) = \frac{1}{2}(e^2 - 1)e^{-2t}$ for $t > 1$.

29.(a) Multiplying the equation by $e^{\int_{t_0}^t p(s)\,ds}$, we have

$$\left(e^{\int_{t_0}^t p(s)\,ds}y\right)' = e^{\int_{t_0}^t p(s)\,ds}g(t).$$

Integrating this we obtain

$$e^{\int_{t_0}^t p(s)\,ds}y(t) = y_0 + \int_{t_0}^t e^{\int_{t_0}^s p(r)\,dr}g(s)\,ds,$$

which implies

$$y(t) = y_0 e^{-\int_{t_0}^t p(s)\,ds} + \int_{t_0}^t e^{-\int_s^t p(r)\,dr}g(s)\,ds.$$

(b) Assume $p(t) \geq p_0 > 0$ for all $t \geq t_0$ and $|g(t)| \leq M$ for all $t \geq t_0$. Therefore,

$$\int_{t_0}^t p(s)\,ds \geq \int_{t_0}^t p_0\,ds = p_0(t - t_0),$$

which implies

$$e^{-\int_{t_0}^t p(s)\,ds} \leq e^{-\int_{t_0}^t p_0\,ds} = e^{-p_0(t - t_0)} \leq 1 \text{ for } t \geq t_0.$$

Also,

$$\int_{t_0}^t e^{-\int_s^t p(r)\,dr} g(s)\,ds \ \le\ \int_{t_0}^t e^{-\int_s^t p(r)\,dr}|g(s)|\,ds \le \int_{t_0}^t e^{-p_0(t-s)} M\,ds$$

$$\le\ M\left.\frac{e^{-p_0(t-s)}}{p_0}\right|_{t_0}^t = M\left[\frac{1}{p_0} - \frac{e^{-p_0(t-t_0)}}{p_0}\right] \le \frac{M}{p_0}.$$

(c) Let $p(t) = 2t + 1 \ge 1$ for all $t \ge 0$ and let $g(t) = e^{-t^2}$. Therefore, $|g(t)| \le 1$ for all $t \ge 0$. By the answer to part (a),

$$y(t) = e^{-\int_0^t (2s+1)\,ds} + \int_0^t e^{-\int_s^t (2r+1)\,dr} e^{-s^2}\,ds = e^{-(t^2+t)} + e^{-t^2-t}\int_0^t e^s\,ds = e^{-t^2}.$$

We see that y satisfies the property that y is bounded for all time $t \ge 0$.

2.5 Autonomous Equations and Population Dynamics

1.(a) The equation is separable. Using partial fractions, it can be written as

$$\left(\frac{1}{y} + \frac{1/K}{1 - y/K}\right)dy = r\,dt.$$

Integrating both sides and using the initial condition $y_0 = K/3$, we know the solution y satisfies

$$\ln\left|\frac{y}{1 - y/K}\right| = rt + \ln\left|\frac{K}{2}\right|.$$

To find the time τ such that $y = 2y_0 = 2K/3$, we substitute $y = 2K/3$ and $t = \tau$ into the equation above. Using the properties of logarithmic functions, we conclude that $\tau = (\ln 4)/r$. If $r = 0.025$, then $\tau \approx 55.452$ years.

(b) Using the analysis from part (a), we know the general solution satisfies

$$\ln\left|\frac{y}{1 - y/K}\right| = rt + c.$$

The initial condition $y_0 = \alpha K$ implies $c = \ln|\alpha K/(1 - \alpha)|$. Therefore,

$$\ln\left|\frac{y}{1 - y/K}\right| = rt + \ln\left|\frac{\alpha K}{1 - \alpha}\right|.$$

In order to find the time T at which $y(T) = \beta K$, we use the equation above. We conclude that

$$T = (1/r)\ln|\beta(1 - \alpha)/\alpha(1 - \beta)|.$$

When $\alpha = 0.1$, $\beta = 0.9$, $r = 0.025$, $\tau \approx 175.78$ years.

3.(a) Using the substitution $u = \ln(y/K)$ and differentiating both sides with respect to t, we conclude that $u' = y'/y$. Substitution into the Gompertz equation yields $u' = -ru$. The solution of this equation is $u = u_0 e^{-rt}$. Therefore,

$$\frac{y}{K} = \exp[\ln(y_0/K)e^{-rt}].$$

(b) For $K = 80.5 \times 10^6$, $y_0/K = 0.25$ and $r = 0.71$, we conclude that $y(2) \approx 57.58 \times 10^6$.

(c) Solving the equation in part (a) for t, we see that

$$t = -\frac{1}{r}\ln\left[\frac{\ln(y/K)}{\ln(y_0/K)}\right].$$

Plugging in the given values, we conclude that $\tau \approx 2.21$ years.

5.(a) The rate of increase of the volume is given by

$$\frac{dV}{dt} = k - \alpha a\sqrt{2gh}.$$

Since the cross-section is constant, $dV/dt = A dh/dt$. Therefore,

$$\frac{dh}{dt} = (k - \alpha a\sqrt{2gh})/A.$$

(b) Setting $dh/dt = 0$, we conclude that the equilibrium height of water is

$$h_e = \frac{1}{2g}\left(\frac{k}{\alpha a}\right)^2.$$

Since $f'(h_e) < 0$, the equilibrium height is stable.

7.(a) The solution of the separable equation is $y(t) = y_0 e^{-\beta t}$.

(b) Using the result from part (a), we see that $dx/dt = -\alpha x y_0 e^{-\beta t}$. This equation is separable with solution $x(t) = x_0 \exp[-\alpha y_0(1 - e^{-\beta t})/\beta]$.

(c) As $t \to \infty$, $y \to 0$ and $x \to x_0 \exp(-\alpha y_0/\beta)$.

9.(a) Since the critical points are $x^* = p, q$, we will look at their stability. Since $f'(x) = -\alpha q - \alpha p + 2\alpha x^2$, we see that $f'(p) = \alpha(2p^2 - q - p)$ and $f'(q) = \alpha(2q^2 - q - p)$. Now if $p > q$, then $-p < -q$, and, therefore, $f'(q) = \alpha(2q^2 - q - p) < \alpha(2q^2 - 2q) = 2\alpha q(q - 1) < 0$ since $0 < q < 1$. Therefore, if $p > q$, $f'(q) < 0$, and, therefore, $x^* = q$ is asymptotically stable. Similarly, if $p < q$, then $x^* = p$ is asymptotically stable, and therefore, we can conclude that $x(t) \to \min\{p, q\}$ as $t \to \infty$.
The equation is separable. It can be solved by using partial fractions as follows. We can rewrite the equation as

$$\left(\frac{1/(q-p)}{p-x} + \frac{1/(p-q)}{q-x}\right)dx = \alpha dt,$$

which implies

$$\ln\left|\frac{p-x}{q-x}\right| = (p-q)\alpha t + C.$$

The initial condition $x_0 = 0$ implies $C = \ln|p/q|$, and, therefore,

$$\ln\left|\frac{q(p-x)}{p(q-x)}\right| = (p-q)\alpha t.$$

Applying the exponential function and simplifying, we conclude that

$$x(t) = \frac{pq(e^{(p-q)\alpha t} - 1)}{pe^{(p-q)\alpha t} - q}.$$

(b) In this case, $x^* = p$ is the only critical point. Since $f(x) = \alpha(p-x)^2$ is concave up, we conclude that $x^* = p$ is semistable. Further, if $x_0 = 0$, we can conclude that $x \to p$ as $t \to \infty$. The phase line is shown below.

This equation is separable. Its solution is given by

$$x(t) = \frac{p^2\alpha t}{p\alpha t + 1}.$$

11.(a) First, for $a < 0$, the only critical point is $y^* = 0$. Second, for $a = 0$, the only critical point is $y^* = 0$. Third, for $a > 0$, the critical points are at $y^* = 0, \pm\sqrt{a}$. Here, $f'(y) = a - 3y^2$. If $a < 0$, then $f'(y) < 0$ for all y, and, therefore, $y^* = 0$ will be asymptotically stable. If $a = 0$, then $f'(0) = 0$. From the behavior on either side of $y^* = 0$, we see that $y^* = 0$ will be asymptotically stable. If $a > 0$, then $f'(0) = a > 0$ which implies that $y^* = 0$ is unstable for $a > 0$. Further, $f'(\pm\sqrt{a}) = -2a < 0$. Therefore, $y^* = \pm\sqrt{a}$ are asymptotically stable for $a > 0$. The phase lines are shown below.

$$a < 0 \qquad\qquad a = 0 \qquad\qquad a > 0$$

(b) Below, we graph solutions in the case $a = -1$, $a = 0$, and $a = 1$, respectively.

(c)

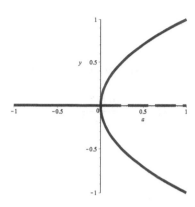

2.6   Exact Equations and Integrating Factors

1.(a) Here $M(x,y) = 2x + 3$ and $N(x,y) = 2y - 2$. Since $M_y = N_x = 0$, the equation is exact.

(b) Since $\psi_x = M = 2x + 3$, to solve for ψ, we integrate M with respect to x. We conclude that $\psi = x^2 + 3x + h(y)$. Then $\psi_y = h'(y) = N = 2y - 2$ implies $h(y) = y^2 - 2y$. Therefore, $\psi(x,y) = x^2 + 3x + y^2 - 2y = c$.

(c)

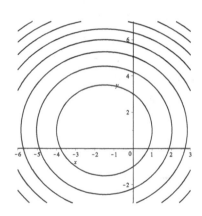

3.(a) Here $M(x,y) = 3x^2 - 2xy + 2$ and $N(x,y) = 6y^2 - x^2 + 3$. Since $M_y = -2x = N_x$, the equation is exact.

(b) Since $\psi_x = M = 3x^2 - 2xy + 2$, to solve for ψ, we integrate M with respect to x. We conclude that $\psi = x^3 - x^2y + 2x + h(y)$. Then $\psi_y = -x^2 + h'(y) = N = 6y^2 - x^2 + 3$ implies $h'(y) = 6y^2 + 3$. Therefore, $h(y) = 2y^3 + 3y$ and $\psi(x,y) = x^3 - x^2y + 2x + 2y^3 + 3y = c$.

(c)

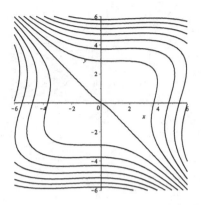

5.(a) Here $M(x,y) = 4x + 2y$ and $N(x,y) = 2x + 3y$. Since $M_y = 2 = N_x$, the equation is exact.

(b) Since $\psi_x = M = 4x + 2y$, to solve for ψ, we integrate M with respect to x. We conclude that $\psi = 2x^2 + 2xy + h(y)$. Then $\psi_y = 2x + h'(y) = N = 2x + 3y$ implies $h'(y) = 3y$. Therefore, $h(y) = 3y^2/2$ and $\psi(x,y) = 2x^2 + 2xy + 3y^2/2 = k$.

(c)

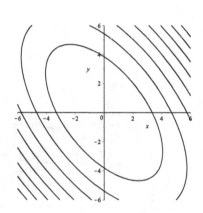

7.(a) Here $M(x,y) = e^x \sin y - 2y \sin x$ and $N(x,y) = e^x \cos y + 2 \cos x$. Since $M_y = e^x \cos y - \sin x = N_x$, the equation is exact.

(b) Since $\psi_x = M = e^x \sin y - 2y \sin x$, to solve for ψ, we integrate M with respect to x. We conclude that $\psi = e^x \sin y + 2y \cos x + h(y)$. Then $\psi_y = e^x \cos y + 2 \cos x + h'(y) = N = e^x \cos y + 2 \cos x$ implies $h'(y) = 0$. Therefore, $h(y) = c$ and $\psi(x,y) = e^x \sin y + 2y \cos x = c$.

(c)

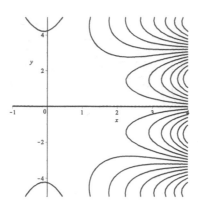

9.(a) Here $M(x,y) = ye^{xy}\cos 2x - 2e^{xy}\sin 2x + 2x$ and $N(x,y) = xe^{xy}\cos 2x - 3$. Since $M_y = e^{xy}\cos 2x + xye^{xy}\cos 2x - 2xe^{xy}\sin 2x = N_x$, the equation is exact.

(b) Since $\psi_x = M = ye^{xy}\cos 2x - 2e^{xy}\sin 2x + 2x$, to solve for ψ, we integrate M with respect to x. We conclude that $\psi = e^{xy}\cos 2x + x^2 + h(y)$. Then $\psi_y = xe^{xy}\cos 2x + h'(y) = N = xe^{xy}\cos 2x - 3$ implies $h'(y) = -3$. Therefore, $h(y) = -3y$ and $\psi(x,y) = e^{xy}\cos 2x + x^2 - 3y = c$.

(c)

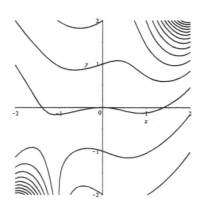

11.(a) Here $M(x,y) = x\ln y + xy$ and $N(x,y) = y\ln x + xy$. Since $M_y = x/y + x$ and $N_x = y/x + y$, we conclude that the equation is not exact.

13. Here $M(x,y) = 2x - y$ and $N(x,y) = 2y - x$. Therefore, $M_y = N_x = -1$ which implies that the equation is exact. Integrating M with respect to x, we conclude that $\psi = x^2 - xy + h(y)$. Then $\psi_y = -x + h'(y) = N = 2y - x$ implies $h'(y) = 2y$. Therefore, $h(y) = y^2$ and we conclude that $\psi = x^2 - xy + y^2 = c$. The initial condition $y(1) = 3$ implies $c = 7$. Therefore, $x^2 - xy + y^2 = 7$. Solving for y, we conclude that $y = (x + \sqrt{28 - 3x^2})/2$. Therefore, the solution is valid for $3x^2 \le 28$, i.e. for $-\sqrt{28/3} < x < \sqrt{28/3}$.

15. Here $M(x,y) = xy^2 + bx^2y$ and $N(x,y) = x^3 + x^2y$. Therefore, $M_y = 2xy + bx^2$ and $N_x = 3x^2 + 2xy$. In order for the equation to be exact, we need $b = 3$. Taking this value for b, we integrating M with respect to x. We conclude that $\psi = x^2y^2/2 + x^3y + h(y)$. Then $\psi_y = x^2y + x^3 + h'(y) = N = x^3 + x^2y$ implies $h'(y) = 0$. Therefore, $h(y) = c$ and $\psi(x,y) = x^2y^2/2 + x^3y = c$. That is, the solution is given implicitly as $x^2y^2 + 2x^3y = k$.

17. We notice that $\psi(x,y) = f(x) + g(y)$. Therefore, $\psi_x = f'(x)$ and $\psi_y = g'(y)$. That is, $\psi_x = M(x, y_0)$, and $\psi_y = N(x_0, y)$. Furthermore, $\psi_{xy} = M_y$ and $\psi_{yx} = N_x$. Based on the hypothesis, $\psi_{xy} = \psi_{yx}$ and $M_y = N_x$.

19.(a) Here $M(x,y) = x^2 y^3$ and $N(x,y) = x + xy^2$. Therefore, $M_y = 3x^2 y^2$ and $N_x = 1 + y^2$. We see that the equation is not exact. Now, multiplying the equation by $\mu(x,y) = 1/xy^3$, the equation becomes $x\,dx + (1+y^2)/y^3\,dy = 0$. Now we see that for this equation $M = x$ and $N = (1+y^2)/y^3$. Therefore, $M_y = 0 = N_x$.

(b) Integrating M with respect to x, we see that $\psi = x^2/2 + h(y)$. Further, $\psi_y = h'(y) = N = (1+y^2)/y^3 = 1/y^3 + 1/y$. Therefore, $h(y) = -1/2y^2 + \ln y$ and we conclude that the solution of the equation is given implicitly by $x^2 - 1/y^2 + 2\ln y = c$ and $y = 0$.

(c)

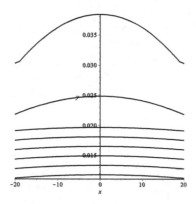

21.(a) We see that $M_y = 1$ while $N_x = 2$. Therefore, $M_y \neq N_x$. However, multiplying the equation by $\mu(x,y) = y$, the equation becomes $y^2\,dx + (2xy - y^2 e^y)\,dy = 0$. Now we see that for this equation $M = y^2$ and $N = 2xy - y^2 e^y$. Therefore, $M_y = 2y = N_x$.

(b) Integrating M with respect to x, we see that $\psi = xy^2 + h(y)$. Further, $\psi_y = 2xy + h'(y) = N = 2xy - y^2 e^y$. Therefore, $h'(y) = -y^2 e^y$ which implies that $h(y) = -e^y(y^2 - 2y + 2)$, and we conclude that the solution of the equation is given implicitly by $xy^2 - e^y(y^2 - 2y + 2) = c$.

(c)

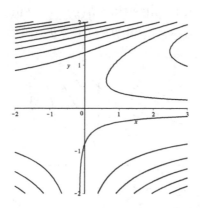

23. Suppose μ is an integrating factor which will make the equation exact. Then multiplying the equation by μ, we have $\mu M\,dx + \mu N\,dy = 0$. Then we need $(\mu M)_y = (\mu N)_x$. That is, we

need $\mu_y M + \mu M_y = \mu_x N + \mu N_x$. Then we rewrite the equation as $\mu(N_x - M_y) = \mu_y M - \mu_x N$. Suppose μ does not depend on x. Then $\mu_x = 0$. Therefore, $\mu(N_x - M_y) = \mu_y M$. Using the assumption that $(N_x - M_y)/M = Q(y)$, we can find an integrating factor μ by choosing μ which satisfies $\mu_y/\mu = Q$. We conclude that $\mu(y) = \exp \int Q(y)\, dy$ is an integrating factor of the differential equation.

25.(a) Since $(M_y - N_x)/N = 3$ is a function of x only, we know that $\mu = e^{3x}$ is an integrating factor for this equation. Multiplying the equation by μ, we have

$$e^{3x}(3x^2 y + 2xy + y^3)dx + e^{3x}(x^2 + y^2)dy = 0.$$

Then $M_y = e^{3x}(3x^2 + 2x + 3y^2) = N_x$. Therefore, this new equation is exact. Integrating M with respect to x, we conclude that $\psi = (x^2 y + y^3/3)e^{3x} + h(y)$. Then $\psi_y = (x^2 + y^2)e^{3x} + h'(y) = N = e^{3x}(x^2 + y^2)$. Therefore, $h'(y) = 0$ and we conclude that the solution is given implicitly by $(3x^2 y + y^3)e^{3x} = c$.

(b)

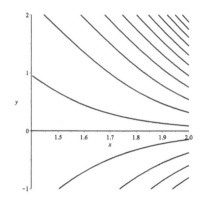

27.(a) Since $(N_x - M_y)/M = 1/y$ is a function of y only, we know that $\mu(y) = e^{\int 1/y\, dy} = y$ is an integrating factor for this equation. Multiplying the equation by μ, we have

$$y dx + (x - y \sin y)dy = 0.$$

Then for this equation, $M_y = 1 = N_x$. Therefore, this new equation is exact. Integrating M with respect to x, we conclude that $\psi = xy + h(y)$. Then $\psi_y = x + h'(y) = N = x - y \sin y$. Therefore, $h'(y) = -y \sin y$ which implies that $h(y) = -\sin y + y \cos y$, and we conclude that the solution is given implicitly by $xy - \sin y + y \cos y = c$.

(b)

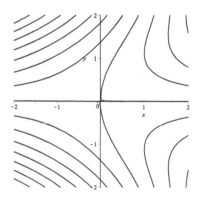

29.(a) Since $(N_x - M_y)/M = \cot y$ is a function of y only, we know that $\mu(y) = e^{\int \cot(y)\,dy} = \sin y$ is an integrating factor for this equation. Multiplying the equation by μ, we have

$$e^x \sin y\,dx + (e^x \cos y + 2y)\,dy = 0.$$

Then for this equation, $M_y = N_x$. Therefore, this new equation is exact. Integrating M with respect to x, we conclude that $\psi = e^x \sin y + h(y)$. Then $\psi_y = e^x \cos y + h'(y) = N = e^x \cos y + 2y$. Therefore, $h'(y) = 2y$ which implies that $h(y) = y^2$, and we conclude that the solution is given implicitly by $e^x siny + y^2 = c$.

(b)

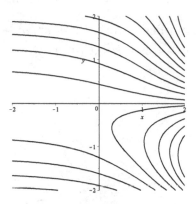

31. Since $(N_x - M_y)/(xM - yN) = 1/xy$ is a function of xy only, we know that $\mu(xy) = e^{\int 1/xy\,dy} = xy$ is an integrating factor for this equation. Multiplying the equation by μ, we have

$$(3x^2 y + 6x)\,dx + (x^3 + 3y^2)\,dy = 0.$$

Then for this equation, $M_y = N_x$. Therefore, this new equation is exact. Integrating M with respect to x, we conclude that $\psi = x^3 y + 3x^2 + h(y)$. Then $\psi_y = x^3 + h'(y) = N = x^3 + 3y^2$. Therefore, $h'(y) = 3y^2$ which implies that $h(y) = y^3$, and we conclude that the solution is given implicitly by $x^3 y + 3x^2 + y^3 = c$.

(b)

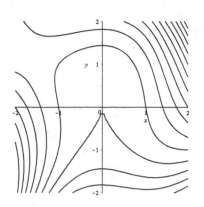

2.7   Substitution Methods

1.(a) $f(x, y) = (x + 1)/y$, thus $f(\lambda x, \lambda y) = (\lambda x + 1)/\lambda y \neq (x + 1)/y$. The equation is not homogeneous.

3.(a) $f(x, y) = (3x^2 y + y^3)/(3x^3 - xy^2)$ satisfies $f(\lambda x, \lambda y) = f(x, y)$. The equation is homogeneous.

(b) The equation is $y' = (3x^2 y + y^3)/(3x^3 - xy^2) = (3(y/x) + (y/x)^3)/(3 - (y/x)^2)$. Let $y = ux$. Then $y' = u'x + u$, thus $u'x = (3u + u^3)/(3 - u^2) - u = 2u^3/(3 - u^2)$. We obtain

$$\int \frac{3 - u^2}{2u^3} \, du = -\frac{3}{4}u^{-2} - \frac{1}{2}\ln|u| = \int \frac{1}{x} \, dx = \ln|x| + c.$$

Therefore, the solution is given implicitly by $-(3/4)x^2/y^2 - (1/2)\ln|y/x| = \ln|x| + c$. Also, $u = 0$ solves the equation, thus $y = 0$ is a solution as well.

(c)

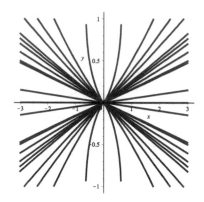

5.(a) $f(x, y) = (\sqrt{x^2 - y^2} + y)/x$ satisfies $f(\lambda x, \lambda y) = f(x, y)$. The equation is homogeneous.

(b) The equation is $y' = (\sqrt{x^2 - y^2} + y)/x = \sqrt{1 - (y/x)^2} + y/x$. Let $y = ux$. Then $y' = u'x + u$, thus $u'x = \sqrt{1 - u^2} + u - u = \sqrt{1 - u^2}$. We obtain

$$\int \frac{1}{\sqrt{1 - u^2}} \, du = \arcsin u = \int \frac{1}{x} \, dx = \ln|x| + c.$$

Therefore, the solution is given implicitly by $\arcsin(y/x) = \ln|x| + c$, thus $y = x\sin(\ln|x| + c)$. Also, $y = x$ and $y = -x$ are solutions.

(c)

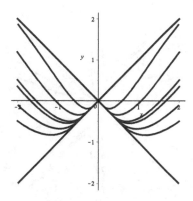

7.(a) $f(x, y) = (4y - 7x)/(5x - y)$ satisfies $f(\lambda x, \lambda y) = f(x, y)$. The equation is homogeneous.

(b) The equation is $y' = (4y - 7x)/(5x - y) = (4y/x - 7)/(5 - y/x)$. Let $y = ux$. Then $y' = u'x + u$, thus $u'x = (4u - 7)/(5 - u) - u = (u^2 - u - 7)/(5 - u)$. We obtain

$$\int \frac{5 - u}{u^2 - u - 7} \, du = \frac{9\sqrt{29} - 29}{58} \ln|1 + \sqrt{29} - 2u| - \frac{29 + 9\sqrt{29}}{58} \ln|-1 + \sqrt{29} + 2u| = \ln|x| + c.$$

The solution is given implicitly by substituting back $u = y/x$. Also, $y = x(1 \pm \sqrt{29})/2$ are solutions.

(c)

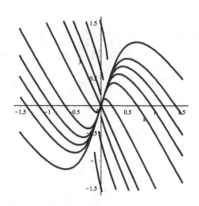

9.(a) $f(x, y) = (y^4 + 2xy^3 - 3x^2y^2 - 2x^3y)/(2x^2y^2 - 2x^3y - 2x^4)$ satisfies $f(\lambda x, \lambda y) = f(x, y)$. The equation is homogeneous.

(b) The equation is $y' = (y^4 + 2xy^3 - 3x^2y^2 - 2x^3y)/(2x^2y^2 - 2x^3y - 2x^4) = ((y/x)^4 + 2(y/x)^3 - 3(y/x)^2 - 2(y/x))/(2(y/x)^2 - 2(y/x) - 2)$. Let $y = ux$. Then $y' = u'x + u$, thus $u'x = (u^4 + 2u^3 - 3u^2 - 2u)/(2u^2 - 2u - 2) - u = (u^4 - u^2)/(2u^2 - 2u - 2)$. We obtain

$$\int \frac{2u^2 - 2u - 2}{u^4 - u^2} \, du = -\frac{2}{u} + 2\ln|u| - \ln|1 - u^2| = \ln|x| + c.$$

The solution is given implicitly by $\ln|1 - y^2/x^2| + 2x/y + \ln|x| = c$. Also, $y = x$, $y = -x$ and $y = 0$ are solutions.

(c)

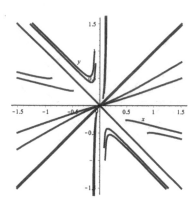

11. The equation is homogeneous. Let $y = ux$; we obtain $y' = u'x + u = 1/u + u$, thus $udu = (1/x)dx$ and we obtain $(y/x)^2 = u^2 = 2(\ln x + c)$. The initial condition gives $1/4 = 2(\ln 2 + c)$, thus $c = 1/8 - \ln 2$ and the solution is $y = x\sqrt{2\ln x + 1/4 - \ln 4}$. The solution exists on the interval $(2e^{-1/8}, \infty)$.

13.(a) $y' + (1/t)y = ty^2$

(b) Here $n = 2$, thus we set $u = y^{-1}$. The equation becomes $u' - (1/t)u = -t$; the integrating factor is $\mu = 1/t$ and we obtain $(u/t)' = -1$. After integration, we get $u/t = -t + c$, thus $u = -t^2 + ct$ and then $y = 1/(ct - t^2)$. Also, $y = 0$ is a solution.

(c)

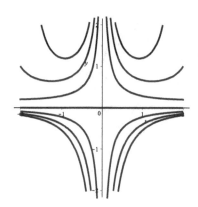

15.(a) $y' + (3/t)y = t^2y^2$

(b) Here $n = 2$, thus we set $u = y^{-1}$. The equation becomes $u' - (3/t)u = -t^2$; the integrating factor is $\mu = 1/t^3$ and we obtain $(u/t^3)' = -1/t$. After integration, we get $u/t^3 = -\ln t + c$, thus $u = -t^3\ln t + ct^3$ and then $y = 1/(ct^3 - t^3\ln t)$. Also, $y = 0$ is a solution.

(c)

17.(a) $y' + (4t/5(1+t^2))y = (4t/5(1+t^2))y^4$

(b) Here $n = 4$, thus we set $u = y^{-3}$. The equation becomes $u' - 12t/5(1+t^2)u = -12t/5(1+t^2)$; the integrating factor is $\mu = (1+t^2)^{-6/5}$ and we obtain $(u\mu)' = -12t(1+t^2)^{-11/5}/5$. After integration, $u = 1 + c(1+t^2)^{6/5}$, thus $y = (1 + c(1+t^2)^{6/5})^{-1/3}$. Also, $y = 0$ is a solution.

(c)

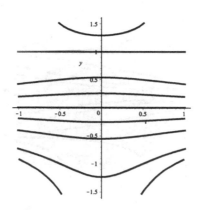

19.(a) $y' - y = y^{1/2}$

(b) Here $n = 1/2$, thus we set $u = y^{1/2}$. The equation becomes $u' - u/2 = 1/2$; the integrating factor is $\mu = e^{-t/2}$ and we obtain $(ue^{-t/2})' = e^{-t/2}/2$. After integration, $ue^{-t/2} = -e^{-t/2} + c$, thus $u = ce^{t/2} - 1$ and then $y = (ce^{t/2} - 1)^2$. Also, $y = 0$ is a solution.

(c)

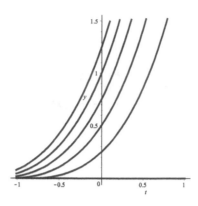

21.(a) $y' - \epsilon y = -\sigma y^3$

(b) Here $n = 3$. Therefore, $u = y^{-2}$ satisfies $u' + 2\epsilon u = 2\sigma$. This equation is linear with integrating factor $e^{2\epsilon t}$. Its solution is given by $u = (\sigma + c\epsilon e^{-2\epsilon t})/\epsilon$. Then, using the fact that $y^2 = 1/u$, we see that $y = \pm\sqrt{\epsilon}/\sqrt{\sigma + c\epsilon e^{-2\epsilon t}}$.

(c) The figure shows the solutions for $\epsilon = 1$, $\sigma = 1$.

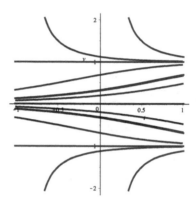

23.(a) Assume y_1 solves the equation: $y_1' = A + By_1 + Cy_1^2$. Let $y = y_1 + v$; we obtain $y_1' + v' = y' = A + By + Cy^2 = A + B(y_1 + v) + C(y_1 + v)^2 = A + By_1 + Bv + Cy_1^2 + 2Cy_1v + Cv^2$. Then $v' = Bv + 2Cy_1v + Cv^2$, i.e. $v' - (B + 2Cy_1)v = Cv^2$ which is a Bernoulli equation with $n = 2$.

(b) If $y_1 = 4t$, then $y_1' = 4$ and $4 + 3t \cdot 4t = 4 - 4t^2 + (4t)^2$. Using the previous idea, let $y = y_1 + v$; we obtain $4 + v' = y_1' + v' = y' = 4 - 4t^2 + y^2 - 3ty = 4 - 4t^2 + (4t + v)^2 - 3t(4t + v)$, i.e. $v' = -4t^2 + 16t^2 + 8tv + v^2 - 12t^2 - 3tv = 5tv + v^2$. Let $u = v^{-1}$, then we obtain $u' + 5tu = -1$. The integrating factor is $\mu = e^{5t^2/2}$, and we obtain $u = -e^{-5t^2/2}\int_0^t e^{5s^2/2}\,ds$. Thus $y = 4t - (e^{-5t^2/2}\int_0^t e^{5s^2/2}\,ds)^{-1}$.

25.(a) Linear.

(b) Consider the equation so that $x = x(y)$. Then $dx/dy = -2x + 3e^y$; the integrating factor is $\mu = e^{2y}$, we obtain $(e^{2y}x)' = 3e^{3y}$. After integration, $e^{2y}x = e^{3y} + c$, thus $x = e^y + ce^{-2y}$.

27.(a) Linear.

(b) The integrating factor is $\mu = e^{x + \ln x} = xe^x$; the equation turns into $(xe^x y)' = xe^x$, after integration $xe^x y = xe^x - e^x + c$, and then $y = 1 - 1/x + ce^{-x}/x$.

29.(a) Separable, linear.

(b) Separation of variables gives $dy/y = dx/\sqrt{x}$; after integration, we get $\ln y = 2\sqrt{x} + c$ and then $y = ce^{2\sqrt{x}}$.

31.(a) Exact, Bernoulli.

(b) Write $(y^2 + 1 + \ln x)dx + 2xy \, dy = 0$. We need $\psi(x, y)$ so that $\psi_x = y^2 + 1 + \ln x$; thus $\psi(x, y) = y^2 x + x \ln x + h(y)$. Now $\psi_y = 2xy + h'(y) = 2xy$, thus we obtain that the solution is given implicitly as $y^2 x + x \ln x = c$.

33.(a) Separable, autonomous (if viewed as dx/dy).

(b) $dy/dx = -x/\ln x$, thus after integration, $y = -\ln \ln x + C$.

35.(a) Bernoulli, homogeneous.

(b) Let $u = y^2$. Then $u' = 2yy' = 4x + (5/2x)y^2 = 4x + (5/2x)u$; we get the linear equation $u' - (5/2x)u = 4x$. The integrating factor is $\mu = x^{-5/2}$, and the equation turns into $(ux^{-5/2})' = 4x^{-3/2}$. After integration, we get $u = y^2 = -8x^2 + cx^{5/2}$.

Chapter 3

Systems of Two First Order Equations

3.1 Systems of Two Linear Algebraic Equations

1.(a)

$$A = \begin{pmatrix} 2 & 3 \\ -3 & 1 \end{pmatrix}, \quad \text{thus} \quad A^{-1} = \frac{1}{11} \begin{pmatrix} 1 & -3 \\ 3 & 2 \end{pmatrix}.$$

Therefore,

$$\mathbf{x} = A^{-1}\mathbf{b} = \frac{1}{11} \begin{pmatrix} 1 & -3 \\ 3 & 2 \end{pmatrix} \begin{pmatrix} 7 \\ -5 \end{pmatrix} = \begin{pmatrix} 2 \\ 1 \end{pmatrix}.$$

That is, $x_1 = 2, x_2 = 1$.

(b) The lines are intersecting, as shown below.

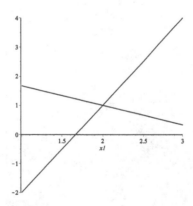

3.(a)

$$A = \begin{pmatrix} 1 & 3 \\ 2 & -1 \end{pmatrix}, \quad \text{thus} \quad A^{-1} = \frac{1}{7} \begin{pmatrix} 1 & 3 \\ 2 & -1 \end{pmatrix}.$$

Therefore,

$$\mathbf{x} = A^{-1}\mathbf{b} = \frac{1}{7} \begin{pmatrix} 1 & 3 \\ 2 & -1 \end{pmatrix} \begin{pmatrix} 0 \\ 0 \end{pmatrix} = \begin{pmatrix} 0 \\ 0 \end{pmatrix}.$$

That is, $x_1 = 0$, $x_2 = 0$.

(b) The lines are intersecting, as shown below.

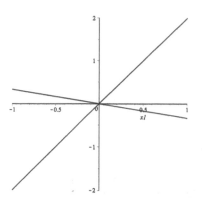

5.(a)

$$A = \begin{pmatrix} 2 & -3 \\ 1 & 2 \end{pmatrix}, \quad \text{thus} \quad A^{-1} = \frac{1}{7}\begin{pmatrix} 2 & 3 \\ -1 & 2 \end{pmatrix}.$$

Therefore,

$$\mathbf{x} = A^{-1}\mathbf{b} = \frac{1}{7}\begin{pmatrix} 2 & 3 \\ -1 & 2 \end{pmatrix}\begin{pmatrix} 4 \\ -5 \end{pmatrix} = \begin{pmatrix} -1 \\ -2 \end{pmatrix}.$$

That is, $x_1 = -1$, $x_2 = -2$.

(b) The lines are intersecting, as shown below.

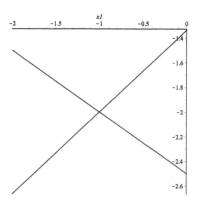

7.(a)

$$A = \begin{pmatrix} 2 & -3 \\ -4 & 6 \end{pmatrix},$$

$\det(A) = 0$, therefore, there are no solutions or an infinite number of solutions. Multiplying the first equation by -2, we see the two equations are equations for the same line. Therefore, there is an infinite number of solutions. If we set $x_1 = c$, then $x_2 = (2c - 6)/3$. This means that any pair (x_1, x_2) such that $x_2 = (2x_1 - 6)/3$ will satisfy the system of equations.

(b) The lines are coincident, as shown below.

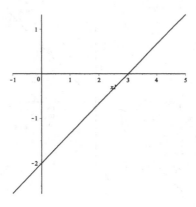

9.(a)

$$A = \begin{pmatrix} 1 & 4 \\ 4 & 1 \end{pmatrix}, \quad \text{thus} \quad A^{-1} = -\frac{1}{15} \begin{pmatrix} 1 & -4 \\ -4 & 1 \end{pmatrix}.$$

Therefore,

$$\mathbf{x} = A^{-1}\mathbf{b} = -\frac{1}{15} \begin{pmatrix} 1 & -4 \\ -4 & 1 \end{pmatrix} \begin{pmatrix} 10 \\ 10 \end{pmatrix} = \begin{pmatrix} 2 \\ 2 \end{pmatrix}.$$

That is, $x_1 = 2$, $x_2 = 2$.

(b) The lines are intersecting, as shown below.

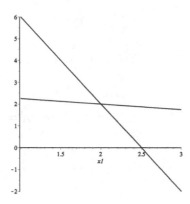

11.(a)

$$A = \begin{pmatrix} 4 & -3 \\ -2 & 5 \end{pmatrix}, \quad \text{thus} \quad A^{-1} = \frac{1}{14} \begin{pmatrix} 5 & 3 \\ 2 & 4 \end{pmatrix}.$$

Therefore,

$$\mathbf{x} = A^{-1}\mathbf{b} = \frac{1}{14} \begin{pmatrix} 5 & 3 \\ 2 & 4 \end{pmatrix} \begin{pmatrix} 0 \\ 0 \end{pmatrix} = \begin{pmatrix} 0 \\ 0 \end{pmatrix}.$$

That is, $x_1 = 0$, $x_2 = 0$.

(b) The lines are intersecting, as shown below.

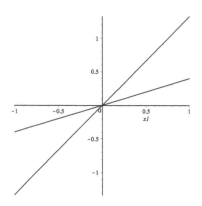

13. $\det(A - \lambda I) = \lambda^2 - \lambda - 2 = 0$, thus $\lambda = 2, -1$. First, $\lambda_1 = 2$ implies

$$A - \lambda_1 I = \begin{pmatrix} 1 & -2 \\ 2 & -4 \end{pmatrix}.$$

Therefore,

$$\mathbf{x}_1 = \begin{pmatrix} 2 \\ 1 \end{pmatrix}$$

is an eigenvector for λ_1. Second, $\lambda_2 = -1$ implies

$$A - \lambda_2 I = \begin{pmatrix} 4 & -2 \\ 2 & -1 \end{pmatrix}.$$

Therefore,

$$\mathbf{x}_2 = \begin{pmatrix} 1 \\ 2 \end{pmatrix}$$

is an eigenvector for λ_2.

15. $\det(A - \lambda I) = \lambda^2 - 2\lambda + 1 = 0$, thus $\lambda = 1$. Now, $\lambda = 1$ implies

$$A - \lambda_1 I = \begin{pmatrix} 2 & -4 \\ 1 & -2 \end{pmatrix}.$$

Therefore,

$$\mathbf{x}_1 = \begin{pmatrix} 2 \\ 1 \end{pmatrix}$$

is an eigenvector for λ.

17. $\det(A - \lambda I) = \lambda^2 + 2\lambda + 5 = 0$, thus $\lambda = -1 + 2i, -1 - 2i$. First, $\lambda_1 = -1 + 2i$ implies

$$A - \lambda_1 I = \begin{pmatrix} -2i & -4 \\ 1 & -2i \end{pmatrix}.$$

Therefore,

$$\mathbf{x}_1 = \begin{pmatrix} 2i \\ 1 \end{pmatrix}$$

is an eigenvector for λ_1. Second, since $\lambda_2 = -1 - 2i$ is the complex conjugate of λ_1,

$$\mathbf{x}_2 = \overline{\mathbf{x}}_1 = \begin{pmatrix} -2i \\ 1 \end{pmatrix}$$

is an eigenvector for λ_2.

19. $\det(A - \lambda I) = \lambda^2 + 2\lambda + 1 = 0$, thus $\lambda = -1$. Now, $\lambda = -1$ implies

$$A - \lambda_1 I = \begin{pmatrix} -1/2 & 1 \\ -1/4 & 1/2 \end{pmatrix}.$$

Therefore,

$$\mathbf{x}_1 = \begin{pmatrix} 1 \\ 1/2 \end{pmatrix}$$

is an eigenvector for λ.

21. $\det(A - \lambda I) = \lambda^2 + 1 = 0$, thus $\lambda = i, -i$. First, $\lambda_1 = i$ implies

$$A - \lambda_1 I = \begin{pmatrix} 2 - i & -5 \\ 1 & -2 - i \end{pmatrix}.$$

Therefore,

$$\mathbf{x}_1 = \begin{pmatrix} 2 + i \\ 1 \end{pmatrix}$$

is an eigenvector for λ_1. Second, since λ_2 is the complex conjugate of λ_1,

$$\mathbf{x}_2 = \overline{\mathbf{x}}_1 = \begin{pmatrix} 2 - i \\ 1 \end{pmatrix}$$

is an eigenvector for λ_2.

23. $\det(A - \lambda I) = \lambda^2 + \lambda - 6 = 0$, thus $\lambda = 2, -3$. First, $\lambda_1 = 2$ implies

$$A - \lambda_1 I = \begin{pmatrix} -1 & 1 \\ 4 & -4 \end{pmatrix}.$$

Therefore,

$$\mathbf{x}_1 = \begin{pmatrix} 1 \\ 1 \end{pmatrix}$$

is an eigenvector for λ_1. Second, $\lambda_2 = -3$ implies

$$A - \lambda_2 I = \begin{pmatrix} 4 & 1 \\ 4 & 1 \end{pmatrix}.$$

Therefore,

$$\mathbf{x}_2 = \begin{pmatrix} 1 \\ -4 \end{pmatrix}$$

is an eigenvector for λ_2.

25. $\det(A - \lambda I) = \lambda^2 + \lambda + 1/4 = 0$, thus $\lambda = -1/2$. Now, $\lambda = -1/2$ implies

$$A - \lambda_1 I = \begin{pmatrix} -5/2 & 5/2 \\ -5/2 & 5/2 \end{pmatrix}.$$

Therefore,

$$\mathbf{x}_1 = \begin{pmatrix} 1 \\ 1 \end{pmatrix}$$

is an eigenvector for λ.

27. $\det(A - \lambda I) = \lambda^2 + 2\lambda = 0$, thus $\lambda = 0, -2$. First, $\lambda_1 = 0$ implies

$$A - \lambda_1 I = \begin{pmatrix} 1 & 4/3 \\ -9/4 & -3 \end{pmatrix}.$$

Therefore,

$$\mathbf{x}_1 = \begin{pmatrix} -4/3 \\ 1 \end{pmatrix}$$

is an eigenvector for λ_1. Second, $\lambda_2 = -2$ implies

$$A - \lambda_2 I = \begin{pmatrix} 3 & 4/3 \\ -9/4 & -1 \end{pmatrix}.$$

Therefore,

$$\mathbf{x}_2 = \begin{pmatrix} -4/9 \\ 1 \end{pmatrix}$$

is an eigenvector for λ_2.

29. $\det(A - \lambda I) = \lambda^2 + 9 = 0$, thus $\lambda = \pm 3i$. First, $\lambda_1 = 3i$ implies

$$A - \lambda_1 I = \begin{pmatrix} 1 - 3i & 2 \\ -5 & -1 - 3i \end{pmatrix}.$$

Therefore,

$$\mathbf{x}_1 = \begin{pmatrix} 1 \\ \frac{-1+3i}{2} \end{pmatrix}$$

is an eigenvector for λ_1. Second, since λ_2 is the complex conjugate of λ_1,

$$\mathbf{x}_2 = \overline{\mathbf{x}}_1 = \begin{pmatrix} 1 \\ \frac{-1-3i}{2} \end{pmatrix}$$

is an eigenvector for λ_2.

31. $\det(A - \lambda I) = \lambda^2 - \frac{10}{4}\lambda + 1 = 0$, thus $\lambda = 2, 1/2$. First, $\lambda_1 = 2$ implies

$$A - \lambda_1 I = \begin{pmatrix} -3/4 & 3/4 \\ 3/4 & -3/4 \end{pmatrix}.$$

Therefore,

$$\mathbf{x}_1 = \begin{pmatrix} 1 \\ 1 \end{pmatrix}$$

is an eigenvector for λ_1. Second, $\lambda_2 = 1/2$ implies

$$A - \lambda_2 I = \begin{pmatrix} 3/4 & 3/4 \\ 3/4 & 3/4 \end{pmatrix}.$$

Therefore,

$$\mathbf{x}_2 = \begin{pmatrix} 1 \\ -1 \end{pmatrix}$$

is an eigenvector for λ_2.

33.(a) $\det(A - \lambda I) = \lambda^2 + \lambda - 6 - \alpha = 0$, thus $\lambda = (-1 \pm \sqrt{25 + 4\alpha})/2$.

(b) If $25 + 4\alpha > 0$, then there will be two real eigenvalues. If $25 + 4\alpha = 0$, then there will be one real eigenvalue. If $25 + 4\alpha < 0$, then there will be two eigenvalues with non-zero imaginary part (in particular, they will be a complex conjugate pair).

35.(a) $\det(A - \lambda I) = \lambda^2 - (\alpha + 1)\lambda + \alpha - 6 = 0$, thus $\lambda = (\alpha + 1 \pm \sqrt{\alpha^2 - 2\alpha + 25})/2$.

(b) Now $\alpha^2 - 2\alpha + 25 > 0$ for all α since $\alpha^2 - 2\alpha + 25 = (\alpha - 1)^2 + 24$. Therefore, there will be two real eigenvalues no matter what the value of α is.

37. Let

$$A = \begin{pmatrix} a_{11} & a_{12} \\ a_{21} & a_{22} \end{pmatrix},$$

and let

$$A^{-1} = \begin{pmatrix} b & c \\ d & e \end{pmatrix}.$$

Now $A^{-1}A = I$ implies

$$\begin{pmatrix} b & c \\ d & e \end{pmatrix} \begin{pmatrix} a_{11} & a_{12} \\ a_{21} & a_{22} \end{pmatrix} = \begin{pmatrix} 1 & 0 \\ 0 & 1 \end{pmatrix}.$$

Therefore, we have the following system of equations:

$$\begin{aligned} ba_{11} + ca_{21} &= 1 \\ ba_{12} + ca_{22} &= 0 \\ da_{11} + ea_{21} &= 0 \\ da_{12} + ea_{22} &= 1. \end{aligned}$$

Multiplying the first equation by a_{12}, the second equation by $-a_{11}$ and adding them implies $c(a_{21}a_{12} - a_{11}a_{22}) = a_{12}$. Therefore,

$$c = \frac{1}{a_{21}a_{12} - a_{11}a_{22}} a_{12} = -\frac{1}{\det(A)} a_{12}.$$

Plugging this value for c into the first equation, we are able to solve for b. In particular,

$$ba_{11} - \frac{a_{12}}{\det(A)}a_{21} = 1, \text{ thus } ba_{11} = \frac{a_{11}a_{22}}{\det(A)}.$$

Therefore,

$$b = \frac{a_{22}}{\det(A)}.$$

Similarly, multiplying the third equation by a_{12}, the fourth equation by $-a_{11}$ and adding them, we see that $e(a_{11}a_{22} - a_{21}a_{12}) = a_{11}$. Therefore,

$$e = \frac{a_{11}}{\det(A)}.$$

Plugging this value for e into the third equation, we see that

$$d = -\frac{a_{21}}{\det(A)}.$$

3.2   Systems of Two First Order Linear Differential Equations

1. The system is autonomous, nonhomogeneous. Let $\mathbf{u} = \begin{pmatrix} x \\ y \end{pmatrix}$. Then

$$\mathbf{u}' = \begin{pmatrix} 0 & 1 \\ 1 & 0 \end{pmatrix}\mathbf{u} + \begin{pmatrix} 0 \\ 4 \end{pmatrix}.$$

3. The system is nonautonomous, homogeneous. Let $\mathbf{u} = \begin{pmatrix} x \\ y \end{pmatrix}$. Then

$$\mathbf{u}' = \begin{pmatrix} -2t & 1 \\ 3 & -1 \end{pmatrix}\mathbf{u}.$$

5. The system is autonomous, homogeneous. Let $\mathbf{u} = \begin{pmatrix} x \\ y \end{pmatrix}$. Then

$$\mathbf{u}' = \begin{pmatrix} 3 & -1 \\ 1 & 2 \end{pmatrix}\mathbf{u}.$$

7. The system is nonautonomous, nonhomogeneous. Let $\mathbf{u} = \begin{pmatrix} x \\ y \end{pmatrix}$. Then

$$\mathbf{u}' = \begin{pmatrix} 1 & 1 \\ -2 & \sin t \end{pmatrix}\mathbf{u} + \begin{pmatrix} 4 \\ 0 \end{pmatrix}.$$

9.(a) We differentiate: $x' = 3e^t - 7e^{-t} = 2(3e^t + 7e^{-t}) - (3e^t + 21e^{-t}) = 2x - y$ and $y' = 3e^t - 21e^{-t} = 3(3e^t + 7e^{-t}) - 2(3e^t + 21e^{-t}) = 3x - 2y$. Also, $x(0) = 10$ and $x(0) = 24$.

(b)

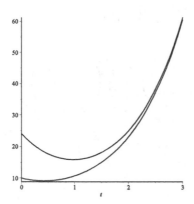

11.(a) We differentiate:

$$\mathbf{x}' = \begin{pmatrix} \cos t - \cos t + t \sin t \\ \sin t + t \cos t \end{pmatrix} = \begin{pmatrix} t \sin t \\ \sin t + t \cos t \end{pmatrix}.$$

Also,

$$\begin{pmatrix} 0 & 1 \\ -1 & 0 \end{pmatrix} \mathbf{x} + \begin{pmatrix} 0 \\ 2 \sin t \end{pmatrix} = \begin{pmatrix} t \sin t \\ \sin t + t \cos t \end{pmatrix}.$$

The initial condition is $\mathbf{x}(0) = \begin{pmatrix} 0 \\ 0 \end{pmatrix}$.

(b)

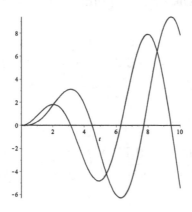

13. The second equation implies that at an equilibrium $u_1 = u_2$. Plugging this into the first equation and solving for u_1 we obtain that the only equilibrium is $u_1 = u_2 = T_a$.

15.(a) The system

$$\begin{aligned} x' &= -x + y + 1 = 0 \\ y' &= x + y - 3 = 0 \end{aligned}$$

can be rewritten in matrix form as

$$\begin{pmatrix} -1 & 1 \\ 1 & 1 \end{pmatrix} \begin{pmatrix} x \\ y \end{pmatrix} = \begin{pmatrix} -1 \\ 3 \end{pmatrix}.$$

Now

$$A = \begin{pmatrix} -1 & 1 \\ 1 & 1 \end{pmatrix}, \text{ thus } A^{-1} = \frac{1}{2} \begin{pmatrix} -1 & 1 \\ 1 & 1 \end{pmatrix}.$$

Therefore,

$$\begin{pmatrix} x \\ y \end{pmatrix} = A^{-1} \begin{pmatrix} -1 \\ 3 \end{pmatrix} = \frac{1}{2} \begin{pmatrix} -1 & 1 \\ 1 & 1 \end{pmatrix} \begin{pmatrix} -1 \\ 3 \end{pmatrix} = \begin{pmatrix} 2 \\ 1 \end{pmatrix}.$$

Therefore, the equilibrium solution is

$$\begin{pmatrix} 2 \\ 1 \end{pmatrix}.$$

(b)

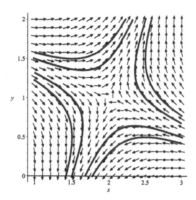

(c) Solutions in the vicinity of the critical point tend away from the critical point.

17.(a) The system

$$\begin{aligned} x' &= -0.25x - 0.75y + 8 = 0 \\ y' &= 0.5x + y - 11.5 = 0 \end{aligned}$$

can be rewritten in matrix form as

$$\begin{pmatrix} -0.25 & -0.75 \\ 0.5 & 1 \end{pmatrix} \begin{pmatrix} x \\ y \end{pmatrix} = \begin{pmatrix} -8 \\ 11.5 \end{pmatrix}.$$

Now

$$A = \begin{pmatrix} -0.25 & -0.75 \\ 0.5 & 1 \end{pmatrix}, \text{ thus } A^{-1} = 8 \begin{pmatrix} 1 & 3/4 \\ -1/2 & -1/4 \end{pmatrix}.$$

Therefore,

$$\begin{pmatrix} x \\ y \end{pmatrix} = A^{-1} \begin{pmatrix} -8 \\ 11.5 \end{pmatrix} = 8 \begin{pmatrix} 1 & 3/4 \\ -1/2 & -1/4 \end{pmatrix} \begin{pmatrix} -8 \\ 11.5 \end{pmatrix} = \begin{pmatrix} 5 \\ 9 \end{pmatrix}.$$

Therefore, the equilibrium solution is

$$\begin{pmatrix} 5 \\ 9 \end{pmatrix}.$$

(b)

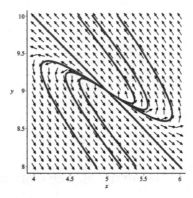

(c) Solutions in the vicinity of the critical point tend away from the critical point.

19.(a) The system

$$x' = x + y - 3 = 0$$
$$y' = -x + y + 1 = 0$$

can be rewritten in matrix form as

$$\begin{pmatrix} 1 & 1 \\ -1 & 1 \end{pmatrix} \begin{pmatrix} x \\ y \end{pmatrix} = \begin{pmatrix} 3 \\ -1 \end{pmatrix}.$$

Now

$$A = \begin{pmatrix} 1 & 1 \\ -1 & 1 \end{pmatrix}, \text{ thus } A^{-1} = \frac{1}{2} \begin{pmatrix} 1 & -1 \\ 1 & 1 \end{pmatrix}.$$

Therefore,

$$\begin{pmatrix} x \\ y \end{pmatrix} = A^{-1} \begin{pmatrix} 3 \\ -1 \end{pmatrix} = \frac{1}{2} \begin{pmatrix} 1 & -1 \\ 1 & 1 \end{pmatrix} \begin{pmatrix} 3 \\ -1 \end{pmatrix} = \begin{pmatrix} 2 \\ 1 \end{pmatrix}.$$

Therefore, the equilibrium solution is

$$\begin{pmatrix} 2 \\ 1 \end{pmatrix}.$$

(b)

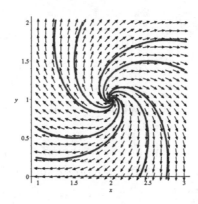

(c) Solutions in the vicinity of the critical point spiral away from the critical point.

21. Let $x_1 = u$ and $x_2 = u'$. Then $x_1' = x_2$ and

$$x_2' = u'' = -2u - 0.5u' = -2x_1 - 0.5x_2.$$

Therefore, we obtain the system of equations

$$\begin{aligned} x_1' &= x_2 \\ x_2' &= -2x_1 - 0.5x_2. \end{aligned}$$

23. First divide the equation by t^2. We arrive at the equation

$$u'' = -\frac{1}{t}u' - \left(1 - \frac{1}{4t^2}\right)u.$$

Now let $x_1 = u$ and $x_2 = u'$. Then $x_1' = x_2$ and

$$x_2' = u'' = -\frac{1}{t}u' - \left(1 - \frac{1}{4t^2}\right)u = -\frac{1}{t}x_2 - \left(1 - \frac{1}{4t^2}\right)x_1.$$

Therefore, we obtain the system of equations

$$\begin{aligned} x_1' &= x_2 \\ x_2' &= -\left(1 - \frac{1}{4t^2}\right)x_1 - \frac{1}{t}x_2. \end{aligned}$$

25. Let $x_1 = u$ and $x_2 = u'$. Then $x_1' = x_2$ and

$$x_2' = u'' = -0.25u' - 4u + 2\cos 3t = -4x_1 - 0.25x_2 + 2\cos 3t.$$

Now $u(0) = 1$ implies $x_1(0) = 1$ and $u'(0) = -2$ implies $x_2(0) = -2$. Therefore, we obtain the system of equations

$$\begin{aligned} x_1' &= x_2 \\ x_2' &= -4x_1 - 0.25x_2 + 2\cos 3t \end{aligned}$$

with initial conditions

$$\begin{aligned} x_1(0) &= 1 \\ x_2(0) &= -2. \end{aligned}$$

27.(a) Taking a *clockwise* loop around each of the paths, it is easy to see that voltage drops are given by $v_1 - v_2 = 0$ and $v_2 - v_3 = 0$.

(b) Consider the right node. The current in is given by $i_1 + i_2$. The current leaving the node is $-i_3$. Therefore, the current passing through the node is $(i_1 + i_2) - (-i_3)$. Based on Kirchoff's first law, $i_1 + i_2 + i_3 = 0$.

(c) In the capacitor, $Cv_1' = i_1$. In the resistor, $v_2 = Ri_2$. In the inductor, $Li_3' = v_3$.

(d) Based on part (a), $v_3 = v_2 = v_1$. Based on parts (b) and (c),

$$Cv_1' + \frac{1}{R}v_2 + i_3 = 0.$$

It follows that

$$Cv_1' = -\frac{1}{R}v_1 - i_3 \quad \text{and} \quad Li_3' = v_1.$$

29. Let i_1, i_2, i_3 and i_4 be the current through the resistors, inductor and capacitor, respectively. Assign v_1, v_2, v_3 and v_4 as the respective voltage drops. Based on Kirchoff's second law, the net voltage drops around each loop satisfy

$$v_1 + v_3 + v_4 = 0, \quad v_1 + v_3 + v_2 = 0, \quad \text{and} \quad v_4 - v_2 = 0.$$

Applying Kirchoff's first law to the upper right node, we have

$$i_3 - (i_2 + i_4) = 0.$$

Likewise, in the remaining nodes, we have

$$i_1 - i_3 = 0 \quad \text{and} \quad i_2 + i_4 - i_1 = 0.$$

Combining the above equations, we have

$$v_4 - v_2 = 0, \quad v_1 + v_3 + v_4 = 0, \quad \text{and} \quad i_2 + i_4 - i_3 = 0.$$

Using the current-voltage relations, we have

$$v_1 = R_1 i_1, \quad v_2 = R_2 i_2, \quad Li_3' = v_3, \quad Cv_4' = i_4.$$

Combining these equations, we have

$$R_1 i_3 + Li_3' + v_4 = 0 \quad \text{and} \quad Cv_4' = i_3 - \frac{1}{R_2}v_4.$$

Now set $i_3 = i$ and $v_4 = v$, to obtain the system of equations

$$Li' = -R_1 i - v \quad \text{and} \quad Cv' = i - \frac{1}{R_2}v.$$

31.(a) Let $Q_1(t)$ and $Q_2(t)$ be the amount of salt in the respective tanks at time t. The volume of each tank remains constant. Based on conservation of mass, the rate of increase of salt is given by

$$\text{rate of increase} = \text{rate in} - \text{rate out}.$$

For Tank 1, the rate of salt flowing in is $r_{in} = 3q_1 + Q_2/100$. The rate of flow out of Tank 1 is $r_{out} = Q_1/15$. Therefore,

$$\frac{dQ_1}{dt} = 3q_1 + \frac{Q_2}{100} - \frac{Q_1}{15}.$$

Similarly, for Tank 2,

$$\frac{dQ_2}{dt} = q_2 + \frac{Q_1}{30} - \frac{3Q_2}{100}.$$

The initial conditions are $Q_1(0) = Q_1^0$ and $Q_2(0) = Q_2^0$.

(b) The equilibrium values are obtained by solving $-Q_1/15 + Q_2/100 + 3q_1 = 0$ and $Q_1/30 - 3Q_2/100 + q_2 = 0$. Its solution is given by $Q_1^E = 54q_1 + 6q_2$ and $Q_2^E = 60q_1 + 40q_2$.

(c) We would need to be able to solve the system

$$\begin{aligned} 54q_1 + 6q_2 &= 60 \\ 60q_1 + 40q_2 &= 50. \end{aligned}$$

The solution of this system is $q_1 = 7/6$ and $q_2 = -1/2$ which is not a physically realistic solution. Therefore, it is not possible to have those values as an equilibrium state.

(d) We can write

$$\begin{aligned} q_2 &= -9q_1 + \frac{Q_1^E}{6} \\ q_2 &= -\frac{3}{2}q_1 + \frac{Q_2^E}{40} \end{aligned}$$

which are the equations of two lines in the $q_1 q_2$-plane. The intercepts of the first line are $(Q_1^E/54, 0)$ and $(0, Q_1^E/6)$. The intercepts of the second line are $(Q_2^E/60, 0)$ and $(0, Q_2^E/40)$. Therefore, the system will have a unique solution in the first quadrant as long as $Q_1^E/54 \leq Q_2^E/60$ or $Q_2^E/40 \leq Q_1^E/6$. That is, $10/9 \leq Q_2^E/Q_1^E \leq 20/3$.

3.3   Homogeneous Linear Systems with Constant Coefficients

1. We look for eigenvalues and eigenvectors of

$$A = \begin{pmatrix} 3 & -2 \\ 2 & -2 \end{pmatrix}.$$

We see that $\det(A - \lambda I) = \lambda^2 - \lambda - 2 = (\lambda - 2)(\lambda + 1)$. Therefore, the eigenvalues are given by $\lambda = 2, -1$. First, $\lambda_1 = 2$ implies

$$A - \lambda_1 I = \begin{pmatrix} 1 & -2 \\ 2 & -4 \end{pmatrix}.$$

Therefore,

$$\mathbf{v}_1 = \begin{pmatrix} 2 \\ 1 \end{pmatrix}$$

is an eigenvector associated with λ_1 and

$$\mathbf{x}_1(t) = e^{2t} \begin{pmatrix} 2 \\ 1 \end{pmatrix}$$

is one solution of the system. Second, $\lambda_2 = -1$ implies

$$A - \lambda_2 I = \begin{pmatrix} 4 & -2 \\ 2 & -1 \end{pmatrix}.$$

Therefore,

$$\mathbf{v}_2 = \begin{pmatrix} 1 \\ 2 \end{pmatrix}$$

is an eigenvector associated with λ_2 and

$$\mathbf{x}_2(t) = e^{-t} \begin{pmatrix} 1 \\ 2 \end{pmatrix}$$

is a second solution of the system. Therefore, the general solution is given by

$$\mathbf{x}(t) = c_1 e^{2t} \begin{pmatrix} 2 \\ 1 \end{pmatrix} + c_2 e^{-t} \begin{pmatrix} 1 \\ 2 \end{pmatrix}.$$

If the initial condition is a multiple of $\begin{pmatrix} 1 & 2 \end{pmatrix}^T$, then the solution will tend to the origin along the eigenvector $\begin{pmatrix} 1 & 2 \end{pmatrix}^T$. Otherwise, the solution will grow, following the eigenvector $\begin{pmatrix} 2 & 1 \end{pmatrix}^T$.

3. We look for eigenvalues and eigenvectors of

$$A = \begin{pmatrix} 2 & -1 \\ 3 & -2 \end{pmatrix}.$$

We see that $\det(A - \lambda I) = \lambda^2 - 1 = (\lambda - 1)(\lambda + 1)$. Therefore, the eigenvalues are given by $\lambda = 1, -1$. First, $\lambda_1 = 1$ implies

$$A - \lambda_1 I = \begin{pmatrix} 1 & -1 \\ 3 & -3 \end{pmatrix}.$$

Therefore,

$$\mathbf{v}_1 = \begin{pmatrix} 1 \\ 1 \end{pmatrix}$$

is an eigenvector associated with λ_1 and

$$\mathbf{x}_1(t) = e^t \begin{pmatrix} 1 \\ 1 \end{pmatrix}$$

is one solution of the system. Second, $\lambda_2 = -1$ implies

$$A - \lambda_2 I = \begin{pmatrix} 3 & -1 \\ 3 & -1 \end{pmatrix}.$$

Therefore,

$$\mathbf{v}_2 = \begin{pmatrix} 1 \\ 3 \end{pmatrix}$$

is an eigenvector associated with λ_2 and

$$\mathbf{x}_2(t) = e^{-t} \begin{pmatrix} 1 \\ 3 \end{pmatrix}$$

is a second solution of the system. Therefore, the general solution is given by

$$\mathbf{x}(t) = c_1 e^t \begin{pmatrix} 1 \\ 1 \end{pmatrix} + c_2 e^{-t} \begin{pmatrix} 1 \\ 3 \end{pmatrix}.$$

If the initial condition is a multiple of $\begin{pmatrix} 1 & 3 \end{pmatrix}^T$, then the solution will tend to the origin along the eigenvector $\begin{pmatrix} 1 & 3 \end{pmatrix}^T$. Otherwise, the solution will grow, following the eigenvector $\begin{pmatrix} 1 & 1 \end{pmatrix}^T$.

5. We look for eigenvalues and eigenvectors of

$$A = \begin{pmatrix} 4 & -3 \\ 8 & -6 \end{pmatrix}.$$

We see that $\det(A - \lambda I) = \lambda^2 + 2\lambda = \lambda(\lambda + 2)$. Therefore, the eigenvalues are given by $\lambda = 0, -2$. First, $\lambda_1 = 0$ implies

$$A - \lambda_1 I = \begin{pmatrix} 4 & -3 \\ 8 & -6 \end{pmatrix}.$$

Therefore,

$$\mathbf{v}_1 = \begin{pmatrix} 3 \\ 4 \end{pmatrix}$$

is an eigenvector associated with λ_1 and

$$\mathbf{x}_1(t) = \begin{pmatrix} 3 \\ 4 \end{pmatrix}$$

is one solution of the system. Second, $\lambda_2 = -2$ implies

$$A - \lambda_2 I = \begin{pmatrix} 6 & -3 \\ 8 & -4 \end{pmatrix}.$$

Therefore,

$$\mathbf{v}_2 = \begin{pmatrix} 1 \\ 2 \end{pmatrix}$$

is an eigenvector associated with λ_2 and

$$\mathbf{x}_2(t) = e^{-2t} \begin{pmatrix} 1 \\ 2 \end{pmatrix}$$

is a second solution of the system. Therefore, the general solution is given by

$$\mathbf{x}(t) = c_1 \begin{pmatrix} 3 \\ 4 \end{pmatrix} + c_2 e^{-2t} \begin{pmatrix} 1 \\ 2 \end{pmatrix}.$$

Any solution starting at a point on the line of critical points remains fixed for all time at its starting point. A solution starting at any other point in the plane moves on a line parallel to $\mathbf{v}_2$ toward the point of intersection of this line with the line of critical points.

7. We look for eigenvalues and eigenvectors of

$$A = \begin{pmatrix} 5/4 & 3/4 \\ 3/4 & 5/4 \end{pmatrix}.$$

We see that $\det(A - \lambda I) = \lambda^2 - \frac{5}{2}\lambda + 1 = (\lambda - 2)(\lambda - 1/2)$. Therefore, the eigenvalues are given by $\lambda = 2, 1/2$. First, $\lambda_1 = 2$ implies

$$A - \lambda_1 I = \begin{pmatrix} -3/4 & 3/4 \\ 3/4 & -3/4 \end{pmatrix}.$$

Therefore,

$$\mathbf{v}_1 = \begin{pmatrix} 1 \\ 1 \end{pmatrix}$$

is an eigenvector associated with λ_1 and

$$\mathbf{x}_1(t) = e^{2t} \begin{pmatrix} 1 \\ 1 \end{pmatrix}$$

is one solution of the system. Second, $\lambda_2 = 1/2$ implies

$$A - \lambda_2 I = \begin{pmatrix} 3/4 & 3/4 \\ 3/4 & 3/4 \end{pmatrix}.$$

Therefore,

$$\mathbf{v}_2 = \begin{pmatrix} 1 \\ -1 \end{pmatrix}$$

is an eigenvector associated with λ_2 and

$$\mathbf{x}_2(t) = e^{t/2} \begin{pmatrix} 1 \\ -1 \end{pmatrix}$$

is a second solution of the system. Therefore, the general solution is given by

$$\mathbf{x}(t) = c_1 e^{2t} \begin{pmatrix} 1 \\ 1 \end{pmatrix} + c_2 e^{\frac{t}{2}} \begin{pmatrix} 1 \\ -1 \end{pmatrix}.$$

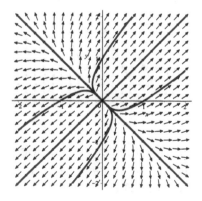

If the initial condition is a multiple of $\begin{pmatrix} 1 & -1 \end{pmatrix}^T$, then the solution will grow following the the eigenvector $\begin{pmatrix} 1 & -1 \end{pmatrix}^T$. Otherwise, the solution will grow, following the eigenvector $\begin{pmatrix} 1 & 1 \end{pmatrix}^T$.

9. We look for eigenvalues and eigenvectors of

$$A = \begin{pmatrix} -1/4 & -3/4 \\ 1/2 & 1 \end{pmatrix}.$$

We see that $\det(A - \lambda I) = \lambda^2 - \dfrac{3}{4}\lambda + \dfrac{1}{8} = \left(\lambda - \dfrac{1}{2}\right)\left(\lambda - \dfrac{1}{4}\right)$. Therefore, the eigenvalues are given by $\lambda = 1/2, 1/4$. First, $\lambda_1 = 1/2$ implies

$$A - \lambda_1 I = \begin{pmatrix} -3/4 & -3/4 \\ 1/2 & 1/2 \end{pmatrix}.$$

Therefore,

$$\mathbf{v}_1 = \begin{pmatrix} 1 \\ -1 \end{pmatrix}$$

is an eigenvector associated with λ_1 and

$$\mathbf{x}_1(t) = e^{t/2}\begin{pmatrix} 1 \\ -1 \end{pmatrix}$$

is one solution of the system. Second, $\lambda_2 = 1/4$ implies

$$A - \lambda_2 I = \begin{pmatrix} -1/2 & -3/4 \\ 1/2 & 3/4 \end{pmatrix}.$$

Therefore,

$$\mathbf{v}_2 = \begin{pmatrix} 3 \\ -2 \end{pmatrix}$$

is an eigenvector associated with λ_2 and

$$\mathbf{x}_2(t) = e^{t/4}\begin{pmatrix} 3 \\ -2 \end{pmatrix}$$

is a second solution of the system. Therefore, the general solution is given by

$$\mathbf{x}(t) = c_1 e^{t/2}\begin{pmatrix} 1 \\ -1 \end{pmatrix} + c_2 e^{t/4}\begin{pmatrix} 3 \\ -2 \end{pmatrix}.$$

If the initial condition is a multiple of $\begin{pmatrix} 3 & -2 \end{pmatrix}^T$, then the solution will stay on the eigenvector $\begin{pmatrix} 3 & -2 \end{pmatrix}^T$. Otherwise, the solution will grow, following the eigenvector $\begin{pmatrix} 1 & -1 \end{pmatrix}^T$.

11. We look for eigenvalues and eigenvectors of

$$A = \begin{pmatrix} -2 & 1 \\ -5 & 4 \end{pmatrix}.$$

We see that $\det(A - \lambda I) = \lambda^2 - 2\lambda - 3 = (\lambda - 3)(\lambda + 1)$. Therefore, the eigenvalues are given by $\lambda = 3, -1$. First, $\lambda_1 = 3$ implies

$$A - \lambda_1 I = \begin{pmatrix} -5 & 1 \\ -5 & 1 \end{pmatrix}.$$

Therefore,

$$\mathbf{v}_1 = \begin{pmatrix} 1 \\ 5 \end{pmatrix}$$

is an eigenvector associated with λ_1 and

$$\mathbf{x_1}(t) = e^{3t} \begin{pmatrix} 1 \\ 5 \end{pmatrix}$$

is one solution of the system. Second, $\lambda_2 = -1$ implies

$$A - \lambda_2 I = \begin{pmatrix} -1 & 1 \\ -5 & 5 \end{pmatrix}.$$

Therefore,

$$\mathbf{v}_2 = \begin{pmatrix} 1 \\ 1 \end{pmatrix}$$

is an eigenvector associated with λ_2 and

$$\mathbf{x_2}(t) = e^{-t} \begin{pmatrix} 1 \\ 1 \end{pmatrix}$$

is a second solution of the system. Therefore, the general solution is given by

$$\mathbf{x}(t) = c_1 e^{3t} \begin{pmatrix} 1 \\ 5 \end{pmatrix} + c_2 e^{-t} \begin{pmatrix} 1 \\ 1 \end{pmatrix}.$$

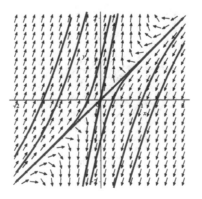

If the initial condition is a multiple of $(\ 1 \ \ 1 \ )^T$, then the solution will tend to the origin along the eigenvector $(\ 1 \ \ 1 \ )^T$. Otherwise, the solution will grow, following the eigenvector $(\ 1 \ \ 5 \ )^T$.

13. From problem 2, the general solution is

$$\mathbf{x}(t) = c_1 e^{-2t} \begin{pmatrix} 2 \\ 3 \end{pmatrix} + c_2 e^{-t} \begin{pmatrix} 1 \\ 1 \end{pmatrix}.$$

The initial condition $\mathbf{x}(0) = \begin{pmatrix} 3 \\ 1 \end{pmatrix}$ implies

$$\begin{pmatrix} 2 & 1 \\ 3 & 1 \end{pmatrix} \begin{pmatrix} c_1 \\ c_2 \end{pmatrix} = \begin{pmatrix} 3 \\ 1 \end{pmatrix}.$$

Therefore,

$$\begin{pmatrix} c_1 \\ c_2 \end{pmatrix} = \begin{pmatrix} -1 & 1 \\ 3 & -2 \end{pmatrix} \begin{pmatrix} 3 \\ 1 \end{pmatrix} = \begin{pmatrix} -2 \\ 7 \end{pmatrix}.$$

Therefore, the solution is

$$\mathbf{x}(t) = -2e^{-2t} \begin{pmatrix} 2 \\ 3 \end{pmatrix} + 7e^{-t} \begin{pmatrix} 1 \\ 1 \end{pmatrix}.$$

 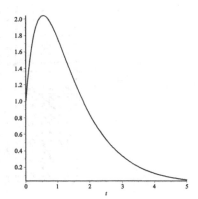

Both components tend to zero as $t \to \infty$.

15. From problem 10, the general solution is

$$\mathbf{x}(t) = c_1 e^{2t} \begin{pmatrix} 1 \\ 3 \end{pmatrix} + c_2 e^{4t} \begin{pmatrix} 1 \\ 1 \end{pmatrix}.$$

The initial condition $\mathbf{x}(0) = \begin{pmatrix} 2 \\ -1 \end{pmatrix}$ implies

$$\begin{pmatrix} 1 & 1 \\ 3 & 1 \end{pmatrix} \begin{pmatrix} c_1 \\ c_2 \end{pmatrix} = \begin{pmatrix} 2 \\ -1 \end{pmatrix}.$$

Therefore,

$$\begin{pmatrix} c_1 \\ c_2 \end{pmatrix} = -\frac{1}{2}\begin{pmatrix} 1 & -1 \\ -3 & 1 \end{pmatrix}\begin{pmatrix} 2 \\ -1 \end{pmatrix} = \begin{pmatrix} -3/2 \\ 7/2 \end{pmatrix}.$$

Therefore, the solution is

$$\mathbf{x}(t) = -\frac{3}{2}e^{2t}\begin{pmatrix} 1 \\ 3 \end{pmatrix} + \frac{7}{2}e^{4t}\begin{pmatrix} 1 \\ 1 \end{pmatrix}.$$

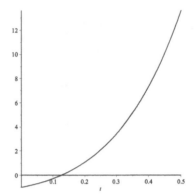

Both components tend to $+\infty$ as $t \to \infty$.

17.(a)

(b)

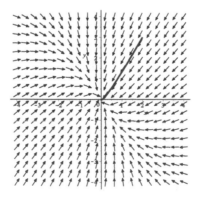

(c) The general solution is given by

$$\mathbf{x}(t) = c_1 e^{-t} \begin{pmatrix} -1 \\ 2 \end{pmatrix} + c_2 e^{-2t} \begin{pmatrix} 1 \\ 2 \end{pmatrix}.$$

The initial condition $\mathbf{x}(0) = \begin{pmatrix} 2 \\ 3 \end{pmatrix}$ implies

$$\begin{pmatrix} c_1 \\ c_2 \end{pmatrix} = \begin{pmatrix} -1/4 \\ 7/4 \end{pmatrix}.$$

Therefore, the solution passing through the initial point $(2, 3)$ is given by

$$\mathbf{x}(t) = -\frac{1}{4} e^{-t} \begin{pmatrix} -1 \\ 2 \end{pmatrix} + \frac{7}{4} e^{-2t} \begin{pmatrix} 1 \\ 2 \end{pmatrix}.$$

The component plots are shown below.

19.(a)

(b)

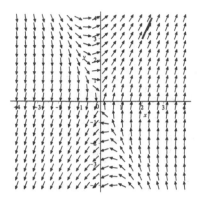

(c) The general solution is given by

$$\mathbf{x}(t) = c_1 e^{-t} \begin{pmatrix} -1 \\ 2 \end{pmatrix} + c_2 e^{2t} \begin{pmatrix} 1 \\ 2 \end{pmatrix}.$$

The initial condition $\mathbf{x}(0) = \begin{pmatrix} 2 \\ 3 \end{pmatrix}$ implies

$$\begin{pmatrix} c_1 \\ c_2 \end{pmatrix} = \begin{pmatrix} -1/4 \\ 7/4 \end{pmatrix}.$$

Therefore, the solution passing through the initial point $(2, 3)$ is given by

$$\mathbf{x}(t) = -\frac{1}{4} e^{-t} \begin{pmatrix} -1 \\ 2 \end{pmatrix} + \frac{7}{4} e^{2t} \begin{pmatrix} 1 \\ 2 \end{pmatrix}.$$

The component plots are shown below.

21.(a)

(b)

(c) The general solution is given by

$$\mathbf{x}(t) = c_1 e^{t/2} \begin{pmatrix} 1 \\ 4 \end{pmatrix} + c_2 e^{-t/2} \begin{pmatrix} 4 \\ 1 \end{pmatrix}.$$

The initial condition $\mathbf{x}(0) = \begin{pmatrix} 2 \\ 3 \end{pmatrix}$ implies

$$\begin{pmatrix} 1 & 4 \\ 4 & 1 \end{pmatrix} \begin{pmatrix} c_1 \\ c_2 \end{pmatrix} = \begin{pmatrix} 2 \\ 3 \end{pmatrix}.$$

Therefore,

$$\begin{pmatrix} c_1 \\ c_2 \end{pmatrix} = -\frac{1}{15} \begin{pmatrix} 1 & -4 \\ -4 & 1 \end{pmatrix} \begin{pmatrix} 2 \\ 3 \end{pmatrix} = \begin{pmatrix} 2/3 \\ 1/3 \end{pmatrix}.$$

Therefore, the solution passing through the initial point $(2, 3)$ is given by

$$\mathbf{x}(t) = \frac{2}{3} e^{t/2} \begin{pmatrix} 1 \\ 4 \end{pmatrix} + \frac{1}{3} e^{-t/2} \begin{pmatrix} 4 \\ 1 \end{pmatrix}.$$

The component plots are shown below.

23.(a)

(b)

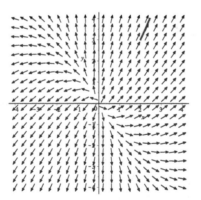

(c) The general solution is given by

$$\mathbf{x}(t) = c_1 e^{3t/10} \begin{pmatrix} 1 \\ -2 \end{pmatrix} + c_2 e^{3t/5} \begin{pmatrix} 1 \\ 3 \end{pmatrix}.$$

The initial condition $\mathbf{x}(0) = \begin{pmatrix} 2 \\ 3 \end{pmatrix}$ implies

$$\begin{pmatrix} 1 & 1 \\ -2 & 3 \end{pmatrix} \begin{pmatrix} c_1 \\ c_2 \end{pmatrix} = \begin{pmatrix} 2 \\ 3 \end{pmatrix}.$$

Therefore,

$$\begin{pmatrix} c_1 \\ c_2 \end{pmatrix} = \frac{1}{5} \begin{pmatrix} 3 & -1 \\ 2 & 1 \end{pmatrix} \begin{pmatrix} 2 \\ 3 \end{pmatrix} = \begin{pmatrix} 3/5 \\ 1 \end{pmatrix}.$$

Therefore, the solution passing through the initial point $(2, 3)$ is given by

$$\mathbf{x}(t) = \frac{7}{5} e^{3t/10} \begin{pmatrix} 2 \\ 1 \end{pmatrix} + \frac{4}{5} e^{3t/5} \begin{pmatrix} -1 \\ 2 \end{pmatrix}.$$

The component plots are shown below.

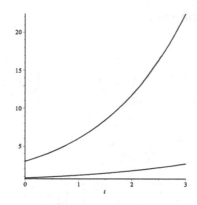

25.(a) Set $x_1 = y$, $x_2 = x_1' = y'$; we obtain

$$\mathbf{x}' = \begin{pmatrix} 0 & 1 \\ -10 & -7 \end{pmatrix} \mathbf{x}.$$

(b) The eigenvalues and corresponding eigenvectors are $\lambda_1 = -5$, $\mathbf{v}_1 = \begin{pmatrix} 1 \\ -5 \end{pmatrix}$, and $\lambda_2 = -2$, $\mathbf{v}_2 = \begin{pmatrix} 1 \\ -2 \end{pmatrix}$, thus the general solution is

$$\mathbf{x} = c_1 e^{-5t} \begin{pmatrix} 1 \\ -5 \end{pmatrix} + c_2 e^{-2t} \begin{pmatrix} 1 \\ -2 \end{pmatrix}.$$

(c) $y(t) = x_1 = c_1 e^{-5t} + c_2 e^{-2t}$.

(d)

(e) $(0,0)$ is an asymptotically stable node.

27.(a) Set $x_1 = y$, $x_2 = x_1' = y'$; we obtain

$$\mathbf{x}' = \begin{pmatrix} 0 & 1 \\ -12 & 7 \end{pmatrix} \mathbf{x}.$$

(b) The eigenvalues and corresponding eigenvectors are $\lambda_1 = 4$, $\mathbf{v}_1 = \begin{pmatrix} 1 \\ 4 \end{pmatrix}$, and $\lambda_2 = 3$, $\mathbf{v}_2 = \begin{pmatrix} 1 \\ 3 \end{pmatrix}$, thus the general solution is

$$\mathbf{x} = c_1 e^{4t} \begin{pmatrix} 1 \\ 4 \end{pmatrix} + c_2 e^{3t} \begin{pmatrix} 1 \\ 3 \end{pmatrix}.$$

(c) $y(t) = x_1 = c_1 e^{4t} + c_2 e^{3t}.$

(d)

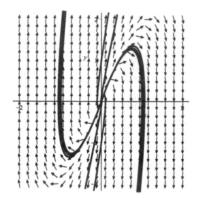

(e) $(0,0)$ is an unstable node.

29.(a) Set $x_1 = y$, $x_2 = x_1' = y'$; we obtain

$$\mathbf{x}' = \begin{pmatrix} 0 & 1 \\ -1/3 & -7/6 \end{pmatrix} \mathbf{x}.$$

(b) The eigenvalues and corresponding eigenvectors are $\lambda_1 = -2/3$, $\mathbf{v}_1 = \begin{pmatrix} 1 \\ -2/3 \end{pmatrix}$, and $\lambda_2 = -1/2$, $\mathbf{v}_2 = \begin{pmatrix} 1 \\ -1/2 \end{pmatrix}$, thus the general solution is

$$\mathbf{x} = c_1 e^{-2t/3} \begin{pmatrix} 1 \\ -2/3 \end{pmatrix} + c_2 e^{-t/2} \begin{pmatrix} 1 \\ -1/2 \end{pmatrix}.$$

(c) $y(t) = x_1 = c_1 e^{-2t/3} + c_2 e^{-t/2}.$

(d)

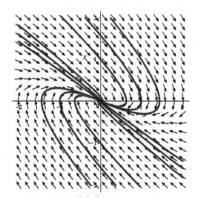

(e) $(0,0)$ is an asymptotically stable node.

31.(a) Let $\mathbf{K} = \begin{pmatrix} -(k_1 + k_2) & k_2 \\ \epsilon k_2 & -\epsilon k_2 \end{pmatrix}$ and $\hat{\mathbf{u}} = \begin{pmatrix} T_a \\ T_a \end{pmatrix}$. If we introduce the new variable $\mathbf{x} = \mathbf{u} - \hat{\mathbf{u}}$, then we can check that

$$\mathbf{x}' = \mathbf{u}' = \mathbf{K}\mathbf{u} + \begin{pmatrix} k_1 T_a \\ 0 \end{pmatrix} = \mathbf{K}(\mathbf{x} + \hat{\mathbf{u}}) + \begin{pmatrix} k_1 T_a \\ 0 \end{pmatrix} = \mathbf{K}\mathbf{x}.$$

This shows the claim.

(b) The characteristic equation of $\mathbf{K}$ is $\lambda^2 + \lambda(k_1 + k_2 + \epsilon k_2) + \epsilon k_1 k_2 = 0$. We supposed that $0 < \epsilon \ll 1$, which implies that $\epsilon^2 \approx 0$. Using the quadratic formula the roots are

$$\lambda_{1,2} = \frac{-(k_1 + k_2 + \epsilon k_2) \pm \sqrt{(k_1 + k_2 + \epsilon k_2)^2 - 4\epsilon k_1 k_2}}{2} =$$

$$= \frac{-(k_1 + k_2 + \epsilon k_2) \pm \sqrt{(k_1 + k_2)^2 + 2\epsilon(k_1 + k_2)k_2 + \epsilon^2 k_2^2 - 4\epsilon k_1 k_2}}{2}.$$

Now using the Taylor series expansion for $\sqrt{x}$ we approximate $\sqrt{x + \delta} \approx \sqrt{x} + \frac{1}{2}\frac{1}{\sqrt{x}}\delta$ when δ is small, so using that $\epsilon^2 \approx 0$ we obtain that

$$\lambda_{1,2} \approx \frac{-(k_1 + k_2 + \epsilon k_2)}{2} \pm \frac{1}{2}\left((k_1 + k_2) + \frac{1}{2}\frac{1}{k_1 + k_2}\epsilon(2(k_1 + k_2)k_2 - 4k_1 k_2)\right).$$

After simplification, we obtain that $\lambda_1 \approx -(k_1 + k_2) - \dfrac{\epsilon k_2^2}{k_1 + k_2}$ and $\lambda_2 \approx -\dfrac{\epsilon k_1 k_2}{k_1 + k_2}$.

(c) Considering the matrices $\mathbf{K} - \lambda_i \mathbf{I}$ for $i = 1, 2$ we obtain that for the first eigenvalue the first row of $\mathbf{K} - \lambda_1 \mathbf{I}$ is $(\epsilon k_2^2/(k_1 + k_2) \quad k_2)$, which gives the first eigenvector $\mathbf{v}_1(\epsilon)$ and for the second eigenvalue the second row of $\mathbf{K} - \lambda_2 \mathbf{I}$ is $(\epsilon k_2 \quad -\epsilon k_2^2/(k_1 + k_2))$, which gives the second eigenvector $\mathbf{v}_2(\epsilon)$.

(d) Using the previous parts, the solution is

$$\mathbf{u} = c_1 e^{\lambda_1 t}\mathbf{v}_1 + c_2 e^{\lambda_2 t}\mathbf{v}_2 + \hat{\mathbf{u}} = \begin{pmatrix} T_a + c_1 e^{(-(k_1+k_2)-\frac{\epsilon k_2^2}{k_1+k_2})t} + c_2 \frac{k_2}{k_1+k_2} e^{-\frac{\epsilon k_1 k_2}{k_1+k_2}t} \\ T_a - c_1 \frac{\epsilon k_2}{k_1+k_2} e^{(-(k_1+k_2)-\frac{\epsilon k_2^2}{k_1+k_2})t} + c_2 e^{-\frac{\epsilon k_1 k_2}{k_1+k_2}t} \end{pmatrix}$$

The sketch of the phase plane is given by the following figure.

(e)The initial condition $u_2(0) = u_{20}$ implies that with the previous part's notations $u_2(0) = T_a - c_1 \frac{\epsilon k_2}{k_1+k_2} + c_2 = u_{20}$; because ϵ is small we get that $c_2 \approx u_{20} - T_a$. Now if the system is in the quasi-steady state, then $u_2 \approx T_a + (u_{20} - T_a)e^{-\frac{\epsilon k_1 k_2}{k_1+k_2}t}$ (the term with the c_1 coefficient is close to 0 at this point in time) and then $u_1 \approx T_a + c_2\frac{k_2}{k_1+k_2}e^{-\frac{\epsilon k_1 k_2}{k_1+k_2}t} = \frac{k_1}{k_1+k_2}T_a + \frac{k_2}{k_1+k_2}u_2$ (again, the term with the c_1 coefficient is close to 0 at this point in time).

(f) The eigenvalue close to 0 corresponds to the slow temporal change in the system, while the eigenvalue close to $-(k_1 + k_2)$ corresponds to the fast temporal change in the system.

33.(a) For the general equation (i), the characteristic equation is given by

$$\lambda^2 + \left(\frac{L + CR_1R_2}{LCR_2}\right)\lambda + \frac{R_1 + R_2}{LCR_2} = 0.$$

The eigenvalues are real and distinct provided that the discriminant is positive. That is,

$$\left(\frac{L + CR_1R_2}{LCR_2}\right)^2 - 4\left(\frac{R_1 + R_2}{LCR_2}\right) > 0,$$

which simplifies to the condition

$$\left(\frac{1}{CR_2} - \frac{R_1}{L}\right)^2 - \frac{4}{LC} > 0.$$

(b) Since the sum of the roots of the characteristic equation is

$$-\frac{L + CR_1R_2}{LCR_2} < 0$$

and the product of the roots is

$$\frac{R_1 + R_2}{LCR_2} > 0,$$

it follows that both roots are negative. Therefore, $i = 0, v = 0$ is a stable node.

3.4 Complex Eigenvalues

1.

$$A = \begin{pmatrix} 3 & -2 \\ 4 & -1 \end{pmatrix}$$

implies

$$\det(A - \lambda I) = \lambda^2 - 2\lambda + 5.$$

Therefore, the eigenvalues are given by $\lambda = 1 \pm 2i$. Now, $\lambda_1 = 1 + 2i$ implies

$$A - \lambda_1 I = \begin{pmatrix} 2 - 2i & -2 \\ 4 & -2 - 2i \end{pmatrix}.$$

Therefore,

$$\mathbf{v}_1 = \begin{pmatrix} 1 \\ 1 - i \end{pmatrix}$$

is an eigenvector for λ_1 and

$$\mathbf{x}_1(t) = e^{(1+2i)t} \begin{pmatrix} 1 \\ 1 - i \end{pmatrix}$$

is a solution of our system. Further, we use the fact that the real and imaginary parts of $\mathbf{x}_1(t)$ are linearly independent solutions of our system. Now

$$\begin{aligned}
\operatorname{Re} \mathbf{x}_1(t) &= e^t \left[\begin{pmatrix} 1 \\ 1 \end{pmatrix} \cos 2t - \begin{pmatrix} 0 \\ -1 \end{pmatrix} \sin 2t \right] = e^t \begin{pmatrix} \cos 2t \\ \cos 2t + \sin 2t \end{pmatrix} \\
\operatorname{Im} \mathbf{x}_1(t) &= e^t \left[\begin{pmatrix} 1 \\ 1 \end{pmatrix} \sin 2t + \begin{pmatrix} 0 \\ -1 \end{pmatrix} \cos 2t \right] = e^t \begin{pmatrix} \sin 2t \\ \sin 2t - \cos 2t \end{pmatrix}.
\end{aligned}$$

Therefore, the general solution is given by

$$\mathbf{x}(t) = c_1 e^t \begin{pmatrix} \cos 2t \\ \cos 2t + \sin 2t \end{pmatrix} + c_2 e^t \begin{pmatrix} \sin 2t \\ \sin 2t - \cos 2t \end{pmatrix}.$$

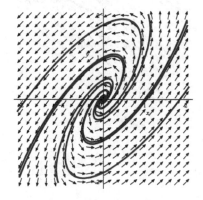

The equilibrium is an unstable spiral point.

3.

$$A = \begin{pmatrix} 2 & -5 \\ 1 & -2 \end{pmatrix}$$

implies

$$\det(A - \lambda I) = \lambda^2 + 1.$$

Therefore, the eigenvalues are given by $\lambda = \pm i$. Now, $\lambda_1 = i$ implies

$$A - \lambda_1 I = \begin{pmatrix} 2 - i & -5 \\ 1 & -2 - i \end{pmatrix}.$$

Therefore,

$$\mathbf{v}_1 = \begin{pmatrix} 2 + i \\ 1 \end{pmatrix}$$

is an eigenvector for λ_1 and

$$\mathbf{x}_1(t) = e^{it} \begin{pmatrix} 2 + i \\ 1 \end{pmatrix}$$

is a solution of our system. Further, we use the fact that the real and imaginary parts of $\mathbf{x}_1(t)$ are linearly independent solutions of our system. Now

$$\text{Re } \mathbf{x}_1(t) = \left[\begin{pmatrix} 2 \\ 1 \end{pmatrix} \cos t - \begin{pmatrix} 1 \\ 0 \end{pmatrix} \sin t \right] = \begin{pmatrix} 2\cos t - \sin t \\ \cos t \end{pmatrix}$$

$$\text{Im } \mathbf{x}_1(t) = \left[\begin{pmatrix} 2 \\ 1 \end{pmatrix} \sin t + \begin{pmatrix} 1 \\ 0 \end{pmatrix} \cos t \right] = \begin{pmatrix} 2\sin t + \cos t \\ \sin t \end{pmatrix}.$$

Therefore, the general solution is given by

$$\mathbf{x}(t) = c_1 \begin{pmatrix} 2\cos t - \sin t \\ \cos t \end{pmatrix} + c_2 \begin{pmatrix} 2\sin t + \cos t \\ \sin t \end{pmatrix}.$$

The equilibrium is a stable center.

5.

$$A = \begin{pmatrix} 1 & -1 \\ 5 & -3 \end{pmatrix}$$

implies

$$\det(A - \lambda I) = \lambda^2 + 2\lambda + 2.$$

Therefore, the eigenvalues are given by $\lambda = -1 \pm i$. Now, $\lambda_1 = -1 - i$ has corresponding eigenvector

$$\mathbf{v}_1 = \begin{pmatrix} 1 \\ 2+i \end{pmatrix}$$

is an eigenvector for λ_1 and

$$\mathbf{x}_1(t) = e^{(-1-i)t} \begin{pmatrix} 1 \\ 2+i \end{pmatrix}$$

is a solution of our system. Further, we use the fact that the real and imaginary parts of $\mathbf{x}_1(t)$ are linearly independent solutions of our system. Now

$$\mathrm{Re}\ \mathbf{x}_1(t) = e^{-t} \begin{pmatrix} \cos t \\ 2\cos t + \sin t \end{pmatrix}$$

$$\mathrm{Im}\ \mathbf{x}_1(t) = e^{-t} \begin{pmatrix} \sin t \\ -\cos t + 2\sin t \end{pmatrix}.$$

Therefore, the general solution is given by

$$\mathbf{x}(t) = c_1 e^{-t} \begin{pmatrix} \cos t \\ 2\cos t + \sin t \end{pmatrix} + c_2 e^{-t} \begin{pmatrix} \sin t \\ -\cos t + 2\sin t \end{pmatrix}.$$

The equilibrium is an asymptotically stable spiral point.

7.

$$A = \begin{pmatrix} -1 & -4 \\ 1 & -1 \end{pmatrix}$$

implies

$$\det(A - \lambda I) = \lambda^2 + 2\lambda + 5.$$

Therefore, the eigenvalues are given by $\lambda = -1 \pm 2i$. Now, $\lambda_1 = -1 + 2i$ implies

$$A - \lambda_1 I = \begin{pmatrix} -2i & -4 \\ 1 & -2i \end{pmatrix}.$$

Therefore,

$$\mathbf{v}_1 = \begin{pmatrix} 2i \\ 1 \end{pmatrix}$$

is an eigenvector for λ_1 and

$$\mathbf{x}_1(t) = e^{(-1+2i)t} \begin{pmatrix} 2i \\ 1 \end{pmatrix}$$

is a solution of our system. Further, we use the fact that the real and imaginary parts of $\mathbf{x}_1(t)$ are linearly independent solutions of our system. Now

$$\text{Re } \mathbf{x}_1(t) = e^{-t}\left[\begin{pmatrix} 0 \\ 1 \end{pmatrix}\cos 2t - \begin{pmatrix} 2 \\ 0 \end{pmatrix}\sin 2t\right] = e^{-t}\begin{pmatrix} -2\sin 2t \\ \cos 2t \end{pmatrix}$$

$$\text{Im } \mathbf{x}_1(t) = e^{-t}\left[\begin{pmatrix} 0 \\ 1 \end{pmatrix}\sin 2t + \begin{pmatrix} 2 \\ 0 \end{pmatrix}\cos 2t\right] = e^{-t}\begin{pmatrix} 2\cos 2t \\ \sin 2t \end{pmatrix}.$$

Therefore, the general solution is given by

$$\mathbf{x}(t) = c_1 e^{-t}\begin{pmatrix} -2\sin 2t \\ \cos 2t \end{pmatrix} + c_2 e^{-t}\begin{pmatrix} 2\cos 2t \\ \sin 2t \end{pmatrix}.$$

The initial condition, $\mathbf{x}(0) = \begin{pmatrix} 4 & -3 \end{pmatrix}^T$ implies

$$c_1\begin{pmatrix} 0 \\ 1 \end{pmatrix} + c_2\begin{pmatrix} 2 \\ 0 \end{pmatrix} = \begin{pmatrix} 4 \\ -3 \end{pmatrix}.$$

Thus $c_1 = -3$, $c_2 = 2$. Therefore, the solution is given by

$$\mathbf{x}(t) = -3e^{-t}\begin{pmatrix} -2\sin 2t \\ \cos 2t \end{pmatrix} + 2e^{-t}\begin{pmatrix} 2\cos 2t \\ \sin 2t \end{pmatrix} = e^{-t}\begin{pmatrix} 4\cos 2t + 6\sin 2t \\ -3\cos 2t + 2\sin 2t \end{pmatrix}.$$

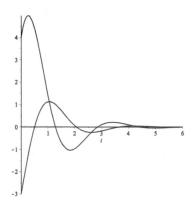

Both components decay to zero as $t \to \infty$.

9.

$$A = \begin{pmatrix} 1 & -5 \\ 1 & -3 \end{pmatrix}$$

implies

$$\det(A - \lambda I) = \lambda^2 + 2\lambda + 2.$$

Therefore, the eigenvalues are given by $\lambda = -1 \pm i$. Now, $\lambda_1 = -1 + i$ implies

$$\mathbf{v}_1 = \begin{pmatrix} 2+i \\ 1 \end{pmatrix}.$$

is an eigenvector for λ_1 and

$$\mathbf{x}_1(t) = e^{(-1+i)t} \begin{pmatrix} 2+i \\ 1 \end{pmatrix}$$

is a solution of our system. Further, we use the fact that the real and imaginary parts of $\mathbf{x}_1(t)$ are linearly independent solutions of our system. Now

$$\text{Re } \mathbf{x}_1(t) = e^{-t} \begin{pmatrix} 2\cos t - \sin t \\ \cos t \end{pmatrix}$$

$$\text{Im } \mathbf{x}_1(t) = e^{-t} \begin{pmatrix} 2\sin t + \cos t \\ \sin t \end{pmatrix}.$$

Therefore, the general solution is given by

$$\mathbf{x}(t) = c_1 e^{-t} \begin{pmatrix} 2\cos t - \sin t \\ \cos t \end{pmatrix} + c_2 e^{-t} \begin{pmatrix} 2\sin t + \cos t \\ \sin t \end{pmatrix}.$$

The initial condition, $\mathbf{x}(0) = \begin{pmatrix} 1 & 1 \end{pmatrix}^T$ implies $c_1 = 1$ and $c_2 = -1$. Therefore, the solution is given by

$$\mathbf{x}(t) = e^{-t} \begin{pmatrix} 2\cos t - \sin t \\ \cos t \end{pmatrix} - e^{-t} \begin{pmatrix} 2\sin t + \cos t \\ \sin t \end{pmatrix} = e^{-t} \begin{pmatrix} \cos t - 3\sin t \\ \cos t - \sin t \end{pmatrix}.$$

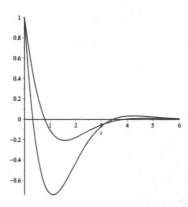

Both components approach zero as $t \to \infty$.

11.(a) The eigenvalues are given by $\lambda = -1/4 \pm i$.

(b) Take $\mathbf{x}(0) = \begin{pmatrix} 1 & 1 \end{pmatrix}^T$. Then the trajectory in the $x_1 x_2$-plane is given by

(c)

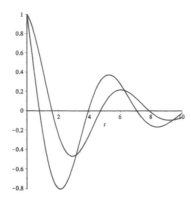

13.(a) The characteristic equation is $\lambda^2 - 2\alpha\lambda + 1 + \alpha^2 = 0$. Therefore, the eigenvalues are $\lambda = \alpha \pm i$.

(b) For $\alpha > 0$, the equilibrium will be an unstable spiral. For $\alpha < 0$, the equilibrium with be a stable spiral. For $\alpha = 0$, the equilibrium will be a center.

(c) Below we show phase portraits for $\alpha = -1/2$ and $\alpha = 1/2$.

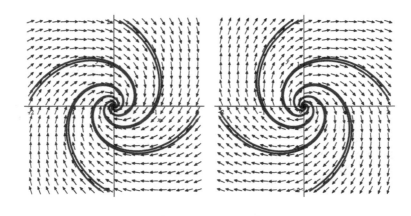

15.(a) The characteristic equation is $\lambda^2 + 5\alpha - 4 = 0$. Therefore, the eigenvalues are $\lambda = \pm\sqrt{4 - 5\alpha}$.

(b) If $4 - 5\alpha < 0$, then the eigenvalues are purely imaginary. In that case, the equilibrium point is a center. If $4 - 5\alpha > 0$, then the eigenvalues are real and distinct. In that case, the equilibrium point is a center. Therefore, the critical value for α is $\alpha = 4/5$.

(c) Below we show phase portraits for $\alpha = 2/5$, and $\alpha = 6/5$.

17.(a) The characteristic equation is $\lambda^2 + 2\lambda + 1 + \alpha = 0$. Therefore, the eigenvalues are $\lambda = -1 \pm \sqrt{-\alpha}$.

(b) If $\alpha \leq 0$, then the eigenvalues are real. If $\alpha < -1$, then the eigenvalues have opposite sign. In this case, the equilibrium point is a saddle. If $-1 < \alpha < 0$, then the eigenvalues are both negative. In this case, the equilibrium is a stable node. If $\alpha > 0$, then the eigenvalues have non-zero imaginary part and negative real part. In this case, the equilibrium point is a stable spiral. Therefore, the critical values for α are $\alpha = 0$ and $\alpha = -1$.

(c) Below we show phase portraits for $\alpha = -1.5$, $\alpha = -0.5$, and $\alpha = 0.5$.

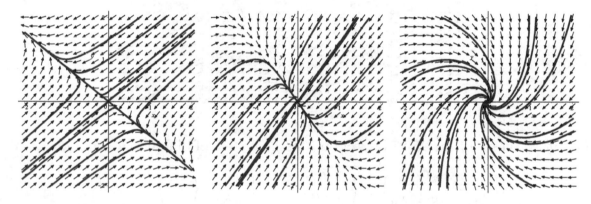

19.(a) The characteristic equation is $\lambda^2 + (4 - \alpha)\lambda + 10 - 4\alpha = 0$. Therefore, the eigenvalues are $\lambda = -2 + \dfrac{\alpha}{2} \pm \sqrt{\alpha^2 + 8\alpha - 24}$.

(b) If $\alpha^2 + 8\alpha - 24 < 0$, the eigenvalues have non-zero imaginary part. Otherwise, the eigenvalues are both real. Therefore, two of the critical values for α are $\alpha = -4 + 2\sqrt{10}$ and $-4 - 2\sqrt{10}$. For $-4 - 2\sqrt{10} < \alpha < -4 + 2\sqrt{10}$, $-2 + \alpha/2 < 0$. Therefore, for $-4 - 2\sqrt{10} < \alpha < -4 + 2\sqrt{10}$, the eigenvalues have non-zero imaginary part with negative real part. In this case, the equilibrium is a stable spiral. Now we consider the case when $\alpha^2 + 8\alpha - 24 \geq 0$. In this case, both eigenvalues are real. We need to determine what the sign of these eigenvalues is. We write the eigenvalues as

$$\lambda = \frac{((\alpha - 4) \pm \sqrt{\alpha^2 + 8\alpha - 24})}{2} = \frac{(\alpha - 4) \pm \sqrt{(\alpha - 4)^2 + 16\alpha - 40}}{2}.$$

We notice that if $16\alpha - 40 < 0$, then both eigenvalues have the same sign. Therefore, if $\alpha < -4 - 2\sqrt{10}$, then both eigenvalues are negative, and, therefore, the equilibrium is a stable node. Also if $-4 + 2\sqrt{10} < \alpha < 5/2$, then the eigenvalues are both negative, and the equilibrium is a stable node. On the other hand, if $5/2 < \alpha$, then the eigenvalues have opposite sign and the equilibrium point is a saddle.

(c) Below we show phase portraits for $\alpha = -8$, $\alpha = 0$, $\alpha = 2.4$ and $\alpha = 3$.

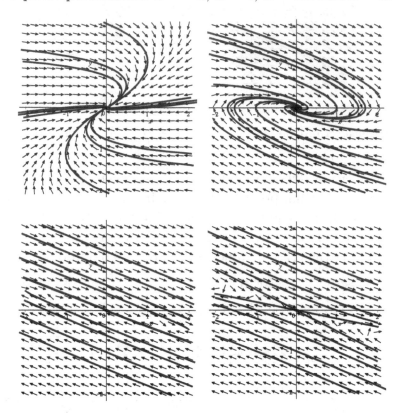

21.(a) Based on problems 25-27 of Section 3.2, the system of differential equations is

$$\frac{d}{dt}\begin{pmatrix} i \\ v \end{pmatrix} = \begin{pmatrix} -R_1/L & -1/L \\ 1/C & -1/CR_2 \end{pmatrix}\begin{pmatrix} i \\ v \end{pmatrix}.$$

Plugging in the given values for R_1, R_2, C, L, we arrive at the given system.

(b) The characteristic equation of the system is $\lambda^2 + \lambda + 1/2 = 0$. Therefore, the eigenvalues are $\lambda = (-1 \pm i)/2$. A corresponding eigenvector for $\lambda_1 = (-1+i)/2$ is

$$\mathbf{v}_1 = \begin{pmatrix} 1 \\ -4i \end{pmatrix}.$$

Therefore, one solution of this equation is

$$\begin{pmatrix} i \\ v \end{pmatrix} = e^{(-1+i)t/2}\begin{pmatrix} 1 \\ -4i \end{pmatrix} = e^{-t/2}\begin{pmatrix} \cos(t/2) \\ 4\sin(t/2) \end{pmatrix} + ie^{-t/2}\begin{pmatrix} \sin(t/2) \\ -4\cos(t/2) \end{pmatrix}.$$

Taking the real and imaginary parts of this solution, we arrive at the general solution,

$$\begin{pmatrix} i \\ v \end{pmatrix} = c_1 e^{-t/2}\begin{pmatrix} \cos(t/2) \\ 4\sin(t/2) \end{pmatrix} + c_2 e^{-t/2}\begin{pmatrix} \sin(t/2) \\ -4\cos(t/2) \end{pmatrix}.$$

(c) For the given initial conditions, we conclude that $c_1 = 2$ and $c_2 = -3/4$. Therefore,

$$\begin{pmatrix} i \\ v \end{pmatrix} = e^{-t/2} \begin{pmatrix} 2\cos(t/2) - \frac{3}{4}\sin(t/2) \\ 8\sin(t/2) + 3\cos(t/2) \end{pmatrix}.$$

(d) Since the eigenvalues have negative real part, all solutions converge to zero.

23.(a) The characteristic equation is $\lambda^2 - (a_{11} + a_{22})\lambda + a_{11}a_{22} - a_{12}a_{21} = 0$. Therefore, the eigenvalues are

$$\lambda = \frac{(a_{11} + a_{22}) \pm \sqrt{(a_{11} + a_{22})^2 - 4(a_{11}a_{22} - a_{12}a_{21})}}{2}.$$

Therefore, from this equation, we see that in order for the eigenvalues to be purely imaginary, we need $a_{11} + a_{22} = 0$ and $a_{11}a_{22} - a_{12}a_{21} > 0$.

(b) Equation (iii) can be rewritten as

$$-(a_{21}x + a_{22}y)dx + (a_{11}x + a_{12}y)dy = 0.$$

To show that this equation is exact, we need to show that for $M = -(a_{21}x + a_{22}y)$ and $N = (a_{11}x + a_{12}y)$, $M_y = N_x$. We see that $M_y = -a_{22}$ and $N_x = a_{11}$. Then, using the fact that $a_{11} + a_{22} = 0$, we see that $N_x = a_{11} = -a_{22} = M_y$. Therefore, the equation is exact.

(c) Since the equation is exact, we begin by integrating M with respect to x. We conclude that $\psi(x, y) = -a_{21}x^2/2 - a_{22}xy + h(y)$. Then, taking the derivative with respect to y, we have $-a_{22}x + h'(y) = N = a_{11}x + a_{12}y$. Therefore, $h'(y) = a_{12}y$ which implies that $h(y) = a_{12}y^2/2$. Therefore, the solution of this exact equation is given implicitly by $-a_{21}x^2/- a_{22}xy + a_{12}y^2/2 = k$ or $a_{21}x^2 = 2a_{22}xy - a_{12}y^2 = k$. The discriminant of the quadratic form is

$$(2a_{22})^2 - 4a_{21}(-a_{12}) = 4a_{22}^2 + 4a_{12}a_{21} = 4(-a_{11}a_{22} + a_{12}a_{21}).$$

Since $a_{11}a_{22} - a_{12}a_{21} > 0$, by equation (ii), we can conclude that the discriminant of the quadratic form is always negative, and, therefore, the graph is always an ellipse.

3.5 Repeated Eigenvalues

1.

$$A - \lambda I = \begin{pmatrix} 3 - \lambda & -4 \\ 1 & -1 - \lambda \end{pmatrix}$$

implies $\det(A - \lambda I) = \lambda^2 - 2\lambda + 1 = (\lambda - 1)^2$. Therefore, $\lambda = 1$ is the only eigenvalue. Now $\lambda = 1$ implies

$$A - \lambda I = \begin{pmatrix} 2 & -4 \\ 1 & -2 \end{pmatrix}.$$

Therefore,

$$\mathbf{v}_1 = \begin{pmatrix} 2 \\ 1 \end{pmatrix}$$

is an eigenvector for $\lambda = 1$ and

$$\mathbf{x}_1(t) = e^t \begin{pmatrix} 2 \\ 1 \end{pmatrix}$$

is one solution. Next, we need to look for a solution $\mathbf{w}$ of

$$(A - I)\mathbf{w} = \begin{pmatrix} 2 \\ 1 \end{pmatrix}.$$

Plugging in for $A - I$, this equation becomes

$$\begin{pmatrix} 2 & -4 \\ 1 & -2 \end{pmatrix} \mathbf{w} = \begin{pmatrix} 2 \\ 1 \end{pmatrix}.$$

We see that

$$\mathbf{w} = \begin{pmatrix} 1 \\ 0 \end{pmatrix}$$

is a solution of this equation. Therefore,

$$\mathbf{x}_2(t) = te^t \begin{pmatrix} 2 \\ 1 \end{pmatrix} + e^t \begin{pmatrix} 1 \\ 0 \end{pmatrix}$$

is also a solution of our original system. Therefore, the general solution of the system is

$$\mathbf{x}(t) = c_1 e^t \begin{pmatrix} 2 \\ 1 \end{pmatrix} + c_2 \left[te^t \begin{pmatrix} 2 \\ 1 \end{pmatrix} + e^t \begin{pmatrix} 1 \\ 0 \end{pmatrix} \right].$$

Nonzero solutions grow as $t \to \infty$.

3.

$$A - \lambda I = \begin{pmatrix} -3/2 - \lambda & 1 \\ -1/4 & -1/2 - \lambda \end{pmatrix}$$

implies $\det(A - \lambda I) = \lambda^2 + 2\lambda + 1 = (\lambda + 1)^2$. Therefore, $\lambda = -1$ is the only eigenvalue. Now $\lambda = -1$ implies

$$A - \lambda I = \begin{pmatrix} -1/2 & 1 \\ -1/4 & 1/2 \end{pmatrix}.$$

Therefore,

$$\mathbf{v}_1 = \begin{pmatrix} 2 \\ 1 \end{pmatrix}$$

is an eigenvector for $\lambda = -1$ and

$$\mathbf{x}_1(t) = e^{-t} \begin{pmatrix} 2 \\ 1 \end{pmatrix}$$

is one solution. Next, we need to look for a solution $\mathbf{w}$ of

$$(A + I)\mathbf{w} = \begin{pmatrix} 2 \\ 1 \end{pmatrix}.$$

Plugging in for $A + I$, this equation becomes

$$\begin{pmatrix} -1/2 & 1 \\ -1/4 & 1/2 \end{pmatrix} \mathbf{w} = \begin{pmatrix} 2 \\ 1 \end{pmatrix}.$$

We see that

$$\mathbf{w} = \begin{pmatrix} 0 \\ 2 \end{pmatrix}$$

is a solution of this equation. Therefore,

$$\mathbf{x}_2(t) = te^{-t} \begin{pmatrix} 2 \\ 1 \end{pmatrix} + e^{-t} \begin{pmatrix} 0 \\ 2 \end{pmatrix}$$

is also a solution of our original system. Therefore, the general solution of the system is

$$\mathbf{x}(t) = c_1 e^{-t} \begin{pmatrix} 2 \\ 1 \end{pmatrix} + c_2 \left[te^{-t} \begin{pmatrix} 2 \\ 1 \end{pmatrix} + e^{-t} \begin{pmatrix} 0 \\ 2 \end{pmatrix} \right].$$

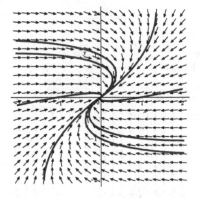

The solutions approach the origin as $t \to \infty$.

5.

$$A - \lambda I = \begin{pmatrix} -1 - \lambda & -1/2 \\ 2 & -3 - \lambda \end{pmatrix}$$

implies $\det(A - \lambda I) = \lambda^2 + 4\lambda + 4 = (\lambda + 2)^2$. Therefore, $\lambda = -2$ is the only eigenvalue. Now $\lambda = -2$ implies

$$A - \lambda I = \begin{pmatrix} 1 & -1/2 \\ 2 & -1 \end{pmatrix}.$$

Therefore,

$$\mathbf{v}_1 = \begin{pmatrix} 1 \\ 2 \end{pmatrix}$$

is an eigenvector for $\lambda = -2$ and

$$\mathbf{x}_1(t) = e^{-2t} \begin{pmatrix} 1 \\ 2 \end{pmatrix}$$

is one solution. Next, we need to look for a solution $\mathbf{w}$ of

$$(A + 2I)\,\mathbf{w} = \begin{pmatrix} 1 \\ 2 \end{pmatrix}.$$

Plugging in for $A + 2I$, this equation becomes

$$\begin{pmatrix} 1 & -1/2 \\ 2 & -1 \end{pmatrix} \mathbf{w} = \begin{pmatrix} 1 \\ 2 \end{pmatrix}.$$

We see that

$$\mathbf{w} = \begin{pmatrix} 1 \\ 0 \end{pmatrix}$$

is a solution of this equation. Therefore,

$$\mathbf{x}_2(t) = te^{-2t} \begin{pmatrix} 1 \\ 2 \end{pmatrix} + e^{-2t} \begin{pmatrix} 1 \\ 0 \end{pmatrix}$$

is also a solution of our original system. Therefore, the general solution of the system is

$$\mathbf{x}(t) = c_1 e^{-2t} \begin{pmatrix} 1 \\ 2 \end{pmatrix} + c_2 \left[te^{-2t} \begin{pmatrix} 1 \\ 2 \end{pmatrix} + e^{-2t} \begin{pmatrix} 1 \\ 0 \end{pmatrix} \right].$$

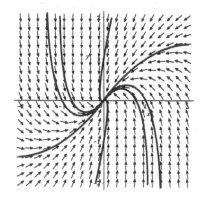

The solutions tend to the origin as $t \to \infty$.

7. The characteristic equation is $(\lambda + 3)^2 = 0$. Therefore, the only eigenvalue is $\lambda = -3$. A corresponding eigenvector is given by

$$\mathbf{v}_1 = \begin{pmatrix} 1 \\ 1 \end{pmatrix}$$

and

$$\mathbf{x}_1(t) = e^{-3t} \begin{pmatrix} 1 \\ 1 \end{pmatrix}$$

is one solution of the system. Next, we need to look for a solution $\mathbf{w}$ of

$$(A + 3I)\,\mathbf{w} = \begin{pmatrix} 1 \\ 1 \end{pmatrix}.$$

Plugging in for $A + 3I$, this equation becomes

$$\begin{pmatrix} 4 & -4 \\ 4 & -4 \end{pmatrix} \mathbf{w} = \begin{pmatrix} 1 \\ 1 \end{pmatrix}.$$

We see that

$$\mathbf{w} = \begin{pmatrix} 1/4 \\ 0 \end{pmatrix}$$

is a solution of this equation. Therefore,

$$\mathbf{x}_2(t) = te^{-3t} \begin{pmatrix} 1 \\ 1 \end{pmatrix} + e^{-3t} \begin{pmatrix} 1/4 \\ 0 \end{pmatrix}$$

is also a solution of our original system. Therefore, the general solution of the system is

$$\mathbf{x}(t) = c_1 e^{-3t} \begin{pmatrix} 1 \\ 1 \end{pmatrix} + c_2 \left[te^{-3t} \begin{pmatrix} 1 \\ 1 \end{pmatrix} + e^{-3t} \begin{pmatrix} 1/4 \\ 0 \end{pmatrix} \right].$$

The initial condition implies that $c_1 = 2$ and $c_2 = 4$. Therefore, the solution of the IVP is

$$\mathbf{x}(t) = \begin{pmatrix} 3 + 4t \\ 2 + 4t \end{pmatrix} e^{-3t}.$$

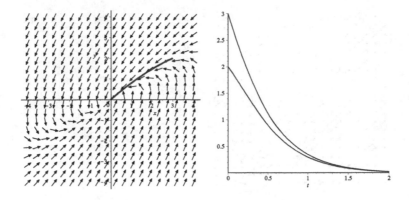

9. The characteristic equation is $(\lambda - 1/2)^2 = 0$. Therefore, the only eigenvalue is $\lambda = 1/2$. A corresponding eigenvector is given by

$$\mathbf{v}_1 = \begin{pmatrix} -1 \\ 1 \end{pmatrix}$$

and

$$\mathbf{x}_1(t) = e^{t/2} \begin{pmatrix} -1 \\ 1 \end{pmatrix}$$

is one solution of the system. Next, we need to look for a solution $\mathbf{w}$ of

$$\left(A - \frac{1}{2}I\right)\mathbf{w} = \begin{pmatrix} -1 \\ 1 \end{pmatrix}.$$

Plugging in for $A - \frac{1}{2}I$, this equation becomes

$$\begin{pmatrix} 3/2 & 3/2 \\ -3/2 & -3/2 \end{pmatrix}\mathbf{w} = \begin{pmatrix} -1 \\ 1 \end{pmatrix}.$$

We see that

$$\mathbf{w} = \begin{pmatrix} -2/3 \\ 0 \end{pmatrix}$$

is a solution of this equation. Therefore,

$$\mathbf{x}_2(t) = te^{t/2} \begin{pmatrix} -1 \\ 1 \end{pmatrix} + e^{t/2} \begin{pmatrix} -2/3 \\ 0 \end{pmatrix}$$

is also a solution of our original system. Therefore, the general solution of the system is

$$\mathbf{x}(t) = c_1 e^{t/2} \begin{pmatrix} -1 \\ 1 \end{pmatrix} + c_2 \left[te^{t/2} \begin{pmatrix} -1 \\ 1 \end{pmatrix} + e^{t/2} \begin{pmatrix} -2/3 \\ 0 \end{pmatrix} \right].$$

The initial condition implies that $c_1 = -2$ and $c_2 = -3/2$. Therefore, the solution of the IVP is

$$\mathbf{x}(t) = \begin{pmatrix} 3 + \frac{3}{2}t \\ -2 - \frac{3}{2}t \end{pmatrix} e^{t/2}.$$

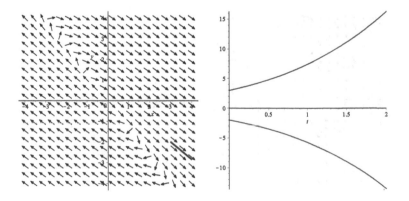

11. From our answer to problem 4, we see that the general solution of this system is given by

$$\mathbf{x}(t) = c_1 e^{-t/2} \begin{pmatrix} 1 \\ 1 \end{pmatrix} + c_2 \left[te^{-t/2} \begin{pmatrix} 1 \\ 1 \end{pmatrix} + e^{-t/2} \begin{pmatrix} -2/5 \\ 0 \end{pmatrix} \right].$$

Now our initial condition $\mathbf{x}(0) = \begin{pmatrix} 3 \\ 1 \end{pmatrix}$ implies

$$\mathbf{x}(0) = c_1 \begin{pmatrix} 1 \\ 1 \end{pmatrix} + c_2 \begin{pmatrix} -2/5 \\ 0 \end{pmatrix} = \begin{pmatrix} 3 \\ 1 \end{pmatrix}.$$

Therefore, $c_1 = 1$ and $c_2 = -5$. Therefore,

$$\mathbf{x}(t) = e^{-t/2} \begin{pmatrix} 1 \\ 1 \end{pmatrix} - 5 \left[te^{-t/2} \begin{pmatrix} 1 \\ 1 \end{pmatrix} + e^{-t/2} \begin{pmatrix} -2/5 \\ 0 \end{pmatrix} \right] = \begin{pmatrix} 3 - 5t \\ 1 - 5t \end{pmatrix} e^{-t/2}.$$

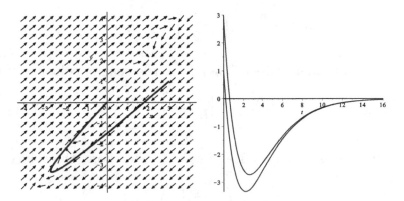

13.(a) The eigenvalues of this system are given by

$$\lambda = -\frac{1}{2RC} \pm \frac{\sqrt{L^2 - 4R^2CL}}{2RCL}.$$

The eigenvalues will be real and equal if the discriminant vanishes. The discriminant will vanish when $L = 4R^2C$.

(b) For the given conditions, the system is

$$\frac{d}{dt} \begin{pmatrix} i \\ v \end{pmatrix} = \begin{pmatrix} 0 & 1/4 \\ -1 & -1 \end{pmatrix} \begin{pmatrix} i \\ v \end{pmatrix}.$$

The characteristic equation of this system is $\lambda^2 + \lambda + 1/4 = 0$. Therefore, the only eigenvalue is $\lambda = -1/2$. An associated eigenvector is given by

$$\mathbf{v}_1 = \begin{pmatrix} 1 \\ -2 \end{pmatrix}.$$

Therefore, one solution of this system is

$$\mathbf{x}_1(t) = e^{-t/2} \begin{pmatrix} 1 \\ -2 \end{pmatrix}.$$

Next, we need to look for a solution $\mathbf{w}$ of

$$\left(A + \frac{1}{2}I\right)\mathbf{w} = \begin{pmatrix} 1 \\ -2 \end{pmatrix}.$$

Plugging in for $A + \frac{1}{2}I$, this equation becomes

$$\begin{pmatrix} 1/2 & 1/4 \\ -1 & -1/2 \end{pmatrix}\mathbf{w} = \begin{pmatrix} 1 \\ -2 \end{pmatrix}.$$

We see that

$$\mathbf{w} = \begin{pmatrix} 0 \\ 4 \end{pmatrix}$$

is a solution of this equation. Therefore,

$$\mathbf{x}_2(t) = te^{-t/2}\begin{pmatrix} 1 \\ -2 \end{pmatrix} + e^{-t/2}\begin{pmatrix} 0 \\ 4 \end{pmatrix}$$

is also a solution of our original system. Therefore, the general solution of the system is

$$\mathbf{x}(t) = c_1 e^{-t/2}\begin{pmatrix} 1 \\ -2 \end{pmatrix} + c_2\left[te^{-t/2}\begin{pmatrix} 1 \\ -2 \end{pmatrix} + e^{-t/2}\begin{pmatrix} 0 \\ 4 \end{pmatrix}\right].$$

The initial condition implies that $c_1 = 1$ and $c_2 = 1$. Therefore, the solution of the IVP is

$$\mathbf{x}(t) = \begin{pmatrix} 1+t \\ 2-2t \end{pmatrix} e^{-t/2}.$$

15. The characteristic equation is $\lambda^2 - p\lambda + q = 0$. The eigenvalues are

$$\lambda = \frac{p \pm \sqrt{p^2 - 4q}}{2} = \frac{p \pm \sqrt{\Delta}}{2}.$$

The results can be verified using Table 3.5.2.

3.6 A Brief Introduction to Nonlinear Systems

1.(a) $\dfrac{dx}{dt} = -x$, $\dfrac{dy}{dt} = -2y$, thus $\dfrac{dy}{dx} = \dfrac{2y}{x}$, and then

$$\int \frac{1}{y}\,dy = 2\int \frac{1}{x}\,dx,$$

thus $\ln|y| = 2\ln|x| + C$. Applying the exponential function to this equation, we conclude that $y = Cx^2$ for any solution of the system. That is, $H(x,y) = y/x^2$.

(b)

(c)

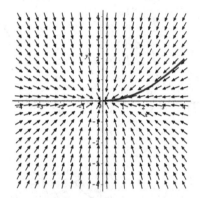

As t increases, the x and y values are decreasing.

3.(a) $\dfrac{dx}{dt} = -x$, $\dfrac{dy}{dt} = 2y$, thus $\dfrac{dy}{dx} = -\dfrac{2y}{x}$, and then

$$\int \frac{1}{y}\, dy = -2 \int \frac{1}{x}\, dx,$$

thus $\ln|y| = -2\ln|x| + C$. Applying the exponential function to this equation, we conclude that $y = C/x^2$ ' for any solution of the system. That is, $H(x,y) = yx^2$.

(b)

(c)

As t increases, the x values decrease and the y values do not change.

5.(a) $\dfrac{dx}{dt} = 2y$, $\dfrac{dy}{dt} = 8x$, thus $\dfrac{dy}{dx} = \dfrac{4x}{y}$, and then

$$\int y \, dy = \int 4x \, dx,$$

thus $y^2/2 = 2x^2 + C$. Therefore, $y^2 - 4x^2 = C$ for any solution of the system. That is, $H(x, y) = y^2 - 4x^2$.

(b)

(c)

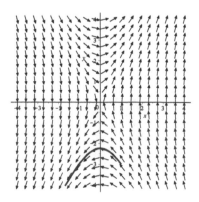

As t increases, the x and y values are decreasing.

7.(a) To find all critical points, we need to solve the following system of equations.

$$\begin{aligned} x' &= 2x - y = 0 \\ y' &= x - 2y = 0. \end{aligned}$$

From the second equation, $x = 2y$ implies $4y - y = 0$ which implies the only critical point is $(0,0)$.

(b) From our equations for dx/dt and dy/dt, we see that

$$\frac{dy}{dx} = \frac{x - 2y}{2x - y},$$

thus $(2x - y)\,dy = (x - 2y)\,dx$, thus $(2y - x)\,dx + (2x - y)\,dy = 0$. We notice that this is an exact equation, which can be rewritten as

$$\left(2xy - \frac{x^2}{2} - \frac{y^2}{2}\right)' = 0.$$

Therefore,

$$H(x, y) = 2xy - \frac{x^2}{2} - \frac{y^2}{2}.$$

(c)

Solutions lying on curves passing through the second quadrant travel to the left. In particular, the x values and y values decrease. Solutions lying on curves solely in the first quadrant will travel to the right along the bottom portion of the curve. Solutions lying on curves solely in the third quadrant will travel to the left along the top portion of the curve. Solutions lying on curves passing through the fourth quadrant will travel to the right with both x and y values increasing.

(d) Solutions lying along the part of the level curve $H(x, y) = 0$ such that $y = (2 + \sqrt{3})x$ will approach the origin. All other solutions will tend away from the critical point as $t \to \infty$.

9.(a) To find all critical points, we need to solve the following system of equations.

$$\begin{aligned} x' &= 2x - 4y = 0 \\ y' &= 2x - 2y = 0. \end{aligned}$$

Solving the two equations, we see that the only critical point is $(0, 0)$.

(b) From our equations for dx/dt and dy/dt, we see that

$$\frac{dy}{dx} = \frac{2x - 2y}{2x - 4y},$$

thus $(2x - 4y)\, dy = (2x - 2y)\, dx$, thus $(2x - 2y)\, dx + (4y - 2x)\, dy = 0$. We notice that this is an exact equation, which can be rewritten as

$$\left(x^2 - 2xy + 2y^2\right)' = 0.$$

Therefore, $H(x, y) = x^2 - 2xy + 2y^2$.

(c)

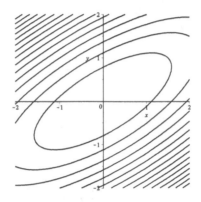

Solutions travel in the counter-clockwise direction along the trajectories shown above.

(d) Solutions orbit the critical point. They do not tend towards or away from the critical point.

11.(a) To find all critical points, we need to solve the following system of equations.

$$\begin{aligned} x' &= 2x^2y - 3x^2 - 4y = 0 \\ y' &= -2xy^2 + 6xy = 0. \end{aligned}$$

Solving the second equation, we see that we must have $x = 0$, $y = 0$ or $y = 3$. Plugging these values into the first equation, we see that the critical points are: $(0, 0)$, $(2, 3)$ and $(-2, 3)$.

(b) From our equations for dx/dt and dy/dt, we see that

$$\frac{dy}{dx} = \frac{-2xy^2 + 6xy}{2x^2y - 3x^2 - 4y}.$$

Rewriting this equation as $(2xy^2 - 6xy)dx + (2x^2y - 3x^2 - 4y)dy = 0$, we see that this is an exact equation. For $M = 2xy^2 - 6xy$, we integrate M with respect to x. We conclude that the level curves are given by $H(x, y) = C$ where $H(x, y) = x^2y^2 - 3x^2y + h(y)$. Differentiating with respect to y, we have $H_y = 2\cdot{}^2y - 3x^2 + h'(y) = N = 2x^2y - 3x^2 - 4y$. Therefore, $h'(y) = -4y$ which implies that $h(y) = -2y^2$. Therefore, $H(x, y) = x^2y^2 - 3x^2y - 2y^2$.

(c)

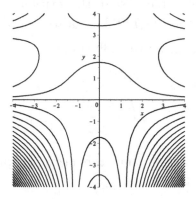

The direction of motion along each trajectory can be seen from the direction field shown below:

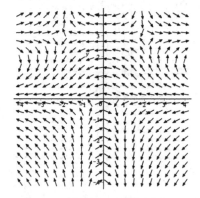

(d) From the direction field above, we see that all critical points are saddle points.

13.(a) The critical points are given by the solutions of the equations

$$
\begin{aligned}
x' &= x(1-y) = 0 \\
y' &= y(1+2x) = 0.
\end{aligned}
$$

The solutions of the first equation are $x = 0$ and $y = 1$. Plugging these values into the second equation, we conclude that the critical points are $(0,0)$, $(-1/2, 1)$.

(b)

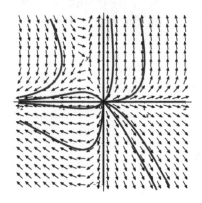

(c) Based on the phase portrait, we conclude that all trajectories starting near the origin diverge. Therefore, $(0,0)$ is unstable. Near the critical point $(-1/2, 1)$, the trajectories behave like a saddle, and, therefore, $(-1/2, 1)$ is also unstable.

15.(a) The critical points are given by the solutions of the equations

$$
\begin{aligned}
x' &= x(1 - x - y) = 0 \\
y' &= y\left(\frac{1}{2} - \frac{1}{4}y - \frac{3}{4}x\right) = 0.
\end{aligned}
$$

Solving these two equations, we see that the critical points are given by $(0,0)$, $(0,2)$, $(1/2, 1/2)$ and $(1, 0)$.

(b)

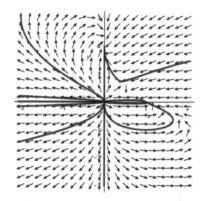

(c) Based on the phase portrait, we conclude that all trajectories starting near the origin diverge. Therefore, $(0,0)$ is unstable. Near the critical point $(0,2)$, the trajectories approach the critical point. Similarly, near $(1,0)$, trajectories approach the critical point. Therefore, both of those critical points are asymptotically stable nodes. Near the critical point $(1/2, 1/2)$, trajectories behave like a saddle.

17.(a) The critical points are given by the solutions of the equations

$$
\begin{aligned}
x' &= y(2 - x - y) = 0 \\
y' &= -x - y - 2xy = 0.
\end{aligned}
$$

The solutions of the first equation are $y = 0$ and $x + y = 2$. Plugging these values into the second equation, we conclude that the critical points are $(0,0)$, $(1 - \sqrt{2}, 1 + \sqrt{2})$, and $(1 + \sqrt{2}, 1 - \sqrt{2})$.

(b)

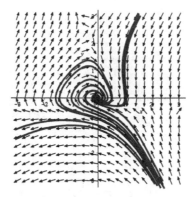

(c) Based on the phase portrait, we conclude that $(0,0)$ is an asymptotically stable spiral point, while $(1 - \sqrt{2}, 1 + \sqrt{2})$ and $(1 + \sqrt{2}, 1 - \sqrt{2})$ are saddle points.

19.(a) The critical points are given by the solutions of the equations

$$\begin{aligned} x' &= -x(1 - 2y) = 0 \\ y' &= y - x^2 - y^2 = 0. \end{aligned}$$

The solutions of the first equation are $x = 0$ and $y = 1/2$. Plugging these values into the second equation, we conclude that the critical points are $(0,0)$, $(0,1)$, $(1/2, 1/2)$ and $(-1/2, 1/2)$.

(b)

(c) Based on the phase portrait, we conclude that $(0,0)$ and $(0,1)$ are saddle points, while $(1/2, 1/2)$ and $(-1/2, 1/2)$ are centers.

21.(a)

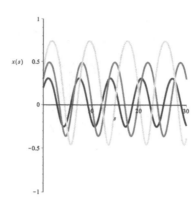

(b) For the first solution, the period is approximately 6 seconds and the amplitude is approximately 0.6. For the second solution, the period is approximately 7 and the amplitude is approximately 1. For the third solution, the period is approximately 8 seconds and the amplitude is approximately 1.5. The period decreases as the amplitude decreases.

23.(a) $A(-1,0)$ and $C(1,0)$ are asymptotically stable spiral points, $B(0,0)$ is a saddle point.

(b) (i) C (ii) A (iii) A (iv) C

25. The critical points are given by the solutions of the equations

$$
\begin{aligned}
x' &= x(1-x-y)=0 \\
y' &= y(3/4-y-x/2)=0.
\end{aligned}
$$

The solutions of the first equation are $x=0$ and $x=1+y$. Plugging these values into the second equation, we conclude that the critical points are $(0,0)$, $(0,3/4)$, $(1,0)$ and $(1/2,1/2)$.

The limiting behavior of the solutions shows that the populations co-exist.

Chapter 4

Second Order Linear Equations

4.1 Definitions and Examples

1. The differential equation is linear, homogeneous.

3. The differential equation is linear, homogeneous.

5. The differential equation is nonlinear, due to the $y^2 y'$ term.

7. The differential equation is linear; nonhomogeneous when $d \neq 0$, homogeneous when $d = 0$.

9. We use the equation $mg - kL = 0$. Here the mass is 10 kg. The force due to gravity is $g = 9.8$ m/s^2. Therefore, $mg = 98$ N. The mass stretches the spring 0.7 meters. Therefore, $k = 98/0.7 = 140$ N/m.

11. The spring constant is $k = 0.98/0.2 = 4.9$ N/m. The mass $m = 0.1$ kg. The equation of motion is

$$0.1y'' + 4.9y = 0$$

or $y'' + 49y = 0$, with initial conditions $y(0) = 0$ m, $y'(0) = 0.05$ m/sec.

13. The inductance $L = 1$ henry. The resistance $R = 0$. The capacitance $C = 0.25 \times 10^{-6}$ farads. Therefore, the equation for charge Q is $Q'' + (4 \times 10^6)Q = 0$, with initial conditions $Q(0) = 10^{-6}$ coulombs, $Q'(0) = 0$ coulombs/sec.

15. The spring constant is $k = 16/(1/4) = 64$ lb/ft. The mass $m = 1/2$ lb·s^2/ft (slug). The damping coefficient is $\gamma = 2$ lb-sec/ft. Therefore, the equation of motion is

$$\frac{1}{2}y'' + 2y' + 64y = 0$$

or $y'' + 4y' + 128y = 0$, with initial conditions $y(0) = 0$ ft, $y'(0) = 1/4$ ft/sec.

17. The inductance $L = 0.2$ henry. The resistance $R = 3 \times 10^2$ ohms. The capacitance $C = 10^{-5}$ farads. Therefore, the equation for charge Q is $0.2Q'' + 300Q' + 10^5 Q = 0$ or $Q'' + (1.5 \times 10^3)Q' + (5 \times 10^5)q = 0$, with initial conditions $Q(0) = 10^{-6}$ coulombs, $Q'(0) = 0$ coulombs/sec.

19.(a) We know that the net force is given by $F_{\text{net}} = mx''(t)$. Here, we are assuming there are no damping of external forces present. Also, we are assuming there is no force due to gravity. Therefore, $F_{\text{net}} = F_s$, i.e. $mx'' = -kx - \epsilon x^3$, so $mx'' + kx + \epsilon x^3 = 0$.

(b) If we assume that the maximum displacement is small, then $x^3 \approx 0$, in which case the linearized equation is $mx'' + kx = 0$.

21. Let $u(t)$ be the depth of the block into the water. Let m be the mass of the block and $F_B(t)$ be the upward buoyant force exerted by the fluid. The equation of motion is given by $mu''(t) = F_B(t)$. Since the density of the cubic block is ρ, the mass $m = \rho l^3$. Since the buoyant force is equal to the weight of the displaced fluid, we see that $F_B = -\rho_0 u(t) l^2 g$. Therefore, $\rho l^3 u'' = -\rho_0 u(t) l^2 g$, or $\rho l u'' + \rho_0 g u = 0$.

23.(a)

(b)

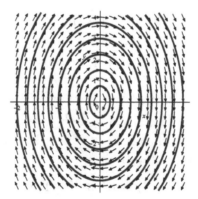

(c) The critical point is stable.

25.(a)

(b)

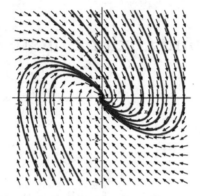

(c) The critical point is asymptotically stable.

27.(a)

(b)

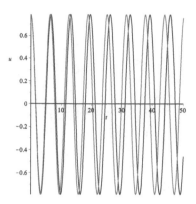

The frequency of the linear pendulum is higher than the frequency of the nonlinear pendulum. The difference between frequencies increases with the amplitude of oscillations.

4.2   Theory of Second Order Linear Homogeneous Equations

1. Write the initial value problem as

$$y'' + \frac{3}{t}y' = 1.$$

Since the function $p(t) = 3/t$ is continuous for all $t > 0$ and $t_0 = 1 > 0$, the initial value problem is guaranteed to have a unique solution for all $t > 0$. The longest interval of existence is $(0, \infty)$.

3. Write the initial value problem as

$$y'' + \frac{3t}{t(t-4)}y' + \frac{4}{t(t-4)}y = \frac{2}{t(t-4)}.$$

Since the coefficient functions are continuous for all t such that $t \neq 0, 4$ and $t_0 = 3$, the initial value problem is guaranteed to have a unique solution for all t such that $0 < t < 4$.

5. Write the initial value problem as

$$y'' + \frac{x}{x+3}y' + \frac{\ln|x|}{x+3}y = 0.$$

Since the coefficient functions are continuous for all x such that $x \neq 0, -3$ and $x_0 = 1$, the initial value problem is guaranteed to have a unique solution for all $x > 0$. The longest interval of existence is $(0, \infty)$.

7. Write the initial value problem as

$$y'' - \frac{2x}{1-x^2}y' + \left(\frac{\alpha(\alpha+1)}{1-x^2} + \frac{\mu^2}{(1-x^2)^2}\right)y = 0.$$

Since the coefficient functions are continuous for all x such that $x \neq -1, 1$ and $x_0 = 0$, the initial value problem is guaranteed to have a unique solution for $-1 < x < 1$.

9.
$$W(e^{2t}, e^{-3t/2}) = \begin{vmatrix} e^{2t} & e^{-3t/2} \\ 2e^{2t} & -\frac{3}{2}e^{-3t/2} \end{vmatrix} = -\frac{7}{2}e^{t/2}.$$

11.
$$W(e^{-2t}, te^{-2t}) = \begin{vmatrix} e^{-2t} & te^{-2t} \\ -2e^{-2t} & (1-2t)e^{-2t} \end{vmatrix} = e^{-4t}.$$

13.
$$W(e^t \sin t, e^t \cos t) = \begin{vmatrix} e^t \sin t & e^t \cos t \\ e^t(\sin t + \cos t) & e^t(\cos t - \sin t) \end{vmatrix} = -e^{2t}.$$

15. If $y_1 = t^2$, then $y_1'' = 2$. Therefore, $t^2 y_1'' - 2y_1 = t^2(2) - 2t^2 = 0$. If $y_2 = t^{-1}$, then $y_2' = 2t^{-3}$. Therefore, $t^2 y_2'' - 2y_2 = t^2(2t^{-3}) - 2t^{-1} = 0$. Since the equation is linear, the function $y_3 = c_1 t^2 + c_2 t^{-1}$ will also be a solution.

17. No. Substituting $y = \sin(t^2)$ into the differential equation, we have

$$-4t^2 \sin(t^2) + 2\cos(t^2) + 2t\cos(t^2)p(t) + \sin(t^2)q(t) = 0.$$

The left side is 2 at $t = 0$, the right side is 0 at $t = 0$, thus the function does not satisfy the equation.

19. $W(t, g(t)) = tg'(t) - g(t) = t^2 e^t$. Dividing both sides of the equation by t, we have $g' - g/t = te^t$. Therefore, $g(t) = te^t + ct$.

21. $W(f, g) = fg' - f'g = t\cos t - \sin t$ and $W(u, v) = (f + 3g)(f - g)' - (f + 3g)'(f - g) = -4fg' + 4f'g$. Therefore, $W(u, v) = -4t\cos t + 4\sin t$.

23. For $y_1 = e^t$, $y_1' = y_1'' = e^t$. Therefore, $y_1'' - 2y_1' + y_1 = e^t - 2e^t + e^t = 0$. For $y_2 = te^t$, $y_2' = (1 + t)e^t$ and $y_2'' = (2 + t)e^t$. Therefore, $y_2'' - 2y_2' + y_2 = (2 + t)e^t - 2(1 + t)e^t + te^t = 0$. Further, $W(e^t, te^t) = e^{2t}$. Therefore, the solutions form a fundamental set.

25. For $y_1 = x$, $y_1' = 1$ and $y_1'' = 0$. Therefore, $(1 - x\cot x)y_1'' - xy_1' + y_1 = -x + x = 0$. For $y_2 = \sin x$, $y_2' = \cos x$ and $y_2'' = -\sin x$. Therefore, $(1 - x\cot x)y_2'' - xy_2' + y_2 = -(1 - x\cot x)\sin x - x\cos x + \sin x = 0$. Further, $W(x, \sin x) = x\cos x - \sin x$ which is nonzero for $0 < x < \pi$. Therefore, the functions form a fundamental set of solutions.

27. We compute:

$$\begin{aligned} K[c_1\mathbf{x}_1 + c_2\mathbf{x}_2] &= (c_1\mathbf{x}_1 + c_2\mathbf{x}_2)' - \mathbf{P}(t)(c_1\mathbf{x}_1 + c_2\mathbf{x}_2) \\ &= c_1(\mathbf{x}_1' - \mathbf{P}(t)\mathbf{x}_1) + c_2(\mathbf{x}_2' - \mathbf{P}(t)\mathbf{x}_2) = c_1 K[\mathbf{x}_1] + c_2 K[\mathbf{x}_2]. \end{aligned}$$

This shows that Theorem 4.2.4 is true. Corollary 4.2.5 is an immediate consequence as in this case the right hand side of this identity is 0, and then so is the left hand side.

29. Following the reduction of order technique given, $y_1 = t^2$, $p(t) = -4/t$, so the equation for v is $t^2 v'' + (4t - 4t^2/t)v' = t^2 v'' = 0$, so we obtain that $v' = c$ and then $v = ct$. (We use the integration constant 0 in the second integration because when we produce the second

solution $y = vy_1$ the second integration constant just adds the first solution y_1.) The second solution in this case is $y = ct^3$. The equation is linear, so we can say $y_2 = t^3$ is the second solution.

31. Following the reduction of order technique given, $y_1 = 1/t$, $p(t) = 3/t$, so the equation for v is $v''/t + v'/t^2 = 0$. After separating the variables the equation becomes $v''/v' = -1/t$, so $\ln v' = -\ln t + c$. We obtain that $v' = c/t$ and then $v = c \ln t$. Thus the second solution is $y_2 = \ln t/t$.

33. Following the reduction of order technique given, $y_1 = \sin(x^2)$, $p(x) = -1/x$, so the equation for v is $\sin(x^2)v'' + (4x \cos(x^2) - \sin(x^2)/x)v' = 0$. After separating the variables the equation becomes $v''/v' = 1/x - 4x \cot(x^2)$, so $\ln v' = \ln x - 2\ln(\sin(x^2))$. We obtain that $v' = x/\sin^2(x^2)$ and then $v = -\cot(x^2)/2$. Thus the second solution is $y_2 = \cos(x^2)$.

35. Following the reduction of order technique given, $y_1 = x^{1/4}e^{2\sqrt{x}}$, $p(x) = 0$, so the equation for v is $x^{1/4}e^{2\sqrt{x}}v'' + 2e^{2\sqrt{x}}(x^{-3/4}/4 + x^{-1/4})v' = 0$. After separating the variables the equation becomes $v''/v' = -1/(2x) - 2/\sqrt{x}$, so $\ln v' = -\ln x/2 - 4\sqrt{x}$. We obtain that $v' = x^{-1/2}e^{-4\sqrt{x}}$ and then $v = -e^{-4\sqrt{x}}/2$. Thus the second solution is $y_2 = x^{1/4}e^{-2\sqrt{x}}$.

37.(a) When $y_1 = e^x$, then $y_1' = y_1'' = e^x$ and

$$xy_1'' - (x+N)y_1' + Ny_1 = xe^x - (x+N)e^x + Ne^x = 0.$$

(b) Let $y_2 = ce^x \int x^N e^{-x}dx$. Then $y_2' = ce^x \int x^N e^{-x}dx + ce^x x^N e^{-x} = y_2 + cx^N$ and $y_2'' = (y_2')' = (y_2 + cx^N)' = y_2' + cNx^{N-1} = y_2 + cx^N + cNx^{N-1}$. This shows that

$$xy_2'' - (x+N)y_2' + Ny_2 = x(y_2 + cx^N + cNx^{N-1}) - (x+N)(y_2 + cx^N) + Ny_2 =$$

$$= xy_2 - xy_2 - Ny_2 + Ny_2 + cx^{N+1} + cNx^N - cx^{N+1} - Ncx^N = 0.$$

When $N = 1$, then $y_2 = c(-1-x)$ and when $N = 2$, then $y_2 = c(-2 - 2x - x^2)$ by direct integration. The generic formula can be found by induction.

4.3   Linear Homogeneous Equations with Constant Coefficients

1.(a) The characteristic equation is given by $\lambda^2 + 2\lambda - 3 = 0$. Therefore, the two distinct roots are $\lambda = -3, 1$. Thus, the general solution is given by

$$y(t) = c_1 e^t + c_2 e^{-3t}.$$

(b) The critical point $(0,0)$ is a saddle point, therefore, unstable.

(c) For y above, we see that $y' = c_1 e^t - 3c_2 e^{-3t}$. Therefore, we can rewrite our solution as the two parameter family

$$\begin{pmatrix} x_1 \\ x_2 \end{pmatrix} = \begin{pmatrix} y \\ y' \end{pmatrix} = \begin{pmatrix} c_1 e^t + c_2 e^{-3t} \\ c_1 e^t - 3c_2 e^{-3t} \end{pmatrix} = c_1 \begin{pmatrix} 1 \\ 1 \end{pmatrix} e^t + c_2 \begin{pmatrix} 1 \\ -3 \end{pmatrix} e^{-3t}.$$

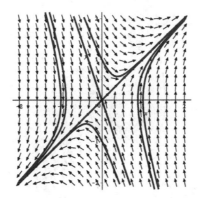

3.(a) The characteristic equation is given by $\lambda^2 - 4\lambda + 4 = 0$. Therefore, we have one repeated root $\lambda = 2$. Therefore, the general solution is given by

$$y(t) = c_1 e^{2t} + c_2 t e^{2t}.$$

(b) The critical point $(0,0)$ is an unstable improper node.

(c) For y above, we see that $y' = 2c_1 e^{2t} + c_2(1 + 2t)e^{2t}$. Therefore, we can rewrite our solution as the two parameter family

$$\begin{pmatrix} x_1 \\ x_2 \end{pmatrix} = \begin{pmatrix} y \\ y' \end{pmatrix} = \begin{pmatrix} c_1 e^{2t} + c_2 t e^{2t} \\ 2c_1 e^{2t} + c_2(1 + 2t)e^{2t} \end{pmatrix} = c_1 \begin{pmatrix} 1 \\ 2 \end{pmatrix} e^{2t} + c_2 \begin{pmatrix} t \\ 1 + 2t \end{pmatrix} e^{2t}.$$

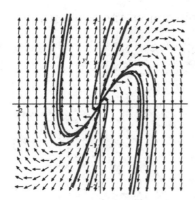

5.(a) The characteristic equation is $\lambda^2 - 2\lambda + 2 = 0$. Therefore, the roots are $\lambda = 1 \pm i$. Therefore, one complex solution is

$$y(t) = e^{(1+i)t} = e^t(\cos t + i \sin t).$$

Considering the real and imaginary parts of this function, we arrive at the general solution

$$y(t) = c_1 e^t \cos t + c_2 e^t \sin t.$$

(b) The critical point $(0,0)$ is an unstable spiral point.

(c) From our solution in part (a), we see that $y'(t) = c_1 e^t(\cos t - \sin t) + c_2 e^t(\sin t + \cos t)$. Therefore,

$$\mathbf{x} = \begin{pmatrix} y \\ y' \end{pmatrix} = c_1 e^t \begin{pmatrix} \cos t \\ \cos t - \sin t \end{pmatrix} + c_2 e^t \begin{pmatrix} \sin t \\ \sin t + \cos t \end{pmatrix}.$$

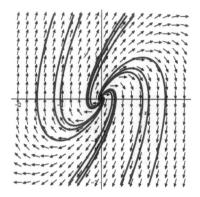

7.(a) The characteristic equation is given by $4\lambda^2 - 4\lambda + 1 = 0$. Therefore, we have one repeated root $\lambda = 1/2$. Therefore, the general solution is given by

$$y(t) = c_1 e^{t/2} + c_2 t e^{t/2}.$$

(b) The critical point $(0,0)$ is an unstable improper node.

(c) For y above, we see that

$$y' = \frac{c_1}{2} e^{t/2} + c_2 \left(1 + \frac{t}{2} \right) e^{t/2}.$$

Therefore, we can rewrite our solution as the two parameter family

$$\begin{pmatrix} x_1 \\ x_2 \end{pmatrix} = \begin{pmatrix} y \\ y' \end{pmatrix} = \begin{pmatrix} c_1 e^{t/2} + c_2 t e^{t/2} \\ \frac{c_1}{2} e^{t/2} + c_2 \left(1 + \frac{t}{2} \right) e^{t/2} \end{pmatrix} = c_1 \begin{pmatrix} 1 \\ \frac{1}{2} \end{pmatrix} e^{t/2} + c_2 \begin{pmatrix} t \\ 1 + \frac{t}{2} \end{pmatrix} e^{t/2}.$$

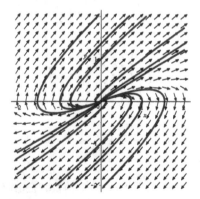

9.(a) The characteristic equation is given by $6\lambda^2 - \lambda - 1 = 0$. Therefore, the two distinct roots are $\lambda = 1/2, -1/3$. Therefore, the general solution is given by

$$y(t) = c_1 e^{t/2} + c_2 e^{-t/3}.$$

(b) The critical point $(0,0)$ is a saddle point, and, therefore, unstable.

(c) For y above, we see that

$$y' = \frac{c_1}{2} e^{t/2} - \frac{c_2}{3} e^{-t/3}.$$

Therefore, we can rewrite our solution as the two parameter family

$$\begin{pmatrix} x_1 \\ x_2 \end{pmatrix} = \begin{pmatrix} y \\ y' \end{pmatrix} = \begin{pmatrix} c_1 e^{t/2} + c_2 e^{-t/3} \\ \frac{c_1}{2} e^{t/2} - \frac{c_2}{3} e^{-t/3} \end{pmatrix} = c_1 \begin{pmatrix} 1 \\ \frac{1}{2} \end{pmatrix} e^{t/2} + c_2 \begin{pmatrix} 1 \\ -\frac{1}{3} \end{pmatrix} e^{-t/3}.$$

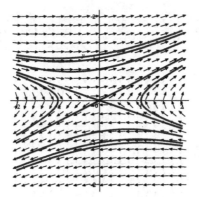

11.(a) The characteristic equation is $\lambda^2 + 2\lambda - 8 = 0$. Therefore, the roots are $\lambda = -4, 2$. Therefore, the general solution is

$$y(t) = c_1 e^{-4t} + c_2 e^{2t}$$

(b) The critical point $(0, 0)$ is an unstable saddle point.

(c) From our solution in part (a), we see that

$$y'(t) = -4c_1 e^{-4t} + 2c_2 e^{2t}.$$

Therefore,

$$\mathbf{x} = \begin{pmatrix} y \\ y' \end{pmatrix} = c_1 e^{-4t} \begin{pmatrix} 1 \\ -4 \end{pmatrix} + c_2 e^{2t} \begin{pmatrix} 1 \\ 2 \end{pmatrix}.$$

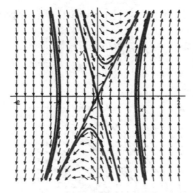

13.(a) The characteristic equation is given by $\lambda^2 + 5\lambda = 0$. Therefore, the two distinct roots are $\lambda = 0, -5$. Therefore, the general solution is given by

$$y(t) = c_1 + c_2 e^{-5t}.$$

(b) Every point of the form $(x_0, 0)$ is a critical point. They are each non-isolated, stable points.

(c) For y above, we see that $y' = -5c_2 e^{-5t}$. Therefore, we can rewrite our solution as the two parameter family

$$\begin{pmatrix} x_1 \\ x_2 \end{pmatrix} = \begin{pmatrix} y \\ y' \end{pmatrix} = \begin{pmatrix} c_1 + c_2 e^{-5t} \\ -5c_2 e^{-5t} \end{pmatrix} = c_1 \begin{pmatrix} 1 \\ 0 \end{pmatrix} + c_2 \begin{pmatrix} 1 \\ -5 \end{pmatrix} e^{-5t}.$$

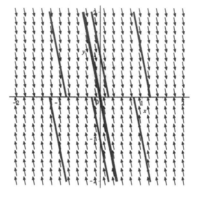

15.(a) The characteristic equation is given by $25\lambda^2 - 20\lambda + 4 = 0$. Therefore, we have one repeated root $\lambda = 2/5$. Therefore, the general solution is given by

$$y(t) = c_1 e^{2t/5} + c_2 t e^{2t/5}.$$

(b) The critical point $(0, 0)$ is an unstable improper node.

(c) For y above, we see that

$$y' = \frac{2c_1}{5} e^{2t/5} + c_2 \left(1 + \frac{2t}{5} \right) e^{2t/5}.$$

Therefore, we can rewrite our solution as the two parameter family

$$\begin{pmatrix} x_1 \\ x_2 \end{pmatrix} = \begin{pmatrix} y \\ y' \end{pmatrix} = \begin{pmatrix} c_1 e^{2t/5} + c_2 t e^{2t/5} \\ \frac{2c_1}{5} e^{2t/5} + c_2 \left(1 + \frac{2t}{5} \right) e^{2t/5} \end{pmatrix} = c_1 \begin{pmatrix} 1 \\ \frac{2}{5} \end{pmatrix} e^{2t/5} + c_2 \begin{pmatrix} t \\ 1 + \frac{2t}{5} \end{pmatrix} e^{2t/5}.$$

17.(a) The characteristic equation is $\lambda^2 + 6\lambda + 13 = 0$. Therefore, the roots are $\lambda = -3 \pm 2i$. Therefore, one complex solution is

$$y(t) = e^{(-3+2i)t} = e^{-3t}(\cos 2t + i\sin 2t).$$

Considering the real and imaginary parts of this function, we arrive at the general solution

$$y(t) = c_1 e^{-3t}\cos 2t + c_2 e^{-3t}\sin 2t.$$

(b) The critical point $(0,0)$ is an asymptotically stable spiral point.

(c) From our solution in part (a), we see that

$$y'(t) = c_1 e^{-3t}(-3\cos 2t - 2\sin 2t) + c_2 e^{-3t}(-3\sin 2t + 2\cos 2t).$$

Therefore,

$$\mathbf{x} = \begin{pmatrix} y \\ y' \end{pmatrix} = c_1 e^{-3t}\begin{pmatrix} \cos 2t \\ -3\cos 2t - 2\sin 2t \end{pmatrix} + c_2 e^{-3t}\begin{pmatrix} \sin 2t \\ -3\sin 2t + 2\cos 2t \end{pmatrix}.$$

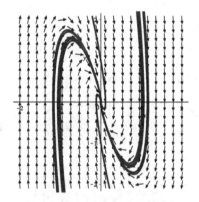

19.(a) The characteristic equation is given by $\lambda^2 - 9\lambda + 9 = 0$. Therefore, the two distinct roots are $\lambda = (9 + 3\sqrt{5})/2, (9 - 3\sqrt{5})/2$. Therefore, the general solution is given by

$$y(t) = c_1 e^{(9+3\sqrt{5})t/2} + c_2 e^{(9-3\sqrt{5})t/2}.$$

(b) The critical point $(0,0)$ is an unstable node.

(c) For y above, we see that

$$y' = \frac{(9+3\sqrt{5})c_1}{2}e^{(9+3\sqrt{5})t/2} + \frac{(9-3\sqrt{5})c_2}{2}e^{(9-3\sqrt{5})t/2}.$$

Therefore, we can rewrite our solution as the two parameter family

$$\begin{pmatrix} x_1 \\ x_2 \end{pmatrix} = \begin{pmatrix} y \\ y' \end{pmatrix} = \begin{pmatrix} c_1 e^{(9+3\sqrt{5})t/2} + c_2 e^{(9-3\sqrt{5})t/2} \\ \frac{(9+3\sqrt{5})c_1}{2}e^{(9+3\sqrt{5})t/2} + \frac{(9-3\sqrt{5})c_2}{2}e^{(9-3\sqrt{5})t/2} \end{pmatrix}$$

$$= c_1 \begin{pmatrix} 1 \\ \frac{9+3\sqrt{5}}{2} \end{pmatrix} e^{(9+3\sqrt{5})t/2} + c_2 \begin{pmatrix} 1 \\ \frac{9-3\sqrt{5}}{2} \end{pmatrix} e^{(9-3\sqrt{5})t/2}.$$

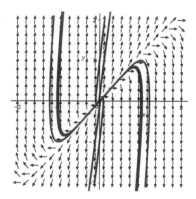

21.(a) The characteristic equation is given by $\lambda^2 + 4\lambda + 4 = 0$. Therefore, we have one repeated root $\lambda = -2$. Therefore, the general solution is given by

$$y(t) = c_1 e^{-2t} + c_2 t e^{-2t}.$$

(b) The critical point $(0,0)$ is an improper node, and it is asymptotically stable.

(c) For y above, we see that

$$y' = -2c_1 e^{-2t} + c_2 e^{-2t} - 2c_2 t e^{-2t}.$$

Therefore, we can rewrite our solution as the two parameter family

$$\begin{pmatrix} x_1 \\ x_2 \end{pmatrix} = \begin{pmatrix} y \\ y' \end{pmatrix} = \begin{pmatrix} c_1 e^{-2t} + c_2 t e^{-2t} \\ -2c_1 e^{-2t} + c_2 e^{-2t} - 2c_2 t e^{-2t} \end{pmatrix} = c_1 \begin{pmatrix} 1 \\ -2 \end{pmatrix} e^{-2t} + c_2 \begin{pmatrix} t \\ 1 - 2t \end{pmatrix} e^{-2t}.$$

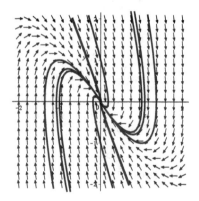

23.(a) The characteristic equation is $4\lambda^2 + 9 = 0$. Therefore, the roots are $\lambda = \pm\dfrac{3}{2}i$. Therefore, one complex solution is

$$y(t) = e^{3it/2} = \cos(3t/2) + i\sin(3t/2).$$

Considering the real and imaginary parts of this function, we arrive at the general solution

$$y(t) = c_1 \cos(3t/2) + c_2 \sin(3t/2).$$

(b) The critical point $(0,0)$ is a stable center.

(c) From our solution in part (a), we see that

$$y'(t) = -\frac{3c_1}{2}\sin(3t/2) + \frac{3c_2}{2}\cos(3t/2).$$

Therefore,

$$\mathbf{x} = \begin{pmatrix} y \\ y' \end{pmatrix} = c_1 \begin{pmatrix} \cos(3t/2) \\ -\frac{3}{2}\sin(3t/2) \end{pmatrix} + c_2 \begin{pmatrix} \sin(3t/2) \\ \frac{3}{2}\cos(3t/2) \end{pmatrix}.$$

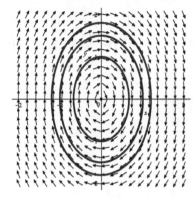

25.(a) The characteristic equation is $\lambda^2 + \lambda + 1.25 = 0$. Therefore, the roots are $\lambda = -\frac{1}{2} \pm i$. Therefore, one complex solution is

$$y(t) = e^{(-\frac{1}{2}+i)t} = e^{-t/2}(\cos t + i\sin t).$$

Considering the real and imaginary parts of this function, we arrive at the general solution

$$y(t) = c_1 e^{-t/2}\cos t + c_2 e^{-t/2}\sin t.$$

(b) The critical point $(0,0)$ is an asymptotically stable spiral point.

(c) From our solution in part (a), we see that

$$y'(t) = c_1 e^{-t/2}\left(-\frac{1}{2}\cos t - \sin t\right) + c_2 e^{-t/2}\left(-\frac{1}{2}\sin t + \cos t\right).$$

Therefore,

$$\mathbf{x} = \begin{pmatrix} y \\ y' \end{pmatrix} = c_1 e^{-t/2}\begin{pmatrix} \cos t \\ -\frac{1}{2}\cos t - \sin t \end{pmatrix} + c_2 e^{-t/2}\begin{pmatrix} \sin t \\ -\frac{1}{2}\sin t + \cos t \end{pmatrix}.$$

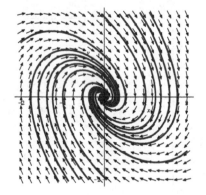

27. The characteristic equation is given by $\lambda^2 + \lambda - 2 = 0$. Therefore, the two distinct roots are $\lambda = -2, 1$. Therefore, the general solution is given by

$$y(t) = c_1 e^{-2t} + c_2 e^t.$$

Therefore,

$$y'(t) = -2c_1 e^{-2t} + c_2 e^t.$$

Now using the initial conditions, we need $c_1 + c_2 = 1$ and $-2c_1 + c_2 = 1$. The solution of this system of equations is $c_1 = 0$ and $c_2 = 1$. Therefore, the specific solution is $y(t) = e^t$.

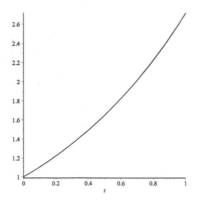

The solution $y \to \infty$ as $t \to \infty$.

29. The characteristic equation is given by $9\lambda^2 - 12\lambda + 4 = 0$. Therefore, there is one re   root, $\lambda = 2/3$. Therefore, the general solution is given by

$$y(t) = c_1 e^{2t/3} + c_2 t e^{2t/3}.$$

Therefore,

$$y'(t) = \frac{2c_1}{3} e^{2t/3} + c_2 \left(1 + \frac{2t}{3} \right) e^{2t/3}.$$

Now using the initial conditions, we need $c_1 = 2$ and $\frac{2c_1}{3} + c_2 = -1$. The solution of this system of equations is $c_1 = 2$ and $c_2 = -7/3$. Therefore, the specific solution is $y(t) = 2e^{2t/3} - \frac{7}{3} t e^{2t/3}.$

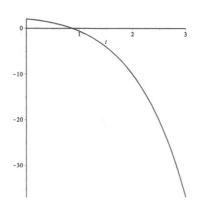

The solution $y \to -\infty$ as $t \to \infty$.

31. The characteristic equation is $\lambda^2 + 4\lambda + 5 = 0$, which has roots $\lambda = -2 \pm i$. Therefore, the general solution is $y(t) = c_1 e^{-2t} \cos t + c_2 e^{-2t} \sin t$. The derivative of y is $y'(t) = c_1 e^{-2t}(-2 \cos t - \sin t) + c_2 e^{-2t}(-2 \sin t + \cos t)$. Using the initial conditions, we have $c_1 = 1$ and $-2c_1 + c_2 = 0$. Therefore, $c_1 = 1$ and $c_2 = 2$, and we conclude that the solution is

$$y(t) = e^{-2t} \cos t + 2e^{-2t} \sin t.$$

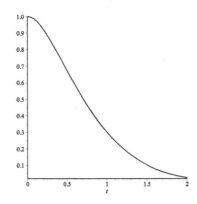

The solution $y \to 0$ as $t \to \infty$.

33. The characteristic equation is given by $\lambda^2 + 6\lambda + 9 = 0$. Therefore, there is one repeated root, $\lambda = -3$. Therefore, the general solution is given by

$$y(t) = c_1 e^{-3t} + c_2 t e^{-3t}.$$

Therefore,

$$y'(t) = -3c_1 e^{-3t} + c_2(1 - 3t)e^{-3t}.$$

Now using the initial conditions, we need $c_1 = 0$ and $-3c_1 + c_2 = 2$. The solution of this system of equations is $c_1 = 0$ and $c_2 = 2$. Therefore, the specific solution is $y(t) = 2te^{-3t}$.

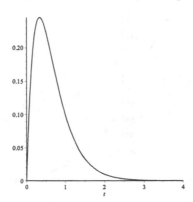

The solution $y \to 0$ as $t \to \infty$.

35. The characteristic equation is given by $\lambda^2 + 3\lambda = 0$. Therefore, the two distinct real roots are $\lambda = 0, -3$. Therefore, the general solution is given by

$$y(t) = c_1 + c_2 e^{-3t}.$$

Therefore,

$$y'(t) = -3c_2 e^{-3t}.$$

Now using the initial conditions, we need $c_1 + c_2 = -2$ and $-3c_2 = 3$. The solution of this system of equations is $c_1 = -1$ and $c_2 = -1$. Therefore, the specific solution is $y(t) = -1 - e^{-3t}$.

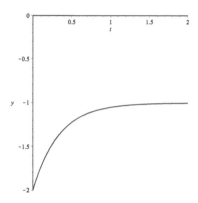

The solution $y \to -1$ as $t \to \infty$.

37. The characteristic equation is given by $\lambda^2 + 4\lambda + 4 = 0$. Therefore, there is one repeated root, $\lambda = -2$. Therefore, the general solution is given by

$$y(t) = c_1 e^{-2t} + c_2 t e^{-2t}.$$

Therefore,

$$y'(t) = -2c_1 e^{-2t} + c_2 (1 - 2t) e^{-2t}.$$

Now using the initial conditions, we need $c_1 e^2 - c_2 e^2 = 2$ and $-2c_1 e^2 + 3c_2 e^2 = 1$. The solution of this system of equations is $c_1 = 7e^{-2}$ and $c_2 = 5e^{-2}$. Therefore, the specific solution is $y(t) = 7e^{-2(1+t)} + 5te^{-2(1+t)}$.

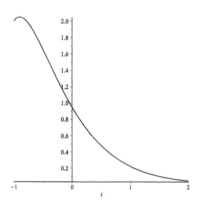

The solution $y \to 0$ as $t \to \infty$.

39. The characteristic equation is $\lambda^2 + \lambda + 1.25 = 0$, which has roots $\lambda = -\dfrac{1}{2} \pm i$. Therefore, the general solution is $y(t) = c_1 e^{-t/2} \cos t + c_2 e^{-t/2} \sin t$. The derivative of y is $y'(t) = c_1 e^{-t/2} \left(-\frac{1}{2} \cos t - \sin t \right) + c_2 e^{-t/2} \left(-\frac{1}{2} \sin t + \cos t \right)$. Using the initial conditions, we have

$c_1 = 3$ and $-c_1/2 + c_2 = 1$. Therefore, $c_1 = 3$ and $c_2 = 5/2$, and we conclude that the solution is

$$y(t) = 3e^{-t/2}\cos t + \frac{5}{2}e^{-t/2}\sin t.$$

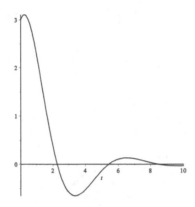

The solution y oscillates as it decays to zero as $t \to \infty$.

41. The characteristic equation is given by $\lambda^2 + 8\lambda - 9 = 0$. Therefore, the two distinct real roots are $\lambda = -9, 1$. Therefore, the general solution is given by

$$y(t) = c_1 e^{-9t} + c_2 e^t.$$

Therefore,

$$y'(t) = -9c_1 e^{-9t} + c_2 e^t.$$

Now using the initial conditions, we need $c_1 e^{-9} + c_2 e = 1$ and $-9c_1 e^{-9} + c_2 e = 0$. The solution of this system of equations is $c_1 = \frac{1}{10}e^9$ and $c_2 = \frac{9}{10}e^{-1}$. Therefore, the specific solution is

$$y(t) = \frac{1}{10}e^{9-9t} + \frac{9}{10}e^{-1+t}.$$

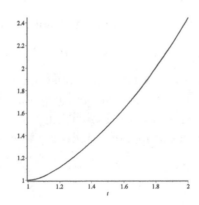

The solution $y \to \infty$ as $t \to \infty$.

43. The characteristic equation is given by $4\lambda^2 - 1 = 0$. Therefore, the two distinct real roots are $\lambda = 1/2, -1/2$. Therefore, the general solution is given by

$$y(t) = c_1 e^{t/2} + c_2 e^{-t/2}.$$

Therefore,

$$y'(t) = \frac{c_1}{2}e^{t/2} - \frac{c_2}{2}e^{-t/2}.$$

Now using the initial conditions, we need $c_1 e^{-1} + c_2 e = 1$ and $\frac{c_1}{2}e^{-1} - \frac{c_2}{2}e = -1$. The solution of this system of equations is $c_1 = -e/2$ and $c_2 = \frac{3}{2}e^{-1}$. Therefore, the specific solution is

$$y(t) = -\frac{1}{2}e^{1+t/2} + \frac{3}{2}e^{-1-t/2}.$$

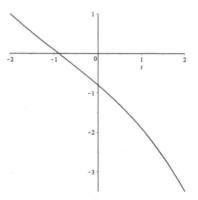

The solution $y \to -\infty$ as $t \to \infty$.

45. We need to find a characteristic equation of degree two with one repeated root $\lambda = -2$. We take $p(\lambda) = (\lambda+2)^2 = \lambda^2 + 4\lambda + 4$. Therefore, the differential equation is $y'' + 4y' + 4y = 0$.

47. The characteristic equation is $\lambda^2 - (2\alpha - 3)\lambda + \alpha(\alpha - 3) = 0$. Solving this equation, we see that the roots are $\lambda = \alpha, \alpha - 3$. Therefore, the general solution is

$$y(t) = c_1 e^{\alpha t} + c_2 e^{(\alpha-3)t}.$$

In order for the solution to tend to zero, we need $\alpha, \alpha - 3 < 0$. Therefore, the solutions will all tend to zero as long as $\alpha < 0$. All solutions will become unbounded if $\alpha > 0$ and $\alpha - 3 > 0$. This will occur exactly when $\alpha > 3$.

49. Suppose the roots are distinct, λ_1, λ_2. Then the solution is

$$y(t) = c_1 e^{\lambda_1 t} + c_2 e^{\lambda_2 t}.$$

Solving the equation $y(t) = 0$, we see that we must have $c_1 e^{\lambda_1 t} = -c_2 e^{\lambda_2 t}$ which implies $e^{(\lambda_1 - \lambda_2)t} = -c_2/c_1$. First, in order to guarantee any solution of this equation, we would need $c_2/c_1 < 0$. Then, applying the natural logarithm function to the equation, we see that $t = \ln(-c_2/c_1)/(\lambda_1 - \lambda_2)$. If the roots are not distinct, then the solution is given by

$$y(t) = c_1 e^{\lambda t} + c_2 t e^{\lambda t}.$$

Therefore, $y(t) = 0$ implies $(c_1 + c_2 t)e^{\lambda t} = 0$. Since $e^{\lambda t} \neq 0$, we must have $c_1 + c_2 t = 0$. Therefore, the solution will be zero only when $t = -c_1/c_2$.

51. The roots of the characteristic equation are

$$\lambda = \frac{-b + \sqrt{b^2 - 4ac}}{2a}.$$

Therefore, the roots will be

(a) real, different and negative if (1) $b^2 - 4ac > 0$ and (2) $b > \sqrt{b^2 - 4ac}$.

(b) real with opposite signs if (1) $b^2 - 4ac > 0$ and (2) $b < \sqrt{b^2 - 4ac}$.

(c) real, different and positive if (1) $b^2 - 4ac > 0$ and (2) $-b > \sqrt{b^2 - 4ac}$.

53. Let $t = \ln(-x)$. Then the chain rule applied to $y(x) = y(t(x))$ gives

$$\frac{dy}{dx} = \frac{dy}{dt}\frac{dt}{dx} = \frac{1}{x}\frac{dy}{dt} \quad \text{and} \quad \frac{d^2y}{dx^2} = \frac{1}{x^2}\frac{d^2y}{dt^2} - \frac{1}{x^2}\frac{dy}{dt},$$

which proves the statement.

55. With $t = \ln x$, the equation transforms into $y'' + 3y' + 2y = 0$. The characteristic roots are $r = -1, -2$. The solution is

$$y = c_1 e^{-t} + c_2 e^{-2t} = c_1 e^{-\ln x} + c_2 e^{-2\ln x} = \frac{c_1}{x} + \frac{c_2}{x^2}.$$

57. With $t = \ln x$, the equation transforms into $y'' - 5y' - 6y = 0$. The characteristic roots are $r = -1, 6$. The solution is

$$y = c_1 e^{-t} + c_2 e^{6t} = c_1 e^{-\ln x} + c_2 e^{6\ln x} = \frac{c_1}{x} + c_2 x^6.$$

59. With $t = \ln x$, the equation transforms into $y'' - 6y' + 9y = 0$. The characteristic root is a double root $r = 3$. The solution is

$$y = c_1 e^{3t} + c_2 t e^{3t} = c_1 e^{3\ln x} + c_2 \ln x e^{3\ln x} = c_1 x^3 + c_2 x^3 \ln x.$$

61. With $t = \ln x$, the equation transforms into $2y'' - 6y' + 6y = 0$. The characteristic roots are $r = (3 \pm \sqrt{3}i)/2$. The solution is

$$y = c_1 e^{3t/2}\cos(\sqrt{3}t/2) + c_2 e^{3t/2}\sin(\sqrt{3}t/2) = c_1 x^{3/2}\cos(\sqrt{3}\ln x/2) + c_2 x^{3/2}\sin(\sqrt{3}\ln x/2).$$

63. With $t = \ln x$, the equation transforms into $4y'' + 4y' + 17y = 0$. The characteristic roots are $r = -\frac{1}{2} \pm 2i$. The solution is

$$y = c_1 e^{-t/2}\cos(2t) + c_2 e^{-t/2}\sin(2t) = c_1 x^{-1/2}\cos(2\ln x) + c_2 x^{-1/2}\sin(2\ln x).$$

The initial conditions state that $c_1 = 1$ and $-c_1/2 + 2c_2 = -3$. This gives us the solution $y = 2x^{-1/2}\cos(2\ln x) - x^{-1/2}\sin(2\ln x)$.

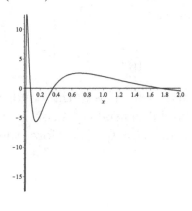

The solution y oscillates with a growing amplitude as $x \to 0$.

65. With $t = \ln x$, the equation transforms into $y'' + 2y' + 5y = 0$. The characteristic roots are $r = -1 \pm 2i$. The solution is

$$y = c_1 e^{-t} \cos 2t + c_2 e^{-t} \sin 2t = c_1 x^{-1} \cos(2 \ln x) + c_2 x^{-1} \sin(2 \ln x).$$

The initial conditions state that $c_1 = 1$ and $-c_1 + 2c_2 = -1$. This gives us the solution $y = x^{-1} \cos(2 \ln x)$.

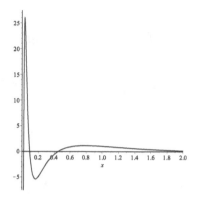

The solution y oscillates with a growing amplitude as $x \to 0$.

4.4   Mechanical and Electrical Vibrations

1. $R \cos \delta = 3$ and $R \sin \delta = 3$ implies $R = \sqrt{18} = 3\sqrt{2}$ and $\delta = \arctan(3/3) = \pi/4$. Therefore,

$$y = 3\sqrt{2} \cos(2t - \frac{\pi}{4}).$$

3. $R \cos \delta = 4$ and $R \sin \delta = -2$ implies $R = \sqrt{20} = 2\sqrt{5}$ and $\delta = \arctan(-1/2) \approx -0.4636$. Therefore,

$$y = 2\sqrt{5} \cos(3t + \arctan \frac{1}{2}).$$

5.(a) The spring constant is $k = 2/(1/2) = 4$ lb/ft. The mass is $m = 2/32 = 1/16$ lb-s^2/ft. Therefore, the equation of motion is

$$\frac{1}{16} y'' + 4y = 0,$$

which can be simplified to $y'' + 64y = 0$. The initial conditions are $y(0) = 1/4$ ft, $y'(0) = 0$ ft/sec. The general solution of the differential equation is $y(t) = A \cos 8t + B \sin 8t$ (ft). The initial condition implies $A = 1/4$ and $B = 0$. Therefore, the solution is

$$y(t) = \frac{1}{4} \cos 8t \text{ (ft)}.$$

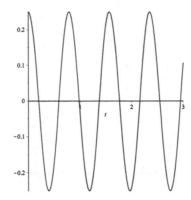

The frequency is $\omega = 8$ rad/sec. The period is $T = \pi/4$ seconds. The amplitude is $R = 1/4$ feet.

(b)

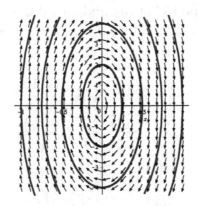

7. The spring constant is $k = 3/(1/4) = 12$ lb/ft. The mass is $3/32$ lb-s^2/ft. Therefore, the equation of motion is

$$\frac{3}{32}y'' + 12y = 0,$$

which can be simplified to $y'' + 128y = 0$. The initial conditions are $y(0) = -1/12$ ft, $y'(0) = 2$ ft/sec. The general solution is $y(t) = A\cos(8\sqrt{2}t) + B\sin(8\sqrt{2}t)$. Considering the initial conditions, we arrive at the solution

$$y(t) = -\frac{1}{12}\cos(8\sqrt{2}t) + \frac{\sqrt{2}}{8}\sin(8\sqrt{2}t).$$

The frequency $\omega_0 = 8\sqrt{2}$ rad/sec. The period is $T = \pi\sqrt{2}/8$ seconds. The amplitude is $R = \sqrt{(1/12)^2 + (\sqrt{2}/8)^2} = \sqrt{11/2}/12$ ft. The phase is $\delta = \pi - \arctan(3/\sqrt{2})$.

9.(a) The spring constant is $k = 0.196/0.05 = 3.92$ N/m. The mass is $m = 0.02$ kg. The damping constant is $\gamma = 400$ dyne-sec/cm $= 0.4$ N-sec/m. Therefore, the equation of motion is

$$0.02y'' + 0.4y' + 3.92y = 0$$

or $y'' + 20y' + 196y = 0$, with initial conditions $y(0) = 0.02$ m, $y'(0) = 0$ m/sec. The solution of this equation is $y(t) = e^{-10t}(A\cos(4\sqrt{6}t) + B\sin(4\sqrt{6}t))$. The initial conditions imply

$$y(t) = e^{-10t}\left[2\cos(4\sqrt{6}t) + (5/\sqrt{6})\sin(4\sqrt{6}t)\right] \quad \text{cm}.$$

The quasi-frequency is $\nu = 4\sqrt{6}$ rad/sec. The quasi- period is $T_d = \pi\sqrt{6}/12$. For the undamped motion, the equation would be $y'' + 196y = 0$, which has the general solution $y(t) = A\cos 14t + B\sin 14t$. Therefore, the period for the undamped motion is $T = \pi/7$. The ratio of the quasi-period to the period of the undamped motion is $T_d/T = 7\sqrt{6}/12 \approx 1.4289$. The solution y will satisfy $|y(\tau)| < 0.05$ for all $\tau > 0.4045$ seconds.

(b)

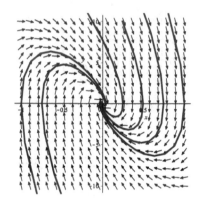

11.(a) The spring constant is $k = 3/0.1 = 30$ N/m. The mass is $m = 2$ kg. The damping coefficient is $\gamma = 3/5$ N-sec/m. Therefore, the equation of motion is

$$2y'' + \frac{3}{5}y' + 30y = 0$$

which can be rewritten as $y'' + 0.3y' + 15y = 0$. The initial conditions are $y(0) = 0.05$ m and $y'(0) = 0.01$ m/sec. The general solution is $y(t) = e^{-0.15t}(A\cos \mu t + B\sin \mu t)$ where $\mu \approx 3.87008$. Based on the initial conditions, we have $A = 0.05$ and $B = 0.00452$. Therefore, the specific solution is

$$y(t) = e^{-0.15t}(0.05\cos \mu t + 0.00452\sin \mu t).$$

The quasi-frequency is $\nu = \mu = 3.87008$ radians/second. The undamped equation is

$$y'' + 15y = 0,$$

which has general solution $y(t) = A\cos \sqrt{15}t + B\sin \sqrt{15}t$. Therefore, the frequency for undamped motion is $\omega_0 = \sqrt{15}$. Therefore, the ratio of ν to ω_0 is $\nu/\omega_0 = 3.87008/\sqrt{15} \approx 0.99925$.

(b)

13. The frequency of the undamped motion is $\omega_0 = 1$. Therefore, the period of the undamped motion is $T = 2\pi$. The quasi-frequency of the damped motion is $\mu = \frac{1}{2}\sqrt{4 - \gamma^2}$. Therefore, the period of the damped motion is $T_d = 4\pi/\sqrt{4 - \gamma^2}$. We want to find γ such that $T_d = 1.5T$. That is, we want γ to satisfy $4\pi/\sqrt{4 - \gamma^2} = 3\pi$. Solving this equation, we have $\gamma = 2\sqrt{5}/3$.

15. Suppose y is a solution of the differential equation

$$my'' + \gamma y' + ky = 0$$

and satisfies the initial conditions $y(t_0) = y_0$, $y'(t_0) = y_1$. Then suppose v is a solution of the same differential equation, but satisfies the initial conditions $v(t_0) = y_0$ and $v'(t_0) = 0$. Let $w = y - v$. Then w satisfies the same differential equation, since the equation is linear, and, further, $w(t_0) = 0$, $w'(t_0) = y_1$. Therefore, $y = v + w$ where v, w satisfy the specified conditions.

17. The spring constant is $k = 8/(1.5/12) = 64$ lb/ft. The mass is $m = 8/32 = 1/4$ lb-s^2/ft. Therefore, the equation of motion is

$$\frac{1}{4}y'' + \gamma y' + 64y = 0.$$

The motion will experience critical damping when $\gamma = 2\sqrt{km} = 8$ lb-s/ft.

19. If the system is critically damped or overdamped, then $\gamma \geq 2\sqrt{km}$. If $\gamma = 2\sqrt{km}$ (critically damped), then the solution is given by

$$y(t) = (A + Bt)e^{-\gamma t/2m}.$$

In this case, if $y = 0$, then we must have $A + Bt = 0$, that is, $t = -A/B$ (assuming $B \neq 0$). If $B = 0$, the solution is never zero (unless $A = 0$). If $\gamma > 2\sqrt{km}$ (overdamped), then the solution is given by

$$y(t) = Ae^{\lambda_1 t} + Be^{\lambda_2 t}$$

where λ_1, λ_2 are given by equation (19) in the text. Assume for the moment that $A, B \neq 0$. Then $y = 0$ implies $Ae^{\lambda_1 t} = -Be^{\lambda_2 t}$, which implies $e^{(\lambda_1 - \lambda_2)t} = -B/A$. There is only one solution to this equation. If $A = 0$ or $B = 0$, then there are no solutions to the equation $y = 0$ (unless they are both zero, in which case, the solution is identically zero).

21.(a) Let $y = Re^{-\gamma t/2m} \cos(\nu t - \delta)$. Then y attains a maximum when $\nu t_k - \delta = 2k\pi$. Therefore, the time between successive maxima is $T_d = t_{k+1} - t_k = \dfrac{1}{\nu}(2(k+1)\pi - 2k\pi) = 2\pi/\nu$.

(b)

$$\frac{y(t_k)}{y(t_{k+1})} = \frac{e^{-\gamma t_k/2m}}{e^{-\gamma t_{k+1}/2m}} = e^{(\gamma t_{k+1} - \gamma t_k)/2m} = e^{\gamma T_d/2m}.$$

(c) From parts (a) and (b), we have $\Delta = \ln[y(t_k)/y(t_{k+1})] = \gamma T_d/2m = 2\pi\gamma/2m\nu = \pi\gamma/m\nu$.

23. For problem 17, the mass is $m = 1/4$ lb-s^2/ft. We suppose that $\Delta = 3$ and $T_d = 0.3$ seconds. From problem 21, we know that $\Delta = \gamma T_d/2m$. Therefore, we have

$$3 = \frac{0.3\gamma}{2(1/4)} = 0.6\gamma.$$

Therefore, the damping coefficient is $\gamma = 5$ lb-sec/ft.

25.(a) The solution of the initial value problem is

$$y(t) = e^{-t/8}\left(2\cos\left(\frac{3}{8}\sqrt{7}\,t\right) + \frac{2\sqrt{7}}{21}\sin\left(\frac{3}{8}\sqrt{7}\,t\right)\right).$$

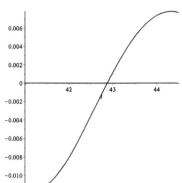

Using the graph above and numerical evidence, we conclude that $\tau \approx 41.715$ seconds.

(b) For $\gamma = 0.5$, $\tau \approx 20.402$; for $\gamma = 1.0$, $\tau \approx 9.168$; for $\gamma = 1.5$, $\tau \approx 7.184$.

(c)

(d) For $\gamma = 1.7$, $\tau \approx 6.767$; for $\gamma = 1.8$, $\tau \approx 5.473$; for $\gamma = 1.9$, $\tau \approx 6.460$. We conclude that τ steadily decreases to about $\tau_{min} \approx 4.873$, corresponding to the critical value $\gamma_0 \approx 1.73$.

(e) We have

$$y(t) = \frac{4e^{-\gamma t/2}}{\sqrt{4 - \gamma^2}} \cos(\nu t - \delta),$$

where $\nu = \frac{1}{2}\sqrt{4 - \gamma^2}$, and $\delta = \tan^{-1}(\gamma/\sqrt{4 - \gamma^2})$. Therefore, $|y(t)| \leq \dfrac{4e^{-\gamma t/2}}{\sqrt{4 - \gamma^2}}$. To find the critical τ, we need to solve $4e^{-\gamma \tau/2}/\sqrt{4 - \gamma^2} = 0.01$. Solving this equation, we see that $\tau = (2/\gamma)\ln(400/\sqrt{4 - \gamma^2})$.

27. Referring to problem 19, the equation of motion is given by $\rho l u'' + \rho_0 g u = 0$. The general solution of this differential equation is

$$u(t) = A\cos\left(\sqrt{\rho_0 g/\rho l}\, t\right) + B\sin\left(\sqrt{\rho_0 g/\rho l}\, t\right).$$

Therefore, the frequency is $\omega_0 = \sqrt{\rho_0 g/\rho l}$ rad/sec, and the period is $T = 2\pi\sqrt{\rho l/\rho_0 g}$ seconds.

29.(a) The general solution is $y(t) = A\cos(\sqrt{2}\,t) + B\sin(\sqrt{2}\,t)$. The initial conditions imply $A = 0$ and $B = \sqrt{2}$. Therefore, the specific solution is $y(t) = \sqrt{2}\sin(\sqrt{2}\,t)$.

(b)

(c)

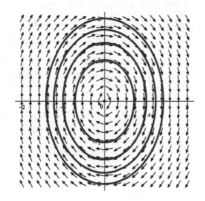

31.(a) The kinetic energy is given by $\frac{1}{2}mv^2$, where v is velocity. Initially, $v = b$, therefore, the kinetic energy initially is $mb^2/2$. The potential energy is given by $\frac{1}{2}ky^2$ where y is the distance the spring is stretched. Therefore, the potential energy initially is $ka^2/2$.

(b) The general solution is

$$y(t) = A\cos(\sqrt{k/m}\,t) + B\sin(\sqrt{k/m}\,t).$$

The initial conditions imply that $A = a$ and $B = b\sqrt{m/k}$. Therefore, the specific solution is

$$y(t) = a\cos(\sqrt{k/m}\,t) + b\sqrt{m/k}\sin(\sqrt{k/m}\,t).$$

(c) By the equation for y, we see that

$$\frac{ky^2}{2} = \frac{k}{2}(A^2\cos^2(\sqrt{k/m}\,t) + 2AB\cos(\sqrt{k/m}\,t)\sin(\sqrt{k/m}\,t) + B^2\sin^2(\sqrt{k/m}\,t)).$$

Further,
$$y'(t) = -A\sqrt{k/m}\sin(\sqrt{k/m}\,t) + B\sqrt{k/m}\cos(\sqrt{k/m}\,t)$$

implies

$$\begin{aligned}
\frac{m(y')^2}{2} &= \frac{m}{2}(A^2(k/m)\sin^2(\sqrt{k/m}\,t) - 2AB(k/m)\sin(\sqrt{k/m}\,t)\cos(\sqrt{k/m}\,t) \\
&+ B^2(k/m)\cos^2(\sqrt{k/m}\,t)).
\end{aligned}$$

Therefore, the total energy satisfies

$$\frac{ky^2}{2} + \frac{m(y')^2}{2} = \frac{k}{2}(A^2 + B^2) = \frac{k}{2}(a^2 + b^2(m/k)) = \frac{ka^2}{2} + \frac{mb^2}{2}.$$

Therefore, energy is conserved.

4.5 Nonhomogeneous Equations; Method of Undetermined Coefficients

1. The characteristic equation for the homogeneous problem is $\lambda^2 - 2\lambda - 3 = 0$, which has roots $\lambda = 3, -1$. Therefore, the solution of the homogeneous problem is $y_h(t) = c_1 e^{3t} + c_2 e^{-t}$. To find a solution of the nonhomogeneous problem, we look for a solution of the form $y_p(t) = Ae^{2t}$. Substituting a function of this form into the differential equation, we have

$$4Ae^{2t} - 4Ae^{2t} - 3Ae^{2t} = 3e^{2t}.$$

Therefore, we need $-3A = 3$, or $A = -1$. Therefore, the general solution of the nonhomogeneous problem is

$$y(t) = c_1 e^{3t} + c_2 e^{-t} - e^{2t}.$$

3. The characteristic equation for the homogeneous problem is $\lambda^2 - 2\lambda - 3 = 0$, which has roots $\lambda = 3, -1$. Therefore, the solution of the homogeneous problem is $y_h(t) = c_1 e^{3t} + c_2 e^{-t}$. Since $y(t) = e^{-t}$ is a solution of the homogeneous problem, to find a solution of the nonhomogeneous problem, we look for a solution of the form $y_p(t) = Ate^{-t} + Bt^2 e^{-t}$. Substituting a function of this form into the differential equation, and equating like terms, we have $-4A + 2B = 0$ and $-8B = -3$. The solution of these equations is $A = 3/16$ and $B = 3/8$. Therefore, the general solution of the nonhomogeneous problem is

$$y(t) = c_1 e^{3t} + c_2 e^{-t} + \frac{3}{16} te^{-t} + \frac{3}{8} t^2 e^{-t}.$$

5. The characteristic equation for the homogeneous problem is $\lambda^2 + 9 = 0$, which has roots $\lambda = \pm 3i$. Therefore, the solution of the homogeneous problem is $y_h(t) = c_1 \cos 3t + c_2 \sin 3t$. To find a solution of the nonhomogeneous problem, we look for a solution of the form $y_p(t) = Ae^{3t} + Bte^{3t} + Ct^2 e^{3t} + D$. Substituting a function of this form into the differential equation, and equating like terms, we have $18A + 6B + 2C = 0$, $18B + 12C = 0$, $18C = 1$ and $9D = 6$. The solution of these equations is $A = 1/162$, $B = -1/27$, $C = 1/18$ and $D = 2/3$. Therefore, the general solution of the nonhomogeneous problem is

$$y(t) = c_1 \cos 3t + c_2 \sin 3t + \frac{1}{162} e^{3t} - \frac{1}{27} te^{3t} + \frac{1}{18} t^2 e^{3t} + \frac{2}{3}.$$

7. The characteristic equation for the homogeneous problem is $\lambda^2 - 5\lambda + 4 = 0$, which has roots $\lambda = 1, 4$. Therefore, the solution of the homogeneous problem is $y_h(t) = c_1 e^t + c_2 e^{4t}$. To find a solution of the nonhomogeneous problem, we look for a solution of the form $y_p(t) = Ate^t$. Substituting a function of this form into the differential equation, and equating like terms, we have $-3A = 2$. Therefore, $A = -2/3$ and the general solution of the nonhomogeneous problem is

$$y(t) = c_1 e^t + c_2 e^{4t} - \frac{2}{3} te^t.$$

9. The characteristic equation for the homogeneous problem is $\lambda^2 + 2\lambda + 1 = 0$, which has the repeated root $\lambda = -1$. Therefore, the solution of the homogeneous problem is $y_h(t) = c_1 e^{-t} + c_2 te^{-t}$. To find a solution of the nonhomogeneous problem, we look for a solution of the form $y_p(t) = At^2 e^{-t}$. Substituting a function of this form into the differential equation, and equating like terms, we have $2A = 3$. Therefore, $A = 3/2$ and the general solution of the nonhomogeneous problem is

$$y(t) = c_1 e^{-t} + c_2 te^{-t} + \frac{3}{2} t^2 e^{-t}.$$

11. The characteristic equation for the homogeneous problem is $2\lambda^2 + 3\lambda + 1 = 0$, which has roots $\lambda = -1, -1/2$. Therefore, the solution of the homogeneous problem is $y_h(t) = c_1 e^{-t} + c_2 e^{-t/2}$. To find a solution of the nonhomogeneous problem, we will first look for a solution of the form $Y_1(t) = A + Bt + Ct^2$ to account for the nonhomogeneous term t^2. Substituting a function of this form into the differential equation, and equating like terms, we have $A + 3B + 4C = 0$, $B + 6C = 0$ and $C = 1$. The solution of these equations is

$A = 14$, $B = -6$, $C = 1$. Therefore, $Y_1 = 14 - 6t + t^2$. Next, we look for a solution of the nonhomogeneous problem of the form $Y_2(t) = D\cos t + E\sin t$. Substituting this function into the differential equation and equating like terms, we find that $D = -9/10$ and $E = -3/10$. Therefore, the general solution of the nonhomogeneous problem is

$$y(t) = c_1 e^{-t} + c_2 e^{-t/2} + 14 - 6t + t^2 - \frac{9}{10}\cos t - \frac{3}{10}\sin t.$$

13. The characteristic equation for the homogeneous problem is $\lambda^2 + \omega_0^2 = 0$, which has roots $\lambda = \pm\omega_0 i$. Therefore, the solution of the homogeneous problem is $y_h(t) = c_1 \cos\omega_0 t + c_2 \sin\omega_0 t$. To find a solution of the nonhomogeneous problem, we look for a solution of the form $y_p(t) = A\cos\omega t + B\sin\omega t$. Substituting a function of this form into the differential equation, and equating like terms, we have $(\omega_0^2 - \omega^2)A = 1$ and $(\omega_0^2 - \omega^2)B = 0$. The solution of these equations is $A = 1/(\omega_0^2 - \omega^2)$ and $B = 0$. Therefore, the general solution of the nonhomogeneous problem is

$$y(t) = c_1 \cos\omega_0 t + c_2 \sin\omega_0 t + \frac{1}{\omega_0^2 - \omega^2}\cos\omega t.$$

15. The characteristic equation for the homogeneous problem is $\lambda^2 + \lambda + 4 = 0$, which has roots $\lambda = (-1 \pm i\sqrt{15})/2$. Therefore, the solution of the homogeneous problem is $y_h(t) = e^{-t/2}(c_1 \cos(\sqrt{15}t/2) + c_2 \sin(\sqrt{15}t/2))$. To find a solution of the nonhomogeneous problem, we look for a solution of the form $y_p(t) = Ae^t + Be^{-t}$. Substituting a function of this form into the differential equation, and equating like terms, we have $6A = 1$ and $4B = -1$. The solution of these equations is $A = 1/6$ and $B = -1/4$. Therefore, the general solution of the nonhomogeneous problem is

$$y(t) = e^{-t/2}(c_1 \cos(\frac{\sqrt{15}\,t}{2}) + c_2 \sin(\frac{\sqrt{15}\,t}{2})) + \frac{1}{6}e^t - \frac{1}{4}e^{-t}.$$

17. The characteristic equation for the homogeneous problem is $\lambda^2 + \lambda - 2 = 0$, which has roots $\lambda = 1, -2$. Therefore, the solution of the homogeneous problem is $y_h(t) = c_1 e^t + c_2 e^{-2t}$. To find a solution of the nonhomogeneous problem, we look for a solution of the form $y_p(t) = At + B$. Substituting a function of this form into the differential equation, and equating like terms, we have $-2A = 2$ and $A - 2B = 0$. The solution of these equations is $A = -1$ and $B = -1/2$. Therefore, the general solution of the nonhomogeneous problem is

$$y(t) = c_1 e^t + c_2 e^{-2t} - t - \frac{1}{2}.$$

The initial conditions imply $c_1 + c_2 - 1/2 = 0$ and $c_1 - 2c_2 - 1 = 1$. Therefore, $c_1 = 1$ and $c_2 = -1/2$, which implies the particular solution of the initial value problem is

$$y(t) = e^t - \frac{1}{2}e^{-2t} - t - \frac{1}{2}.$$

19. The characteristic equation for the homogeneous problem is $\lambda^2 - 2\lambda + 1 = 0$, which has the repeated root $\lambda = 1$. Therefore, the solution of the homogeneous problem is $y_h(t) =$

$c_1 e^t + c_2 t e^t$. First, to find a solution of the nonhomogeneous problem, we look for a solution of the form $Y_1(t) = At^2 e^t + Bt^3 e^t$ to correspond with the term te^t. Substituting a function of this form into the differential equation, and equating like terms, we have $Y_1(t) = t^3 e^t / 6$. Next, considering the nonhomogeneous term 4, we look for a solution of the form $Y_2(t) = C$. Substituting this function into the equation and equating like terms, we have $Y_2 = 4$. Therefore, the general solution of the nonhomogeneous problem is

$$y(t) = c_1 e^t + c_2 t e^t + \frac{1}{6} t^3 e^t + 4.$$

The initial conditions imply $c_1 + 4 = 1$ and $c_1 + c_2 = 1$. Therefore, $c_1 = -3$ and $c_2 = 4$ which implies the particular solution of the initial value problem is

$$y(t) = -3e^t + 4te^t + \frac{1}{6} t^3 e^t + 4.$$

21. The characteristic equation for the homogeneous problem is $\lambda^2 + 4 = 0$, which has roots $\lambda = \pm 2i$. Therefore, the solution of the homogeneous problem is $y_h(t) = c_1 \cos 2t + c_2 \sin 2t$. To find a solution of the nonhomogeneous problem, we look for a solution of the form $y_p(t) = At \cos 2t + Bt \sin 2t$. Substituting a function of this form into the differential equation, and equating like terms, we have $y_p(t) = -\frac{3}{4} t \cos 2t$. Therefore, the general solution of the nonhomogeneous problem is

$$y(t) = c_1 \cos 2t + c_2 \sin 2t - \frac{3}{4} t \cos 2t.$$

The initial conditions imply $c_1 = 2$ and $2c_2 - 3/4 = -1$. Therefore, $c_1 = 2$ and $c_2 = -1/8$ which implies the particular solution of the initial value problem is

$$y(t) = 2 \cos 2t - \frac{1}{8} \sin 2t - \frac{3}{4} t \cos 2t.$$

23.(a) The characteristic equation for the homogeneous problem is $\lambda^2 + 3\lambda = 0$, which has roots $\lambda = 0, -3$. Therefore, the solution of the homogeneous problem is $y_h(t) = c_1 + c_2 e^{-3t}$. Therefore, to find a solution of the nonhomogeneous problem, we look for a solution of the form $Y(t) = t(A_0 t^4 + A_1 t^3 + A_2 t^2 + A_3 t + A_4) + t(B_0 t^2 + B_1 t + B_2)e^{-3t} + D \sin 3t + E \cos 3t$.

(b) Substituting a function of this form into the differential equation, and equating like terms, we have $A_0 = 2/15$, $A_1 = -2/9$, $A_2 = 8/27$, $A_3 = -8/27$, $A_4 = 16/81$, $B_0 = -1/9$, $B_1 = -1/9$, $B_2 = -2/27$, $D = -1/18$, $E = -1/18$ Therefore, the solution of the nonhomogeneous problem is

$$
\begin{aligned}
y(t) &= c_1 + c_2 e^{-3t} + t(\frac{2}{15} t^4 - \frac{2}{9} t^3 + \frac{8}{27} t^2 - \frac{8}{27} t + \frac{16}{81}) + \\
&+ t(-\frac{1}{9} t^2 - \frac{1}{9} t - \frac{2}{27}) e^{-3t} - \frac{1}{18} \sin 3t - \frac{1}{18} \cos 3t.
\end{aligned}
$$

25.(a) The characteristic equation for the homogeneous problem is $\lambda^2 - 5\lambda + 6 = 0$, which has roots $\lambda = 2, 3$. Therefore, the solution of the homogeneous problem is $y_h(t) = c_1 e^{2t} + c_2 e^{3t}$.

Therefore, to find a solution of the nonhomogeneous problem, we look for a solution of the form $Y(t) = e^t(A\cos 2t + B\sin 2t) + (D_0t + D_1)e^{2t}\sin t + (E_0t + E_1)e^{2t}\cos t$.

(b) Substituting a function of this form into the differential equation, and equating like terms, we have $A = -1/20$, $B = -3/20$, $D_0 = -3/2$, $D_1 = -5$, $E_0 = 3/2$, $E_1 = 1/2$. Therefore, the solution of the nonhomogeneous problem is

$$y(t) = c_1e^{2t} + c_2e^{3t} + e^t(-\frac{1}{20}\cos 2t - \frac{3}{20}\sin 2t) + (-\frac{3}{2}t - 5)e^{2t}\sin t + (\frac{3}{2}t + \frac{1}{2})e^{2t}\cos t.$$

27.(a) The characteristic equation for the homogeneous problem is $\lambda^2 - 4\lambda + 4 = 0$, which has the repeated root $\lambda = 2$. Therefore, the solution of the homogeneous problem is $y_h(t) = c_1e^{2t} + c_2te^{2t}$. Therefore, to find a solution of the nonhomogeneous problem, we look for a solution of the form $Y(t) = A_0t^2 + A_1t + A_2 + t^2(B_0t + B_1)e^{2t} + (D_0t + D_1)\sin 2t + (E_0t + E_1)\cos 2t$.

(b) Substituting a function of this form into the differential equation, and equating like terms, we have $A_0 = 1/2$, $A_1 = 1$, $A_2 = 3/4$, $B_0 = 2/3$, $B_1 = 0$, $D_0 = 0$, $D_1 = -1/16$, $E_0 = 1/8$, $E_1 = 1/16$. Therefore, the solution of the nonhomogeneous problem is

$$y(t) = c_1e^{2t} + c_2te^{2t} + \frac{1}{2}t^2 + t + \frac{3}{4} + \frac{2}{3}t^3e^{2t} - \frac{1}{16}\sin 2t + (\frac{1}{8}t + \frac{1}{16})\cos 2t.$$

29.(a) The characteristic equation for the homogeneous problem is $\lambda^2 + 3\lambda + 2 = 0$, which has roots $\lambda = -1, -2$. Therefore, the solution of the homogeneous problem is $y_h(t) = c_1e^{-t} + c_2e^{-2t}$. Therefore, to find a solution of the nonhomogeneous problem, we look for a solution of the form $Y(t) = (A_0t^2 + A_1t + A_2)e^t\sin 2t + (B_0t^2 + B_1t + B_2)e^t\cos 2t + e^{-t}(D\cos t + E\sin t) + Fe^t$.

(b) Substituting a function of this form into the differential equation, and equating like terms, we have $A_0 = 1/52$, $A_1 = 10/169$, $A_2 = -1233/35152$, $B_0 = -5/52$, $B_1 = 73/676$, $B_2 = -4105/35152$, $D = -3/2$, $E = 3/2$, $F = 2/3$ Therefore, the solution of the nonhomogeneous problem is $y(t) = c_1e^{-t} + c_2e^{-2t} + (A_0t^2 + A_1t + A_2)e^t\sin 2t + (B_0t^2 + B_1t + B_2)e^t\cos 2t + e^{-t}(D\cos t + E\sin t) + Fe^t$ with coefficients given above.

31.(a) For $Y = ve^{-t}$, we have $Y' = v'e^{-t} - ve^{-t}$ and $Y'' = v''e^{-t} - 2v'e^{-t} + ve^{-t}$. Then, $Y'' - 3Y' - 4Y = 2e^{-t}$ implies $v''e^{-t} - 2v'e^{-t} + ve^{-t} - 3v'e^{-t} + 3ve^{-t} - 4ve^{-t} = 2e^{-t}$. Simplifying this equation, we obtain $v'' - 5v' = 2$.

(b) We see that $(v')' - 5(v') = 2$. Therefore, letting $w = v'$, we see that w must satisfy $w' - 5w = 2$. This equation is linear with integrating factor $\mu(t) = e^{-5t}$. Therefore, we have $(e^{-5t}w)' = 2e^{-5t}$ which implies that $w = -2/5 + ce^{5t}$.

(c) Integrating w, we see that $v = (-2/5)t + c_1e^{5t} + c_2$. Then using the fact that $Y = ve^{-t}$, we conclude that

$$Y(t) = -\frac{2}{5}te^{-t} + c_1e^{4t} + c_2e^{-t}.$$

33. Letting $t = \ln x$, our equation can be rewritten as $y'' + 6y' + 5y = e^t$. First, we will solve the homogeneous equation. The associated characteristic equation is $\lambda^2 + 6\lambda + 5 = 0$ which

has roots $\lambda = -5, -1$. Therefore, the solution of the homogeneous equation is

$$y(t) = c_1 e^{-5t} + c_2 e^{-t}.$$

Now we solve the nonhomogeneous equation by the method of undetermined coefficients. We look for a solution of the form $y = Ae^t$. If $y = Ae^t$, then $y' = Ae^t$ and $y'' = Ae^t$. Plugging this into the nonhomogeneous equation, we have $Ae^t + 6Ae^t + 5Ae^t = e^t$, which implies $12A = 1$, thus $A = 1/12$. Therefore,

$$y(t) = c_1 e^{-5t} + c_2 e^{-t} + \frac{1}{12} e^t.$$

Rewriting the equation in terms of x, we have

$$y(x) = c_1 x^{-5} + c_2 x^{-1} + \frac{1}{12} x.$$

35. Letting $t = \ln x$, our equation can be rewritten as $y'' + 4y = \sin t$. First, we will solve the homogeneous equation. The associated characteristic equation is $\lambda^2 + 4 = 0$ which has roots $\lambda = \pm 2i$. Therefore, the solution of the homogeneous equation is

$$y(t) = c_1 \cos 2t + c_2 \sin 2t.$$

Now we solve the nonhomogeneous equation by the method of undetermined coefficients. We look for a solution of the form $y = A \sin t + B \cos t$. If $y = A \sin t + B \cos t$, then $y = A \cos t - B \sin t$ and $y'' = -A \sin t - B \cos t$. Plugging this into the nonhomogeneous equation, we get

$$y'' + 4y = -A \sin t - B \cos t + 4(A \sin t + B \cos t) = 3A \sin t + 3B \cos t = \sin t.$$

Equating like coefficients, we obtain $3A = 1$, $3B = 0$. Therefore, $A = 1/3$ and $B = 0$. Thus

$$y(t) = c_1 \cos 2t + c_2 \sin 2t + \frac{1}{3} \sin t.$$

Rewriting the equation in terms of x, we have

$$y(x) = c_1 \cos(2 \ln x) + c_2 \sin(2 \ln x) + \frac{1}{3} \sin(\ln x).$$

37. The solution of the homogeneous problem is $y_h(t) = c_1 \cos t + c_2 \sin t$. To solve the nonhomogeneous initial value problem (starting at $t = 0$), we begin by looking for a solution of the nonhomogeneous equation of the form $Y = A + Bt$. Substituting a function of this form into the differential equation leads to the equation $A + Bt = t$. Therefore, $A = 0$ and $B = 1$, and the solution of the nonhomogeneous problem is $y(t) = c_1 \cos t + c_2 \sin t + t$. Now, we consider the initial conditions. The initial conditions imply that $c_1 = 0$ and $c_2 + 1 = 1$. Therefore, the solution of the nonhomogeneous problem for $0 \leq t \leq \pi$ is given by $y_1(t) = t$. Now we need to solve the nonhomogeneous problem starting at $t = \pi$. We look for a solution of the nonhomogeneous problem of the form $Y(t) = Ce^{-t}$. Substituting a function of this

form into the differential equation leads to the equation $Ae^{-t} + Ae^{-t} = \pi e^{\pi-t}$. Therefore, $2A = \pi e^{\pi}$, or $A = \pi e^{\pi}/2$. Therefore, a solution of the nonhomogeneous problem (starting at $t = \pi$) is given by $y(t) = d_1 \cos t + d_2 \sin t + \pi e^{\pi-t}/2$. Using the solution of the initial value problem starting at $t = 0$, $y_1(t) = t$, we see that at time $t = \pi$, $y_1(\pi) = \pi$ and $y_1'(\pi) = 1$. Using these as our new initial conditions for $t = \pi$, we see that d_1, d_2 must satisfy $-d_1 + \pi/2 = \pi$ and $-d_2 - \pi 2/2 = 1$. The solution of these equations is $d_1 = -\pi/2$, $d_2 = -1 - \pi/2$. Therefore, we conclude that the solution of the nonhomogeneous initial value problem is

$$y(t) = \begin{cases} t & 0 \le t \le \pi \\ -\frac{\pi}{2} \cos t - \left(1 + \frac{\pi}{2}\right) \sin t + \frac{\pi}{2} e^{\pi-t} & t > \pi. \end{cases}$$

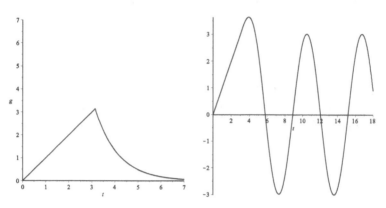

4.6   Forced Vibrations, Frequency Response, and Resonance

1. Consider the trigonometric identities $\cos(\alpha \pm \beta) = \cos\alpha\cos\beta \mp \sin\alpha\sin\beta$. Subtracting these two identities we obtain $\cos(\alpha - \beta) - \cos(\alpha + \beta) = 2\sin\alpha\sin\beta$. Here, our expression is $\cos 11t - \cos 3t$. Therefore, we let $\alpha - \beta = 11t$ and $\alpha + \beta = 3t$. Solving this system of equations, we have $\alpha = 7t$ and $\beta = -4t$. Therefore, we can write $\cos 11t - \cos 3t = 2\sin 7t \sin(-4t) = -2\sin 7t \sin 4t$.

3. Consider the trigonometric identities $\cos(\alpha \pm \beta) = \cos\alpha\cos\beta \mp \sin\alpha\sin\beta$. Adding these two identities we obtain $\cos(\alpha - \beta) + \cos(\alpha + \beta) = 2\cos\alpha\cos\beta$. Here, our expression is $\cos 7\pi t - \cos 2\pi t$. Therefore, we let $\alpha - \beta = 7\pi t$ and $\alpha + \beta = 2\pi t$. Solving this system of equations, we have $\alpha = 9\pi t/2$ and $\beta = -5\pi t/2$. Therefore, we can write $\cos 7\pi t + \cos 2\pi t = 2\cos(9\pi t/2)\cos(5\pi t/2)$.

5. The spring constant is $k = 4/(1.5/12) = 32$ lb/ft. The mass is $m = 4/32 = 1/8$ lb-s^2/ft. Assuming no damping, but an external force, $F(t) = 7\cos 3t$, the equation describing the motion is

$$\frac{1}{8}y'' + 32y = 7\cos 3t,$$

which can be rewritten as $y'' + 256y = 56\cos 3t$. The initial conditions are $y(0) = 12/12 = 1$ (ft) and $y'(0) = 0$ (ft/s).

7.(a) The solution of the homogeneous problem is $y_h(t) = c_1 \cos 16t + c_2 \sin 16t$. To find a solution of the nonhomogeneous problem, we look for a solution of the form $Y(t) = A \cos 3t$ (since there is no first derivative term, we may exclude the $\sin 3t$ function). Looking for a solution of this form, we arrive at the equation

$$-9A \cos 3t + 256A \cos 3t = 56 \cos 3t.$$

Therefore, we need A to satisfy $247A = 56$ or $A = 56/247$. Thus the solution of the nonhomogeneous problem is

$$y(t) = c_1 \cos 16t + c_2 \sin 16t + \frac{56}{247} \cos 3t.$$

The initial conditions are $y(0) = 1$ and $y'(0) = 0$. Therefore, c_1, c_2 must satisfy $c_1 + 56/247 = 1$ and $16c_2 = 0$. Thus we have $c_1 = 191/247$ and $c_2 = 0$. Therefore, the solution is

$$y(t) = \frac{191}{247} \cos 16t + \frac{56}{247} \cos 3t.$$

(b)

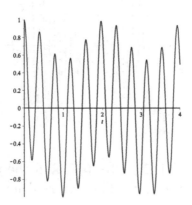

(c) If we replace the given external force with a force $Ae^{i\omega t}$, the nonhomogeneous problem to solve becomes

$$y'' + 256y = Ae^{i\omega t}.$$

The frequency for the homogeneous problem is $\omega_0 = 16$. Therefore, using equation (5), we know that a particular solution of this nonhomogeneous problem will be given by $Y(t) = G(i\omega)Ae^{i\omega t}$ where $G(i\omega) = \dfrac{1}{256 - \omega^2}$ is the frequency response. The gain is given by $|G(i\omega)| = |1/(256 - \omega^2)|$ and the angle

$$\phi(\omega) = \arccos\left(\frac{256 - \omega^2}{\sqrt{(256 - \omega^2)^2}}\right) = \arccos\left(\frac{256 - \omega^2}{|256 - \omega^2|}\right)$$

implies

$$\phi(\omega) = \begin{cases} \arccos(1) = 0 & \omega < 16 \\ \arccos(0) = \pi/2 & \omega = 16 \\ \arccos(-1) = \pi & \omega > 16. \end{cases}$$

Resonance will occur at $\omega_{max} = \omega_0 = 16$.

9. The spring constant is $k = 24$ lb/ft and the mass is $12/32$ lb-s^2/ft. The forcing term is $15\cos 7t$. Therefore, the equation of motion is

$$\frac{12}{32}y'' + 24y = 15\cos 7t$$

which can be simplified to

$$y'' + 64y = 40\cos 7t.$$

The solution of the homogeneous problem is $y_h(t) = c_1\cos 8t + c_2\sin 8t$. Then we look for a particular solution of the form $y_p(t) = A\cos 7t$. (We do not need to include a term of the form $\sin 7t$ since there are no first derivative terms in the equation.) Substituting y_p into the differential equation, we conclude that $A = 40/15 = 8/3$. Therefore, the general solution of this differential equation is $y(t) = c_1\cos 8t + c_2\sin 8t + (8/3)\cos 7t$. The initial conditions $y(0) = 0$ ft and $y'(0) = 0$ ft/s imply that $c_1 = -8/3$ and $c_2 = 0$. Therefore, the solution of this initial value problem is $y(t) = (8/3)(\cos 7t - \cos 8t)$.

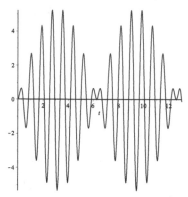

11. The spring constant is $k = 8/(1/2) = 16$ lb/ft and the mass is $8/32 = 1/4$ lb-s^2/ft. The damping constant is $\gamma = 0.25$ lb-sec/ft. The external force is $3\cos 2t$ lbs. Therefore, the equation of motion is

$$\frac{1}{4}y'' + \frac{1}{4}y' + 16y = 3\cos 2t$$

which can be simplified to

$$y'' + y' + 64y = 12\cos 2t.$$

(a) The roots of the characteristic equation are $\lambda = (-1 \pm \sqrt{255}i)/2$. Therefore, the solution of the homogeneous equation will be transient. To find the steady-state solution, we look for a particular solution of the form $y_p(t) = A\cos 2t + B\sin 2t$. Substituting y_p into the differential equation, we conclude that $A = 180/901$ and $B = 6/901$. Therefore, the steady-state response is

$$y(t) = \frac{180}{901}\cos 2t + \frac{6}{901}\sin 2t.$$

(b) With a forcing term of the form $F_0\cos(\omega t)$, the steady-state response can be written as $Y(t) = R\cos(\omega t - \delta)$ where the amplitude

$$R = \frac{F_0}{\sqrt{m^2(\omega_0^2 - \omega^2)^2 + \gamma^2\omega^2}} = \frac{F_0}{\sqrt{(k - m\omega^2)^2 + \gamma^2\omega^2}}.$$

The amplitude will be maximized when the denominator is minimized. This will occur when $k = m\omega^2$; that is, when $m = k/\omega^2 = 16/4 = 4$ slugs.

13. The amplitude of the steady-state response is given by

$$R = \frac{1}{\sqrt{(\omega_0^2 - \omega^2)^2 + 4\delta^2\omega^2}} = \frac{m}{\sqrt{m^2(\omega_0^2 - \omega^2)^2 + 4m^2\delta^2\omega^2}} = \frac{m}{\sqrt{m^2(\omega_0^2 - \omega^2)^2 + \gamma^2\omega^2}}.$$

The amplitude will be a maximum when the denominator is a minimum. Consider the function $f(z) = m^2(\omega_0^2 - z)^2 + \gamma^2 z$. This function has a minimum when $z = \omega_0^2 - \gamma^2/2m^2$. Therefore, the amplitude reaches a maximum at $\omega_{\max}^2 = \omega_0^2 - \gamma^2/2m^2$. Since $\omega_0^2 = k/m$, we can conclude that

$$\omega_{\max}^2 = \omega_0^2\left(1 - \frac{\gamma^2}{2km}\right).$$

Substituting $\omega^2 = \omega_{\max}^2$ into the expression for the amplitude, we have

$$R = \frac{m}{\sqrt{\gamma^4/4m^2 + \gamma^2(\omega_0^2 - \gamma^2/2m^2)}} = \frac{m}{\sqrt{\omega_0^2\gamma^2 - \gamma^4/4m^2}} = \frac{m}{\gamma\omega_0\sqrt{1 - \gamma^2/4mk}}.$$

15. The inductance is $L = 1$ henry. The resistance $R = 5 \cdot 10^3$ ohms. The capacitance is $C = 0.25 \cdot 10^{-6}$ farads. The forcing term is due to the 9-volt battery. Therefore, the equation for charge Q is

$$Q'' + 5000Q' + (4 \cdot 10^6)Q = 9.$$

The initial conditions are $Q(0) = 0$, $Q'(0) = 0$. The solution of the homogeneous problem is $Q_h(t) = c_1 e^{-1000t} + c_2 e^{-4000t}$. Looking for a solution of the nonhomogeneous problem, we find a particular solution of the form $Q_p(t) = (9/4) \cdot 10^{-6}$. Therefore, the general solution is given by $Q(t) = c_1 e^{-1000t} + c_2 e^{-4000t} + (9/4) \cdot 10^{-6}$. Considering our initial conditions, we conclude that $c_1 = -3 \cdot 10^{-6}$ and $c_2 = (3/4) \cdot 10^{-6}$. Therefore, the solution of the initial value problem is $Q(t) = 10^{-6}(-3e^{-1000t} + (3/4)e^{-4000t} + 9/4)$. At $t = 0.001, 0.01$, we have $Q(0.001) \approx 1.1601 \cdot 10^{-6}$ and $Q(0.01) \approx 2.2499 \cdot 10^{-6}$, respectively. From our function Q, we see that $Q(t) \to (9/4) \cdot 10^{-6}$ as $t \to \infty$.

17.(a) The solution of the homogeneous problem is $y_h(t) = c_1 \cos t + c_2 \sin t$. To find a particular solution of the nonhomogeneous problem, we look for a solution of the form $y_p(t) = A\cos\omega t$. Substituting a function of this form into the differential equation, we conclude that $A = 3/(1 - \omega^2)$. Therefore, the general solution of this differential equation is given by

$$y(t) = c_1 \cos t + c_2 \sin t + \frac{3}{1 - \omega^2}\cos\omega t.$$

Applying the initial conditions, we see that $c_1 + 3/(1 - \omega^2) = 0$ and $c_2 = 0$. Therefore, the solution of this initial value problem is

$$y(t) = \frac{3(\cos\omega t - \cos t)}{1 - \omega^2}.$$

(b)

19.

21.

23.(a)

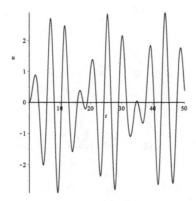

(b) The figures below show the solutions for $\omega = 0.8$, $\omega = 1.5$, and $\omega = 1.9$, respectively.

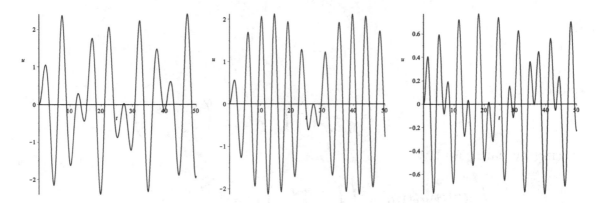

4.7 Variation of Parameters

1.(a)

$$\mathbf{Xu} = \left(\begin{array}{cc} x_{11}(t) & x_{12}(t) \\ x_{21}(t) & x_{22}(t) \end{array} \right) \left(\begin{array}{c} u_1(t) \\ u_2(t) \end{array} \right) = \left(\begin{array}{c} x_{11}(t)u_1(t) + x_{12}(t)u_2(t) \\ x_{21}(t)u_1(t) + x_{22}(t)u_2(t) \end{array} \right).$$

Therefore,

$$(\mathbf{Xu})' = \left(\begin{array}{c} x_{11}'u_1 + x_{11}u_1' + x_{12}'u_2 + x_{12}u_2' \\ x_{21}'u_1 + x_{21}u_1' + x_{22}'u_2 + x_{22}u_2' \end{array} \right).$$

Now

$$\mathbf{X}' = \left(\begin{array}{cc} x_{11}' & x_{12}' \\ x_{21}' & x_{22}' \end{array} \right) \quad \text{and} \quad \mathbf{u}' = \left(\begin{array}{c} u_1' \\ u_2' \end{array} \right).$$

Thus

$$\left(\begin{array}{cc} x_{11}' & x_{12}' \\ x_{21}' & x_{22}' \end{array} \right) \left(\begin{array}{c} u_1 \\ u_2 \end{array} \right) + \left(\begin{array}{cc} x_{11} & x_{12} \\ x_{21} & x_{22} \end{array} \right) \left(\begin{array}{c} u_1' \\ u_2' \end{array} \right) = \left(\begin{array}{c} x_{11}'u_1 + x_{12}'u_2 + x_{11}u_1 + x_{12}u_2' \\ x_{21}'u_1 + x_{22}'u_2 + x_{21}u_1' + x_{22}u_2' \end{array} \right),$$

i.e. $\mathbf{X'u} + \mathbf{Xu'} = (\mathbf{Xu})'$.

(b) Now

$$\mathbf{x}_p(t) = \mathbf{X}(t) \int \mathbf{X}^{-1}(t)\mathbf{g}(t)\, dt.$$

Let

$$\mathbf{u}(t) = \int \mathbf{X}^{-1}(t)\mathbf{g}(t)\, dt.$$

Therefore, $\mathbf{x}_p(t) = \mathbf{X}(t)\mathbf{u}(t)$, which implies

$$\mathbf{x}_p' = (\mathbf{X}\mathbf{u})' = \mathbf{X}'\mathbf{u} + \mathbf{X}\mathbf{u}' = \mathbf{P}(t)\mathbf{X}\mathbf{u} + \mathbf{X}(\mathbf{X}^{-1}\mathbf{g}) = \mathbf{P}(t)\mathbf{x}_p + \mathbf{g}.$$

3. Let $\mathbf{X}(t)$ be the fundamental matrix

$$\mathbf{X}(t) = \begin{pmatrix} e^{-t} & -e^t \\ 0 & e^t \end{pmatrix}.$$

Then

$$\mathbf{X}^{-1}(t) = \begin{pmatrix} e^t & e^t \\ 0 & e^{-t} \end{pmatrix}.$$

Therefore,

$$\mathbf{X}^{-1}(t)\mathbf{g}(t) = \begin{pmatrix} e^t & e^t \\ 0 & e^{-t} \end{pmatrix}\begin{pmatrix} e^{-t} \\ t \end{pmatrix} = \begin{pmatrix} 1 + te^t \\ te^{-t} \end{pmatrix}.$$

Integrating by parts, we have

$$\int \mathbf{X}^{-1}(t)\mathbf{g}(t)\, dt = \int \begin{pmatrix} 1 + te^t \\ te^{-t} \end{pmatrix} dt = \begin{pmatrix} t + te^t - e^t \\ -te^{-t} - e^{-t} \end{pmatrix}.$$

Thus

$$\mathbf{x}_p(t) = \mathbf{X}(t) \int \mathbf{X}^{-1}(t)g(t)\, dt = \begin{pmatrix} e^{-t} & -e^t \\ 0 & e^t \end{pmatrix}\begin{pmatrix} t + te^t - e^t \\ -te^{-t} - e^{-t} \end{pmatrix} = \begin{pmatrix} te^{-t} + 2t \\ -t - 1 \end{pmatrix}.$$

5. Let $\mathbf{X}(t)$ be the fundamental matrix,

$$\mathbf{X}(t) = \begin{pmatrix} \cos t & \sin t \\ -\sin t & \cos t \end{pmatrix}.$$

Then

$$\mathbf{X}^{-1}(t) = \begin{pmatrix} \cos t & -\sin t \\ \sin t & \cos t \end{pmatrix}.$$

Therefore,

$$\mathbf{X}^{-1}(t)\mathbf{g}(t) = \begin{pmatrix} \cos t & -\sin t \\ \sin t & \cos t \end{pmatrix}\begin{pmatrix} \cos t \\ -\sin t \end{pmatrix} = \begin{pmatrix} 1 \\ 0 \end{pmatrix}.$$

Thus

$$\int \mathbf{X}^{-1}(t)\mathbf{g}(t)\, dt = \int \begin{pmatrix} 1 \\ 0 \end{pmatrix} dt = \begin{pmatrix} t \\ 0 \end{pmatrix}.$$

Then

$$\mathbf{x}_p(t) = \mathbf{X}(t) \int \mathbf{X}^{-1}(t) g(t)\, dt = \begin{pmatrix} \cos t & \sin t \\ -\sin t & \cos t \end{pmatrix} \begin{pmatrix} t \\ 0 \end{pmatrix} = \begin{pmatrix} t\cos t \\ -t\sin t \end{pmatrix}.$$

7. The general solution of the equation in problem 3 is

$$\mathbf{x}(t) = c_1 \begin{pmatrix} e^{-t} \\ 0 \end{pmatrix} + c_2 \begin{pmatrix} -e^t \\ e^t \end{pmatrix} + \begin{pmatrix} te^{-t} + 2t \\ -t - 1 \end{pmatrix}.$$

The initial condition $\mathbf{x}(0) = (\begin{matrix} -1 & 1 \end{matrix})^T$ implies

$$\mathbf{x}(0) = c_1 \begin{pmatrix} 1 \\ 0 \end{pmatrix} + c_2 \begin{pmatrix} -1 \\ 1 \end{pmatrix} + \begin{pmatrix} 0 \\ -1 \end{pmatrix} = \begin{pmatrix} -1 \\ 1 \end{pmatrix}.$$

Therefore, $c_1 - c_2 = -1$ and $c_2 - 1 = 1$. Solving this system of equations, we have $c_1 = 1$ and $c_2 = 2$. Thus the solution of the initial value problem is

$$\mathbf{x}(t) = \begin{pmatrix} e^{-t} - 2e^t + te^{-t} + 2t \\ 2e^t - t - 1 \end{pmatrix}.$$

9. The general solution of the equation in problem 5 is

$$\mathbf{x}(t) = c_1 \begin{pmatrix} \cos t \\ -\sin t \end{pmatrix} + c_2 \begin{pmatrix} \sin t \\ \cos t \end{pmatrix} + \begin{pmatrix} t\cos t \\ -t\sin t \end{pmatrix}.$$

The initial condition $\mathbf{x}(0) = (\begin{matrix} 1 & 1 \end{matrix})^T$ implies

$$\mathbf{x}(0) = c_1 \begin{pmatrix} 1 \\ 0 \end{pmatrix} + c_2 \begin{pmatrix} 0 \\ 1 \end{pmatrix} + \begin{pmatrix} 0 \\ 0 \end{pmatrix} = \begin{pmatrix} 1 \\ 1 \end{pmatrix}.$$

Therefore, $c_1 = 1$ and $c_2 = 1$. Thus the solution of the initial value problem is

$$\mathbf{x}(t) = \begin{pmatrix} \cos t \\ -\sin t \end{pmatrix} + \begin{pmatrix} \sin t \\ \cos t \end{pmatrix} + \begin{pmatrix} t\cos t \\ -t\sin t \end{pmatrix}.$$

11. The solution of the homogeneous equation is $y_h(t) = c_1 e^{2t} + c_2 e^{-t}$. The functions $y_1(t) = e^{2t}$ and $y_2(t) = e^{-t}$ form a fundamental set of solutions. The Wronskian of these functions is $W(y_1, y_2) = -3e^t$. Using the method of variation of parameters, a particular solution is given by $Y(t) = u_1(t)y_1(t) + u_2(t)y_2(t)$ where

$$u_1(t) = -\int \frac{e^{-t}(2e^{-t})}{W(t)}\, dt = -\frac{2}{9}e^{-3t} \quad \text{and} \quad u_2(t) = \int \frac{e^{2t}(2e^{-t})}{W(t)}\, dt = -\frac{2t}{3}.$$

Therefore, a particular solution is $Y(t) = -\dfrac{2}{9}e^{-t} - \dfrac{2}{3}te^{-t}$. As $-2e^{-t}/9$ is a solution of the homogeneous equation, we can omit it and conclude that $y_p(t) = -\dfrac{2}{3}te^{-t}$ is a particular solution of the nonhomogeneous problem.

13. The solution of the homogeneous equation is $y_h(t) = c_1 e^{t/2} + c_2 t e^{t/2}$. The functions $y_1(t) = e^{t/2}$ and $y_2(t) = t e^{t/2}$ form a fundamental set of solutions. The Wronskian of these functions is $W(y_1, y_2) = e^t$. Using the method of variation of parameters, a particular solution is given by $Y(t) = u_1(t) y_1(t) + u_2(t) y_2(t)$ where

$$u_1(t) = -\int \frac{t e^{t/2}(4 e^{t/2})}{W(t)}\, dt = -2t^2 \quad \text{and} \quad u_2(t) = \int \frac{e^{t/2}(4 e^{t/2})}{W(t)}\, dt = 4t.$$

Therefore, a particular solution is $Y(t) = -2t^2 e^{t/2} + 4t^2 e^{t/2} = 2t^2 e^{t/2}$.

15. The solution of the homogeneous equation is $y_h(t) = c_1 \cos 2t + c_2 \sin 2t$. The functions $y_1(t) = \cos 2t$ and $y_2(t) = \sin 2t$ form a fundamental set of solutions. The Wronskian of these functions is $W(y_1, y_2) = 2$. Using the method of variation of parameters, a particular solution is given by $Y(t) = u_1(t) y_1(t) + u_2(t) y_2(t)$ where

$$u_1(t) = -\int \frac{\sin 2t (3 \sec^2 2t)}{W(t)}\, dt = -\frac{3}{4} \sec 2t$$

$$u_2(t) = \int \frac{\cos 2t (3 \sec^2 2t)}{W(t)}\, dt = \frac{3}{4} \ln(\sec 2t + \tan 2t).$$

Thus a particular solution is $Y(t) = -3/4 + (3/4)\sin 2t \ln(\sec 2t + \tan 2t)$. Therefore, the general solution is $y(t) = c_1 \cos 2t + c_2 \sin 2t + (3/4)\sin 2t \ln(\sec 2t + \tan 2t) - 3/4$.

17. The solution of the homogeneous equation is $y_h(t) = c_1 \cos 2t + c_2 \sin 2t$. The functions $y_1(t) = \cos 2t$ and $y_2(t) = \sin 2t$ form a fundamental set of solutions. The Wronskian of these functions is $W(y_1, y_2) = 2$. Using the method of variation of parameters, a particular solution is given by $Y(t) = u_1(t) y_1(t) + u_2(t) y_2(t)$ where

$$u_1(t) = -\int \frac{\sin 2t (3 \csc(t/2))}{W(t)}\, dt = -6 \sin(t/2) - 2 \sin(3t/2)$$

$$u_2(t) = \int \frac{\cos 2t (3 \csc(t/2))}{W(t)}\, dt = 6 \cos(t/2) + 2 \cos(3t/2) + 3 \ln \tan(t/4).$$

Thus a particular solution is $Y(t) = (-6 \sin(t/2) - 2 \sin(3t/2)) \cos 2t + (6 \cos(t/2) + 2 \cos(3t/2) + 3 \ln \tan(t/4)) \sin 2t$. Therefore, after simplification, the general solution is $y(t) = c_1 \cos 2t + c_2 \sin 2t + 2 \sin(t/2) + 6 \sin(3t/2) + 3 \ln \tan(t/4) \sin 2t$.

19. The solution of the homogeneous equation is $y_h(t) = c_1 e^t + c_2 t e^t$. The functions $y_1(t) = e^t$ and $y_2(t) = t e^t$ form a fundamental set of solutions. The Wronskian of these functions is $W(y_1, y_2) = e^{2t}$. Using the method of variation of parameters, a particular solution is given by $Y(t) = u_1(t) y_1(t) + u_2(t) y_2(t)$ where

$$u_1(t) = -\int \frac{t e^t (e^t)}{W(t)(1 + t^2)}\, dt = -\frac{1}{2} \ln(1 + t^2)$$

$$u_2(t) = \int \frac{e^t (e^t)}{W(t)(1 + t^2)}\, dt = \arctan t.$$

Thus a particular solution is $Y(t) = -\frac{1}{2} e^t \ln(1 + t^2) + t e^t \arctan t$. Therefore, the general solution is $y(t) = c_1 e^t + c_2 t e^t - \frac{1}{2} e^t \ln(1 + t^2) + t e^t \arctan t$.

21. The solution of the homogeneous equation is $y_h(t) = c_1 \cos 2t + c_2 \sin 2t$. The functions $y_1(t) = \cos 2t$ and $y_2(t) = \sin 2t$ form a fundamental set of solutions. The Wronskian of these functions is $W(y_1, y_2) = 2$. Using the method of variation of parameters, a particular solution is given by $Y(t) = u_1(t)y_1(t) + u_2(t)y_2(t)$ where

$$u_1(t) = -\int \frac{\sin 2s\, g(s)}{W(s)}\, ds = -\frac{1}{2}\int \sin 2s\, g(s)\, ds$$

$$u_2(t) = \int \frac{\cos 2s\, g(s)}{W(s)}\, ds = \frac{1}{2}\int \cos 2s\, g(s)\, ds.$$

Therefore, a particular solution is

$$Y(t) = -\frac{1}{2}\int \cos 2t \sin 2s\, g(s)\, ds + \frac{1}{2}\int \sin 2t \cos 2s\, g(s)\, ds.$$

Using the fact that $\sin 2t \cos 2s - \cos 2t \sin 2s = \sin(2t - 2s)$, we see that the general solution is

$$y(t) = c_1 \cos 2t + c_2 \sin 2t + \frac{1}{2}\int \sin(2t - 2s)\, g(s)\, ds.$$

23. By direct substitution, it can be verified that $y_1(t) = 1+t$ and $y_2(t) = e^t$ are solutions of the homogeneous equations. The Wronskian of these functions is $W(y_1, y_2) = te^t$. Rewriting the equation in standard form, we have

$$y'' - \frac{1+t}{t}y' + \frac{1}{t}y = te^{2t}.$$

Therefore, $g(t) = te^{2t}$. Using the method of variation of parameters, a particular solution is given by $Y(t) = u_1(t)y_1(t) + u_2(t)y_2(t)$ where

$$u_1(t) = -\int \frac{e^t(te^{2t})}{W(t)}\, dt = -\frac{1}{2}e^{2t} \quad \text{and} \quad u_2(t) = \int \frac{(1+t)(te^{2t})}{W(t)}\, dt = te^t.$$

Therefore, a particular solution is $Y(t) = -\frac{1}{2}(1+t)e^{2t} + te^{2t} = \frac{1}{2}(t-1)e^{2t}$.

25. By direct substitution, it can be verified that $y_1(x) = x^{-1/2}\sin x$ and $y_2(x) = x^{-1/2}\cos x$ are solutions of the homogeneous equations. The Wronskian of these functions is $W(y_1, y_2) = -1/x$. Rewriting the equation in standard form, we have

$$y'' + \frac{1}{x}y' + \frac{x^2 - 1/4}{x^2}y = 3\frac{\sin x}{x^{1/2}}.$$

Therefore, $g(x) = 3x^{-1/2}\sin x$. Using the method of variation of parameters, a particular solution is given by $Y(x) = u_1(x)y_1(x) + u_2(x)y_2(x)$ where

$$u_1(x) = -\int \frac{x^{-1/2}\cos x\,(3x^{-1/2}\sin x)}{W(x)}\, dx = -\frac{3}{2}\cos^2 x$$

$$u_2(x) = \int \frac{x^{-1/2}\sin x\,(3x^{-1/2}\sin x)}{W(x)}\, dx = \frac{3}{2}\cos x \sin x - \frac{3x}{2}.$$

Therefore, a particular solution is

$$Y(x) = -\frac{3}{2}x^{-1/2}\cos^2 x \sin x + \left(\frac{3}{2}\cos x \sin x - \frac{3x}{2}\right)x^{-1/2}\cos x = -\frac{3}{2}x^{1/2}\cos x.$$

27. By direct substitution, it can be verified that $y_1(x) = x^{-1/2}\sin x$ and $y_2(x) = x^{-1/2}\cos x$ are solutions of the homogeneous equations. The Wronskian of these functions is $W(y_1, y_2) = -1/x$. Rewriting the equation in standard form, we have

$$y'' + \frac{1}{x}y' + \frac{x^2 - 1/4}{x^2}y = \frac{g(x)}{x^2}.$$

Therefore, the nonhomogeneous term is $g(x)/x^2$. Using the method of variation of parameters, a particular solution is given by $Y(x) = u_1(x)y_1(x) + u_2(x)y_2(x)$ where

$$u_1(x) = -\int \frac{s^{-1/2}\cos s\, g(s)}{s^2 W(s)}\, ds \quad \text{and} \quad u_2(x) = \int \frac{s^{-1/2}\sin s\, g(s)}{s^2 W(s)}\, ds.$$

Therefore, a particular solution is

$$\begin{aligned}
Y(x) &= -x^{-1/2}\sin(x)\int \frac{s^{-1/2}\cos s\, g(s)}{s^2 W(s)}\, ds + x^{-1/2}\cos(x)\int \frac{s^{-1/2}\sin s\, g(s)}{s^2 W(s)}\, ds \\
&= \frac{1}{\sqrt{x}}\int \frac{(\sin x \cos s - \cos x \sin s)g(s)}{s\sqrt{s}}\, ds = \frac{1}{\sqrt{x}}\int \frac{\sin(x-s)g(s)}{s\sqrt{s}}\, ds.
\end{aligned}$$

29. First, we rewrite the equation as

$$y'' - \frac{3}{x}y' + \frac{4}{x^2}y = \ln x.$$

Then, we look for a solution of the homogeneous equation

$$y'' - \frac{3}{x}y' + \frac{4}{x^2}y = 0.$$

We look for a solution of the form x^α. Upon substituting a function of this form into the differential equation, we have

$$\alpha(\alpha - 1)x^{\alpha-2} - 3\alpha x^{\alpha-2} + 4x^{\alpha-2} = 0,$$

which implies $\alpha(\alpha - 1) - 3\alpha + 4 = 0$. This equation can be rewritten as $\alpha^2 - 4\alpha + 4 = 0$. The solution of this equation is $\alpha = 2$. Therefore, $y_1(x) = x^2$ is a solution of the homogeneous equation. Now to solve the nonhomogeneous equation, we look for a solution of the form $y(x) = v(x)x^2$, using the fact that $y_1(x) = x^2$ is a solution of the homogeneous problem. Substituting a function of this form into the differential equation, we see that v must satisfy

$$x^2 v'' + xv' = \ln x.$$

Dividing this equation by x^2, we have

$$v'' + \frac{1}{x}v' = \frac{\ln x}{x^2}.$$

This equation is linear in $w = v'$ with integrating factor $\mu(x) = e^{\int 1/x \, dx} = x$. Therefore, $xv'' + v' = \frac{\ln x}{x}$, which implies $(xv')' = \frac{\ln x}{x}$, thus $xv' = \frac{1}{2}(\ln(x))^2 + c$, and then $v' = \frac{\ln(x)^2}{2x} + \frac{c}{x}$. Integrating v', we conclude that $v = \frac{1}{6}(\ln x)^3 + c_1 \ln x + c_2$. Thus we obtain

$$y(t) = vy_1 = \frac{1}{6}x^2(\ln x)^3 + c_1 x^2 \ln x + c_2 x^2.$$

31. First, we rewrite the equation as

$$y'' + \frac{7}{t}y' + \frac{5}{t^2}y = \frac{1}{t}.$$

Then, we look for a solution of the homogeneous equation

$$y'' + \frac{7}{t}y' + \frac{5}{t^2}y = 0.$$

We look for a solution of the form t^α. Upon substituting a function of this form into the differential equation, we have

$$\alpha(\alpha - 1)t^{\alpha-2} + 7\alpha t^{\alpha-2} + 5t^{\alpha-2} = 0,$$

which implies $\alpha(\alpha - 1) + 7\alpha + 5 = 0$. This equation can be rewritten as $\alpha^2 + 6\alpha + 5 = 0$. The solutions of this equation are $\alpha = -5, -1$. In particular, $y_1(t) = t^{-1}$ is a solution of the homogeneous equation. Now to solve the nonhomogeneous equation, we look for a solution of the form $y(t) = v(t)t^{-1}$, using the fact that $y_1(t) = t^{-1}$ is a solution of the homogeneous problem. Substituting a function of this form into the differential equation, we see that v must satisfy

$$v'' + \frac{5}{t}v' = 1.$$

This equation is linear with integrating factor $\mu(t) = \int e^{5/t} \, dt = t^5$. Therefore, the equation can be written as $t^5 v'' + 5t^4 v' = t^5$ which implies $(t^5 v')' = t^5$. Then integrating, we have $t^5 v' = t^6/6 + c_1$. Therefore, $v' = t/6 + c_1 t^{-5}$. Now integrating this equation, we conclude that $v = t^2/12 + c_1 t^{-4} + c_2$. Thus the general solution is given by

$$y(t) = vy_1 = \frac{t}{12} + c_1 t^{-5} + c_2 t^{-1}.$$

33. We rewrite (27) as

$$Y(t) = -y_1(t)\int_{t_0}^t \frac{y_2(\tau)g(\tau)}{W(y_1, y_2)(\tau)} \, d\tau + y_2(t)\int_{t_0}^t \frac{y_1(\tau)g(\tau)}{W(y_1, y_2)(\tau)} \, d\tau.$$

Bringing the terms $y_1(t)$ and $y_2(t)$ inside the integrals and using the fact that $W(y_1, y_2)(\tau) = y_1(\tau)y_2'(\tau) - y_1'(\tau)y_2(\tau)$, the desired result holds. Then, using the product rule and the Fundamental Theorem of Calculus, we see that

$$
\begin{aligned}
Y'(t) &= \frac{y_1(t)y_2(t) - y_1(t)y_2(t)}{y_1(t)y_2'(t) - y_1'(t)y_2(t)}g(t) + \int_{t_0}^{t} \frac{y_1(\tau)y_2'(t) - y_1'(t)y_2(\tau)}{y_1(\tau)y_2'(\tau) - y_1'(\tau)y_2(\tau)}g(\tau)\,d\tau \\
&= \int_{t_0}^{t} \frac{y_1(\tau)y_2'(t) - y_1'(t)y_2(\tau)}{y_1(\tau)y_2'(\tau) - y_1'(\tau)y_2(\tau)}g(\tau)\,d\tau.
\end{aligned}
$$

Differentiating again, we obtain that

$$
\begin{aligned}
Y''(t) &= \frac{y_1(t)y_2'(t) - y_1'(t)y_2(t)}{y_1(t)y_2'(t) - y_1'(t)y_2(t)}g(t) + \int_{t_0}^{t} \frac{y_1(\tau)y_2''(t) - y_1''(t)y_2(\tau)}{y_1(\tau)y_2'(\tau) - y_1'(\tau)y_2(\tau)}g(\tau)\,d\tau \\
&= g(t) + \int_{t_0}^{t} \frac{y_1(\tau)y_2''(t) - y_1''(t)y_2(\tau)}{y_1(\tau)y_2'(\tau) - y_1'(\tau)y_2(\tau)}g(\tau)\,d\tau.
\end{aligned}
$$

Now substituting Y back into our equation, we get

$$
\begin{aligned}
L[Y] &= Y'' + p(t)Y' + q(t)Y \\
&= g(t) + \int_{t_0}^{t} \frac{y_1(\tau)(y_2''(t) - p(t)y_2'(t) + q(t)y_2(t)) - y_2(\tau)(y_1''(t) + p(t)y_1'(t) + q(t)y_1(t))}{y_1(\tau)y_2'(\tau) - y_1'(\tau)y_2(\tau)}g(\tau)\,d\tau \\
&= g(t).
\end{aligned}
$$

Further, $Y(t_0) = 0$ and $Y'(t_0) = 0$ since the upper and lower limits of integration are t_0.

35. The given linear operator $L[y] = (D-a)(D-b)y$ can be written as $L[y] = y'' - (a+b)y' + aby$. Therefore, $p(t) = -(a+b)$ and $q(t) = ab$. To solve the given nonhomogeneous problem, we first need to solve the associated homogeneous problem. In particular, we need to look for the general solution of $y'' - (a+b)y' + aby = 0$. The roots of the associated equation are $\lambda = a, b$. Therefore, the general solution of the homogeneous problem is $y(t) = c_1 e^{at} + c_2 e^{bt}$. Letting $y_1(t) = e^{at}$ and $y_2(t) = e^{bt}$, we now use the result in problem 33. In particular, a solution of the indicated problem will be given by

$$
Y(t) = \int_{t_0}^{t} \frac{e^{a\tau}e^{bt} - e^{at}e^{b\tau}}{be^{a\tau}e^{b\tau} - ae^{a\tau}e^{b\tau}}g(\tau)\,d\tau = \int_{t_0}^{t} \frac{1}{b-a}(e^{b(t-\tau)} - e^{a(t-\tau)})g(\tau)\,d\tau.
$$

37. The given linear operator $L[y] = (D-a)^2 y$ can be written as $L[y] = y'' - 2ay' + a^2 y$. Therefore, $p(t) = -2a$ and $q(t) = a^2$. To solve the given nonhomogeneous problem, we first need to solve the associated homogeneous problem. In particular, we need to look for the general solution of $y'' - 2ay' + a^2 y = 0$. The repeated root of the associated equation is $\lambda = a$. Therefore, the general solution of the homogeneous problem is $y(t) = c_1 e^{at} + c_2 t e^{at}$. Letting $y_1(t) = e^{at}$ and $y_2(t) = te^{at}$, we now use the result in problem 33. In particular, a solution of the indicated problem will be given by

$$
Y(t) = \int_{t_0}^{t} e^{a(t-\tau)}(t - \tau)g(\tau)\,d\tau.
$$

39. Let $y(t) = v(t)y_1(t)$ where y_1 is a solution of the homogeneous equation. Substituting a function of this form into the differential equation, we have

$$v''y_1 + 2v'y_1' + vy_1'' + p(t)[v'y_1 + vy_1'] + q(t)vy_1 = g(t).$$

By assumption, $y_1'' + p(t)y_1 + q(t)y_1 = 0$. Therefore, v must be a solution of the differential equation

$$v''y_1 + [2y_1' + p(t)y_1]v' = g(t).$$

41. Write the equation as $y'' + t(1-t)^{-1}y' - (1-t)^{-1}y = 2(1-t)e^{-t}$. Using the fact that $y_1(t) = e^t$ is a solution of the homogeneous equation, we consider a function of the form $y(t) = e^t v(t)$. From the method in problem 39, we know that y will be a solution of the given differential equation as long as v satisfies

$$e^t v'' + \left[2e^t + t(1-t)^{-1}e^t\right]v' = 2(1-t)e^{-t}.$$

We can simplify this equation to

$$v'' + \frac{2-t}{1-t}v' = 2(1-t)e^{-2t}.$$

Using the fact that this equation is linear in v' and has an integrating factor $\mu(t) = e^t/(t-1)$, we find that $v' = (t-1)(2e^{-2t} + c_1 e^{-t})$. Integrating this equation, we conclude that $v(t) = (1/2 - t)e^{-2t} - c_1 t e^{-t} + c_2$. Therefore, the solution of the original differential equation is

$$y(t) = \left(\frac{1}{2} - t\right)e^{-t} - c_1 t + c_2 e^t.$$

Chapter 5

The Laplace Transform

5.1 Definition of the Laplace Transform

1.

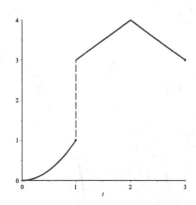

The function is piecewise continuous.

3.

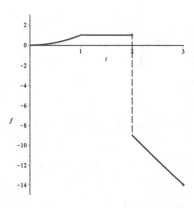

The function is piecewise continuous.

5. The function is of exponential order. We can take $K = 2$, $a = 3$ and $M = 0$.

7. Since $|f(t)| = |e^{5t}\sin(3t)| \le e^{5t}$, the function is of exponential order. We can take $K = 1$, $a = 5$ and $M = 0$.

9. $\cosh t^2 = (e^{t^2} + e^{-t^2})/2$ is not of exponential order, since for all $a > 0$, there exists an M such that $e^{t^2} \geq e^{at}$ for all $t \geq M$.

11. Since $1/(1+t) \leq 1$ for all $t \geq 0$, the function is of exponential order. We can take $K = 1$, $a = 0$ and $M = 0$.

13.(a) For $f(t) = t$, the Laplace transform is

$$F(s) = \int_0^\infty e^{-st} t \, dt = \left. \frac{-e^{-st}t - e^{-st}}{s^2} \right|_{t=0}^\infty = \frac{1}{s^2}.$$

(b) For $f(t) = t^2$, the Laplace transform is

$$F(s) = \int_0^\infty e^{-st} t^2 \, dt = \left. -\frac{e^{-st}s^2t^2 + 2e^{-st}st + 2e^{-st}}{s^3} \right|_{t=0}^\infty = \frac{2}{s^3}.$$

(c) For $f(t) = t^n$, using induction, the Laplace transform is

$$F(s) = \int_0^\infty e^{-st} t^n \, dt = \frac{n!}{s^{n+1}}.$$

15. For

$$f(t) = \begin{cases} 0 & 0 \leq t \leq 1 \\ 1 & 1 < t \leq 2 \\ 0 & 2 < t, \end{cases}$$

the Laplace transform is given by

$$\begin{aligned} F(s) &= \int_0^1 e^{-st} \cdot 0 \, dt + \int_1^2 e^{-st} \, dt + \int_2^\infty e^{-st} \cdot 0 \, dt = \int_1^2 e^{-st} \, dt = \left. \frac{e^{-st}}{-s} \right|_1^2 \\ &= \frac{e^{-2s}}{-s} - \frac{e^{-s}}{-s} = \frac{e^{-s} - e^{-2s}}{s}. \end{aligned}$$

17. For

$$f(t) = \begin{cases} t^2 & 0 \leq t \leq 1 \\ 5 - t & 1 < t \leq 2 \\ 6 & 2 < t \end{cases}$$

the Laplace transform is given by

$$F(s) = \int_0^1 e^{-st} t^2 \, dt + \int_1^2 e^{-st} (5 - t) \, dt + \int_2^\infty 6e^{-st} \, dt.$$

For the first integral on the right-hand side, we proceed as follows:

$$\int_0^1 e^{-st} t^2 \, dt = \left. \frac{t^2 e^{-st}}{-s} \right|_0^1 - \int_0^1 \frac{e^{-st}}{-s} \cdot 2t \, dt = \frac{e^{-s}}{-s} + \frac{2}{s} \int_0^1 te^{-st} \, dt.$$

By the solution to problem 14, we know that

$$\int_0^1 te^{-st} \, dt = \frac{e^{-s}}{-s} - \frac{e^{-s}}{s^2} + \frac{1}{s^2}.$$

Therefore,

$$\int_0^1 e^{-st}t^2\, dt = \frac{e^{-s}}{-s} + \frac{2}{s}\left[\frac{e^{-s}}{-s} - \frac{e^{-s}}{s^2} + \frac{1}{s^2}\right] = \frac{e^{-s}}{-s} - \frac{2e^{-s}}{s^2} - \frac{2e^{-s}}{s^3} + \frac{2}{s^3}.$$

For the second integral on the right-hand side, we begin by separating it into two pieces.

$$\int_1^2 e^{-st}(5-t)\, dt = 5\int_1^2 e^{-st}\, dt - \int_1^2 te^{-st}\, dt \equiv A + B.$$

Now, for A, we have

$$5\int_1^2 e^{-st}\, dt = \left.\frac{5e^{-st}}{-s}\right|_1^2 = \frac{5e^{-s}}{s} - \frac{5e^{-2s}}{s}.$$

For term B, we have

$$-\int_1^2 te^{-st}\, dt = -\left[\left.\frac{te^{-st}}{-s}\right|_1^2 - \int_1^2 \frac{e^{-st}}{-s}\, dt\right] = -\left[\frac{2e^{-2s}}{-s} - \frac{e^{-s}}{-s} + \frac{1}{s}\int_1^2 e^{-st}\, dt\right]$$

$$= -\left[\frac{2e^{-2s}}{-s} + \frac{e^{-s}}{s} = \frac{1}{s}\cdot\left.\frac{e^{-st}}{-s}\right|_1^2\right] = -\left[\frac{2e^{-2s}}{-s} + \frac{e^{-s}}{s} - \frac{e^{-2s}}{s^2} + \frac{e^{-s}}{s^2}\right].$$

Now combining the results for A and B, we have

$$\int_1^2 e^{-st}(5-t)\, dt = \frac{5e^{-s}}{s} - \frac{5e^{-2s}}{s} - \left[\frac{2e^{-2s}}{-s} + \frac{e^{-s}}{s} - \frac{e^{-2s}}{s^2} + \frac{e^{-s}}{s^2}\right].$$

Finally, for the third integral above, we have

$$\int_2^\infty 6e^{-st}\, dt = \left.\frac{6e^{-st}}{-s}\right|_2^\infty = 0 - \frac{6e^{-2s}}{-s} = \frac{6e^{-2s}}{s}.$$

Combining these integrals, we have

$$F(s) = \frac{e^{-s}}{-s} - \frac{2e^{-s}}{s^2} - \frac{2e^{-s}}{s^3} + \frac{2}{s^3} + \frac{5e^{-s}}{s} - \frac{5e^{-2s}}{s} + \frac{2e^{-2s}}{s} - \frac{e^{-s}}{s} + \frac{e^{-2s}}{s^2} - \frac{e^{-s}}{s^2} + \frac{6e^{-2s}}{s}.$$

Simplifying, we conclude that the Laplace transform is given by

$$F(s) = \frac{3e^{-s}}{s} - \frac{3e^{-s}}{s^2} + \frac{3e^{-2s}}{s} + \frac{e^{-2s}}{s^2} - \frac{2e^{-s}}{s^3} + \frac{2}{s^3}.$$

19. The Laplace transform is given by

$$\int_0^a (\sinh bt)e^{-st}\, dt = \frac{1}{2}\int_0^a e^{(b-s)t}\, dt - \frac{1}{2}\int_0^a e^{-(b+s)t}\, dt = \frac{1}{2}\left[\frac{1 - e^{(b-s)a}}{s-b}\right] - \frac{1}{2}\left[\frac{1 - e^{-(b+s)a}}{s+b}\right].$$

Taking the limit as $a \to \infty$, we conclude that

$$\int_0^\infty (\sinh bt)e^{-st}\, dt = \frac{1}{2}\left[\frac{1}{s-b}\right] - \frac{1}{2}\left[\frac{1}{s+b}\right] = \frac{b}{s^2 - b^2}.$$

This function is valid as long as $s > |b|$.

21. We note that $e^{at} \sinh bt = (e^{(a+b)t} - e^{(a-b)t})/2$. Therefore,

$$\int_0^A (e^{at} \sinh bt) e^{-st}\, dt = \frac{1}{2}\left[\frac{1 - e^{(a+b-s)A}}{s-a+b}\right] - \frac{1}{2}\left[\frac{1 - e^{-(b-a+s)A}}{s+b-a}\right].$$

Taking the limit as $A \to \infty$,

$$\int_0^\infty (e^{at}\sinh bt) e^{-st}\, dt = \frac{1}{2}\left[\frac{1}{s-a+b}\right] - \frac{1}{2}\left[\frac{1}{s+b-a}\right] = \frac{b}{(s-a)^2 - b^2}.$$

This limit exists as long as $s - a > |b|$.

23. Since the Laplace transform is a linear operator, we have

$$\mathcal{L}[e^{at}\sin bt] = \frac{1}{2i}\mathcal{L}[e^{(a+ib)t}] - \frac{1}{2i}\mathcal{L}[e^{(a-ib)t}].$$

Now

$$\int_0^\infty e^{(a\pm ib)t} e^{-st}\, dt = \frac{1}{s - a \mp ib}.$$

Therefore (for $s > a$),

$$\mathcal{L}[e^{at}\sin bt] = \frac{1}{2i}\left[\frac{1}{s-a-ib} - \frac{1}{s-a+ib}\right] = \frac{b}{(s-a)^2 + b^2}.$$

25. Integrating by parts, we have

$$\int_0^A te^{at} \cdot e^{-st}\, dt = -\left.\frac{te^{(a-s)t}}{s-a}\right|_0^A + \int_0^A \frac{1}{s-a}e^{(a-s)t}\, dt = \frac{1 - e^{A(a-s)} + A(a-s)e^{A(a-s)}}{(s-a)^2}.$$

Taking the limit as $A \to \infty$ (when $s > a$),

$$\int_0^\infty e^{at} \cdot e^{-st}\, dt = \frac{1}{(s-a)^2}.$$

27. We will use the fact that $t\cosh at = (te^{at} + te^{-at})/2$. We note that for any value of c,

$$\int_0^A te^{ct} \cdot e^{-st}\, dt = -\left.\frac{te^{(c-s)t}}{s-c}\right|_0^A + \int_0^A \frac{e^{(c-s)t}}{s-c}\, dt = \frac{1 - e^{A(c-s)} + A(c-s)e^{A(c-s)}}{(s-c)^2}.$$

Taking the limit as $A \to \infty$, we have

$$\int_0^\infty te^{ct} \cdot e^{-st}\, dt = \frac{1}{(s-c)^2}.$$

Therefore (for $s > |a|$),

$$\int_0^\infty (t\cosh at) \cdot e^{-st}\, dt = \frac{1}{2}\left[\frac{1}{(s-a)^2} + \frac{1}{(s+a)^2}\right] = \frac{s^2 + a^2}{(s-a)^2(s+a)^2}.$$

29. We will use the fact that $\sin at = (e^{iat} - e^{-iat})/2i$. Therefore,

$$\int_0^A (t^2 \sin at)e^{-st}\, dt = \frac{1}{2i}\left[\int_0^A t^2 e^{(ia-s)t}\, dt - \int_0^A t^2 e^{-(ia+s)t}\, dt\right].$$

Now for any value of c, we note that

$$\begin{aligned}\int_0^A t^2 e^{(c-s)t}\, dt &= \left.\frac{e^{(c-s)t}(c-s)^2 t^2 - 2e^{(c-s)t}(c-s)t + 2e^{(c-s)t}}{(c-s)^3}\right|_0^A \\ &= \frac{e^{(c-s)A}(c-s)^2 A^2 - 2e^{(c-s)A}(c-s)A + 2e^{(c-s)A} - 2}{(c-s)^3}.\end{aligned}$$

Taking the limit as $A \to \infty$, we have

$$\int_0^\infty t^2 e^{(c-s)t}\, dt = -\frac{2}{(c-s)^3}.$$

Therefore (for $s > 0$),

$$\int_0^\infty (t^2 \sin at)e^{-st}\, dt = \frac{1}{2i}\left[-\frac{2}{(ia-s)^3} + \frac{2}{(ia+s)^3}\right] = \frac{2a(3s^2 - a^2)}{(s^2 + a^2)^3}.$$

31. First, we note that

$$\int_0^A (t^2 + 1)^{-1}\, dt = \arctan t\big|_0^A = \arctan A - \arctan 0 = \arctan A.$$

Therefore, as $A \to \infty$, we have

$$\int_0^\infty (t^2 + 1)^{-1}\, dt = \lim_{A\to\infty} \arctan A = \frac{\pi}{2}.$$

Therefore, the integral converges.

33. Based on a series expansion, we see that for $t > 0$,

$$e^t > 1 + t + \frac{t^2}{2} + \frac{t^3}{6} + \frac{t^4}{24} > \frac{t^4}{24}.$$

Therefore, $t^{-4}e^t > 1/24$. Thus

$$\int_1^A t^{-4} e^t\, dt > \int_1^A \frac{1}{24}\, dt = \frac{A-1}{24}.$$

As $A \to \infty$, we see that $(A-1)/24 \to \infty$. Therefore, the integral diverges.

35.(a) We see that

$$\begin{aligned}|F(s)| &= \left|\int_0^{M+1} e^{-st} f(t)\, dt + \int_{M+1}^\infty e^{-st} f(t)\, dt\right| \\ &\leq \max_{0\leq t\leq M+1} |f(t)| \int_0^{M+1} e^{-st}\, dt + K\int_{M+1}^\infty e^{-st} e^{at}\, dt \\ &= \max_{0\leq t\leq M+1} |f(t)| \left.\frac{e^{-st}}{-s}\right|_0^{M+1} + K\left.\frac{e^{-(s-a)t}}{-(s-a)}\right|_{M+1}^\infty \\ &= \max_{0\leq t\leq M+1} |f(t)| \frac{1 - e^{-(M+1)t}}{s} + \frac{K}{s-a} e^{-(s-a)(M+1)}.\end{aligned}$$

(b) Since $e^{-(s-a)(M+1)} \to 0$ as $s \to \infty$ and $(s-a)/s \to 1$ as $s \to \infty$, there exists a constant K_1 such that $Ke^{-(s-a)(M+1)} < K_1(s-a)/s$ for s sufficiently large.

(c) Using the results from parts (a) and (b) and the fact that $(1 - e^{-(M+1)t})/s < 1/s$, we conclude that

$$|F(s)| \leq \max_{0 \leq t \leq M+1} |f(t)|\frac{1}{s} + \frac{K_1}{s}$$

for s sufficiently large.

37.(a) By definition,

$$\mathcal{L}(t^p) = \int_0^\infty e^{-st}t^p \, dt.$$

Now letting $x = st$, we see that $dx = s \, dt$. Therefore,

$$\mathcal{L}(t^p) = \int_0^\infty e^{-st}t^p \, dt = \int_0^\infty e^{-st}\left(\frac{x}{s}\right)^p \frac{dx}{s} = \frac{1}{s^{p+1}}\int_0^\infty e^{-st}x^p \, dx = \frac{\Gamma(p+1)}{s^{p+1}}.$$

(b) Using part (a) and the fact that for n a positive integer, $\Gamma(n+1) = n!$, we conclude that $\mathcal{L}(t^n) = n!/s^{n+1}$.

(c) By definition,

$$\mathcal{L}(t^{-1/2}) = \int_0^\infty e^{-st}t^{-1/2} \, dt.$$

Making the change of variables $x = \sqrt{st}$, we see that $x^2 = st$ implies $2x \, dx = s \, dt$. Therefore,

$$\int_0^\infty e^{-st}t^{-1/2} \, dt = \int_0^\infty e^{-x^2}\left(\frac{x^2}{s}\right)^{-1/2} \frac{2x \, dx}{s} = \frac{2}{\sqrt{s}}\int_0^\infty e^{-x^2} \, dx.$$

(d) Using the result from part (a), we see that

$$\mathcal{L}(t^{1/2}) = \frac{\Gamma(3/2)}{s^{3/2}}.$$

Then, using the result from problem 36, we know that $\Gamma(3/2) = \sqrt{\pi}/2$. Therefore,

$$\mathcal{L}(t^{1/2}) = \frac{\sqrt{\pi}}{2s^{3/2}}.$$

5.2   Properties of the Laplace Transform

1. By Example 7 in Section 5.1, for $f(t) = \sin 3t$, $F(s) = \dfrac{3}{s^2 + 9}$. Therefore, by Theorem 5.2.1,

$$\mathcal{L}(e^{-2t}\sin 3t) = F(s+2) = \frac{3}{(s+2)^2 + 9}.$$

3. By Corollary 5.2.5, $\mathcal{L}(t^n) = \dfrac{n!}{s^{n+1}}$. Combining this fact with linearity, we have

$$\mathcal{L}(t^6 - 4t^2 + 5) = \mathcal{L}(t^6) - 4\mathcal{L}(t^2) + 5\mathcal{L}(1) = \frac{6!}{s^7} - 4 \cdot \frac{2!}{s^3} + 5 \cdot \frac{0!}{s^1} = \frac{720}{s^7} - \frac{8}{s^3} + \frac{5}{s}.$$

5. By Theorem 5.2.1, $\mathcal{L}(e^{-2t}(t^3+1)^2) = F(s+2)$ where $F(s) = \mathcal{L}((t^3+1)^2)$. Using linearity and Corollary 5.2.5, we have

$$\mathcal{L}((t^3+1)^2) = \mathcal{L}(t^6 + 2t^3 + 1) = \frac{6!}{s^7} + \frac{12}{s^4} + \frac{1}{s}.$$

Therefore,

$$\mathcal{L}(e^{-2t}(t^3+1)^2) = \frac{6!}{(s+2)^7} + \frac{12}{(s+2)^4} + \frac{1}{s+2}.$$

7. By Theorem 5.2.4, $\mathcal{L}(t^2 \sin bt) = F''(s)$ where $F(s) = \mathcal{L}(\sin bt)$. By example 7 of Section 5.1,

$$F(s) = \mathcal{L}(\sin bt) = \frac{b}{s^2 + b^2}.$$

Differentiating, we get

$$F'(s) = \frac{-2sb}{(s^2+b^2)^2} \quad \text{and} \quad F''(s) = \frac{6bs^2 - 2b^3}{(s^2+b^2)^3}.$$

Therefore,

$$\mathcal{L}(t^2 \sin bt) = \frac{6bs^2 - 2b^3}{(s^2+b^2)^3}.$$

9. By Theorem 5.2.4, $\mathcal{L}(te^{at} \sin bt) = (-1)F'(s)$ where $F(s) = \mathcal{L}(e^{at} \sin bt)$. Further, $\mathcal{L}(e^{at} \sin bt) = G(s-a)$ where $G(s) = \mathcal{L}(\sin bt) = \dfrac{b}{s^2+b^2}$. Therefore,

$$\mathcal{L}(e^{at} \sin(bt)) = \frac{b}{(s-a)^2 + b^2}$$

and

$$\frac{d}{ds}\left[\frac{b}{(s-a)^2+b^2}\right] = \frac{-2b(s-a)}{((s-a)^2+b^2)^2}.$$

Therefore,

$$\mathcal{L}(te^{at} \sin bt) = \frac{2b(s-a)}{((s-a)^2+b^2)^2}.$$

11.(a) Let $g_1(t) = \displaystyle\int_0^t f(t_1)\,dt_1$. Therefore, $g_1'(t) = f(t)$ which implies $\mathcal{L}(g_1'(t)) = \mathcal{L}(f(t)) = F(s)$. Also, by Theorem 5.2.2, $\mathcal{L}(g_1'(t)) = s\mathcal{L}(g_1(t)) - g_1(0) = s\mathcal{L}(g_1(t))$. Therefore, $s\mathcal{L}(g_1(t)) = \mathcal{L}(g_1'(t)) = F(s)$, so

$$\mathcal{L}(g_1(t)) = \frac{F(s)}{s}.$$

That is,

$$\mathcal{L}\left(\int_0^t f(\tau)\,d\tau\right) = \frac{1}{s}F(s).$$

(b) Let

$$g_n(t) = \int_0^t \int_0^{t_n} \cdots \int_0^{t_2} f(t_1)\, dt_1 \cdots dt_n.$$

Therefore,

$$g_n'(t) = \int_0^t \int_0^{t_{n-1}} \cdots \int_0^{t_2} f(t_1)\, dt_1 \cdots dt_{n-1}.$$

Continuing, we conclude that $g_n^{(n)}(t) = f(t)$. Therefore, $\mathcal{L}(g_n^{(n)}(t)) = \mathcal{L}(f(t)) = F(s)$. But also, by Corollary 5.2.3,

$$\mathcal{L}(g_n^{(n)}(t)) = s^n \mathcal{L}(g_n(t)) - s^{n-1} g_n(0) - \cdots - s g_n^{(n-2)}(0) - g_n^{(n-1)}(0) = s^n \mathcal{L}(g_n(t)).$$

Therefore,

$$\mathcal{L}(g_n(t)) = \frac{1}{s^n} \mathcal{L}(g_n^{(n)}(t)) = \frac{1}{s^n} F(s).$$

13. The equation is $9y'' + 12y' + 4y = 0$; applying the Laplace transform we get $9\mathcal{L}(y'') + 12\mathcal{L}(y') + 4\mathcal{L}(y) = 0$. Then $9[s^2\mathcal{L}(y) - sy(0) - y'(0)] + 12[s\mathcal{L}(y) - y(0)] + 4\mathcal{L}(y) = 0$, so $9[s^2\mathcal{L}(y) - 2s + 1] + 12[s\mathcal{L}(y) - 2] + 4\mathcal{L}(y) = 0$. Thus $[9s^2 + 12s + 4]\mathcal{L}(y) - 18s + 9 - 24 = 0$, and then

$$\mathcal{L}(y) = \frac{18s + 15}{9s^2 + 12s + 4}.$$

15. The equation is $6y'' + 5y' + y = 0$; applying the Laplace transform we get $6\mathcal{L}(y'') + 5\mathcal{L}(y') + \mathcal{L}(y) = 0$. Then $6[s^2\mathcal{L}(y) - sy(0) - y'(0)] + 5[s\mathcal{L}(y) - y(0)] + \mathcal{L}(y) = 0$, so $6[s^2\mathcal{L}(y) - 4s] + 5[s\mathcal{L}(y) - 4] + \mathcal{L}(y) = 0$. Thus $[6s^2 + 5s + 1]\mathcal{L}(y) - 24s - 20 = 0$, and then

$$\mathcal{L}(y) = \frac{24s + 20}{6s^2 + 5s + 1}.$$

17. The equation is $y'' - 5y' - 6y = t^2 + 7$; applying the Laplace transform we get $\mathcal{L}(y'') - 5\mathcal{L}(y') - 6\mathcal{L}(y) = \mathcal{L}(t^2 + 7)$. This implies that

$$[s^2\mathcal{L}(y) - sy(0) - y'(0)] - 5[s\mathcal{L}(y) - y(0)] - 6\mathcal{L}(y) = \frac{2!}{s^3} + \frac{7}{s},$$

thus

$$[s^2\mathcal{L}(y) - s] - 5[s\mathcal{L}(y) - 1] - 6\mathcal{L}(y) = \frac{2}{s^3} + \frac{7}{s}.$$

Simplifying, we get

$$[s^2 - 5s - 6]\mathcal{L}(y) - s + 5 = \frac{2}{s^3} + \frac{7}{s},$$

and we obtain

$$\mathcal{L}(y) = \frac{1}{s^2 - 5s - 6}\left[\frac{2}{s^3} + \frac{7}{s} + s - 5 \right].$$

19. The equation is $y'' + 2y' + 5y = t\cos 2t$; applying the Laplace transform we get $\mathcal{L}(y'') + 2\mathcal{L}(y') + 5\mathcal{L}(y) = \mathcal{L}(t\cos 2t)$. Thus $[s^2\mathcal{L}(y) - sy(0) - y'(0)] + 2[s\mathcal{L}(y) - y(0)] + 5\mathcal{L}(y) =$

$(-1)F'(s)$, where $F(s) = \mathcal{L}(\cos 2t) = s/(s^2 + 4)$. Therefore, $F'(s) = (4 - s^2)/(s^2 + 4)^2$ implies

$$\mathcal{L}(t \cos 2t) = \frac{s^2 - 4}{(s^2 + 4)^2}.$$

We obtain

$$[s^2\mathcal{L}(y) - s] + 2[s\mathcal{L}(y) - 1] + 5\mathcal{L}(y) = \frac{s^2 - 4}{(s^2 + 4)^2},$$

which implies that

$$[s^2 + 2s + 5]\mathcal{L}(y) - s - 2 = \frac{s^2 - 4}{(s^2 + 4)^2}.$$

Then

$$\mathcal{L}(y) = \frac{1}{s^2 + 2s + 5}\left[\frac{s^2 - 4}{(s^2 + 4)^2} + s + 2\right].$$

21. The equation is $y'''' - 6y = te^{-t}$; applying the Laplace transform we get $\mathcal{L}(y'''') - 6\mathcal{L}(y) = \mathcal{L}(te^{-t})$. Thus $[s^4\mathcal{L}(y) - s^3y(0) - s^2y'(0) - sy''(0) - y'''(0)] - 6\mathcal{L}(y) = F(s + 1)$, where $F(s) = \mathcal{L}(t) = 1/s^2$. Therefore, $\mathcal{L}(te^{-t}) = 1/(s + 1)^2$, which implies

$$[s^4\mathcal{L}(y) - 9] - 6\mathcal{L}(y) = \frac{1}{(s + 1)^2}.$$

Thus

$$[s^4 - 6]\mathcal{L}(y) - 9 = \frac{1}{(s + 1)^2},$$

and then

$$\mathcal{L}(y) = \frac{1}{s^4 - 6}\left[\frac{1}{(s + 1)^2} + 9\right].$$

23. The equation is $y'' + 16y = f(t)$; applying the Laplace transform we get $\mathcal{L}(y'') + 16\mathcal{L}(y) = \mathcal{L}(f(t))$, so $[s^2\mathcal{L}(y) - sy(0) - y'(0)] + 16\mathcal{L}(y) = \mathcal{L}(f(t))$. Now

$$\mathcal{L}(f(t)) = \int_0^\infty f(t)e^{-st}\,dt = \int_0^\pi e^{-st}\,dt = \frac{1}{s} - \frac{e^{-\pi s}}{s}.$$

Thus

$$[s^2\mathcal{L}(y) - 9s - 2] + 16\mathcal{L}(y) = \frac{1}{s} - \frac{e^{-\pi s}}{s},$$

which implies that

$$[s^2 + 16]\mathcal{L}(y) - 9s - 2 = \frac{1 - e^{-\pi s}}{s},$$

and we obtain

$$\mathcal{L}(y) = \frac{9s^2 + 2s + 1 - e^{-\pi s}}{s(s^2 + 16)}.$$

25. The equation is $y'' + 4y = f(t)$; applying the Laplace transform we get $\mathcal{L}(y'') + 4\mathcal{L}(y) = \mathcal{L}(f(t))$, so $[s^2\mathcal{L}(y) - sy(0) - y'(0)] + 4\mathcal{L}(y) = \mathcal{L}(f(t))$. Now

$$\mathcal{L}(f(t)) = \int_0^\infty f(t)e^{-st}\,dt = \int_0^1 te^{-st}\,dt + \int_1^\infty e^{-st}\,dt = \frac{1 - e^{-s}}{s^2}.$$

Thus

$$s^2 \mathcal{L}(y) + 4\mathcal{L}(y) = \frac{1 - e^{-s}}{s^2},$$

which implies

$$[s^2 + 4]\mathcal{L}(y) = \frac{1 - e^{-s}}{s^2},$$

and we obtain

$$\mathcal{L}(y) = \frac{1 - e^{-s}}{s^2(s^2 + 4)}.$$

27. The equation of motion is given by

$$my'' + \gamma y' + ky = f(t) = \begin{cases} F_0 t & 0 \leq t \leq T \\ 0 & T < t < \infty. \end{cases}$$

Since the mass is initially in the equilibrium state, the initial conditions are $y(0) = 0$ and $y'(0) = 0$. Applying the Laplace transform to the equation $my'' + \gamma y' + ky = f(t)$, we have $m\mathcal{L}(y'') + \gamma\mathcal{L}(y') + k\mathcal{L}(y) = \mathcal{L}(f(t))$, so $m[s^2\mathcal{L}(y) - sy(0) - y'(0)] + \gamma[s\mathcal{L}(y) - y(0)] + k\mathcal{L}(y) = \mathcal{L}(f(t))$. Now

$$\mathcal{L}(f(t)) = \int_0^\infty f(t)e^{-st}\, dt = \int_0^T F_0 t e^{-st}\, dt = F_0 \frac{1 - sTe^{-sT} - e^{-sT}}{s^2}.$$

Thus

$$ms^2\mathcal{L}(y) + \gamma s\mathcal{L}(y) + k\mathcal{L}(y) = F_0\frac{1 - sTe^{-sT} - e^{-sT}}{s^2},$$

which implies

$$[ms^2 + \gamma s + k]\mathcal{L}(y) = F_0\frac{1 - sTe^{-sT} - e^{-sT}}{s^2},$$

and then

$$\mathcal{L}(y) = F_0\frac{1 - sTe^{-sT} - e^{-sT}}{s^2(ms^2 + \gamma s + k)}.$$

29.(a) Taking the Laplace transform of the Airy equation, we have

$$\mathcal{L}(y'') - \mathcal{L}(ty) = 0.$$

Using the differentiation property of the transform, we get

$$\mathcal{L}(y'') + \frac{d}{ds}\mathcal{L}(y) = 0.$$

Therefore,

$$[s^2 Y(s) - sy(0) - y'(0)] + \frac{d}{ds}Y(s) = 0.$$

Using the initial conditions, we obtain

$$[s^2 Y(s) - s] + \frac{d}{ds}Y(s) = 0,$$

which we can rewrite as

$$Y'(s) + s^2 Y(s) = s.$$

(b) Taking the Laplace transform of the Legendre equation, we have

$$\mathcal{L}(y'') - \mathcal{L}(t^2 y'') - 2\mathcal{L}(ty') + \alpha(\alpha+1)\mathcal{L}(y) = 0.$$

Using the differentiation property of the transform, we get

$$\mathcal{L}(y'') - \frac{d^2}{ds^2}\mathcal{L}(y'') + 2\frac{d}{ds}\mathcal{L}(y') + \alpha(\alpha+1)\mathcal{L}(y) = 0.$$

Therefore,

$$[s^2 Y(s) - sy(0) - y'(0)] - \frac{d^2}{ds^2}[s^2 Y(s) - sy(0) - y'(0)] + +2\frac{d}{ds}[sY(s) - y(0)] + \alpha(\alpha+1)Y(s) = 0.$$

Using the initial conditions, we obtain

$$s^2 Y(s) - 1 - \frac{d^2}{ds^2}[s^2 Y(s) - 1] + 2\frac{d}{ds}[sY(s)] + \alpha(\alpha+1)Y(s) = 0.$$

After carrying out the differentiation, the equation simplifies to

$$s^2 Y'' + 2sY' - [s^2 + \alpha(\alpha+1)]Y = -1.$$

5.3 The Inverse Laplace Transform

1. We need to find constants a, b such that

$$a(s-4) + b(s+3) = 3s - 40,$$

which is the same as

$$(a+b)s - 4a + 3b = 3s - 40.$$

Equating like coefficients, we need $a + b = 3$ and $-4a + 3b = -40$. Solving the first equation, we have $a = 3 - b$. Plugging this equation into the second equation, we have $-4(3-b) + 3b = -40$. Therefore, $b = -4$ and $a = 7$.

3. We need a, b, c to satisfy the following equation,

$$a(s^2 + 4) + (bs + 2c)(s + 4) = -3s^2 + 32 - 14s.$$

We can rewrite this equation as

$$(a+b)s^2 + (4b + 2c)s + 4a + 8c = -3s^2 + 32 - 14s.$$

Equating like coefficients, we need $a + b = -3$, $4b + 2c = -14$ and $4a + 8c = 32$. Solving this system of equations, we have $a = 2$, $b = -5$ and $c = 3$.

5. We need a, b, c to satisfy the equation

$$a(s^2 - 2s + 5) + (b(s - 1) + 2c)(s + 1) = 3s^2 - 8s + 5.$$

This equation can be rewritten as

$$(a + b)s^2 + (-2a + 2c)s + (5a - b + 2c) = 3s^2 - 8s + 5.$$

Equating like coefficients, we have $a + b = 3$, $-2a + 2c = -8$ and $5a - b + 2c = 5$. Solving this system of equations, we have $a = 2$, $b = 1$ and $c = -2$.

7. We need $a_1, b_1, a_2, b_2, a_3, b_3$ to satisfy the equation

$$(a_1 s + b_1)(s^2 + 1)^2 + (a_2 s + b_2)(s^2 + 1) + (a_3 s + b_3) = 8s^3 - 15s - s^5.$$

We can rewrite this equation as

$$a_1 s^5 + b_1 s^4 + (2a_1 + a_2)s^3 + (2b_1 + b_2)s^2 + (a_1 + a_2 + a_3)s + (b_1 + b_2 + b_3) = -s^5 + 8s^3 - 15s.$$

Equating like coefficients, we have $a_1 = -1$ and $b_1 = 0$. Then, $2a_1 + a_2 = 8$ implies $a_2 = 10$, while $2b_1 + b_2 = 0$ implies $b_2 = 0$. Finally, $a_1 + a_2 + a_3 = -15$ implies $a_3 = -24$ and $b_1 + b_2 + b_3 = 0$ implies $b_3 = 0$. We conclude that $a_1 = -1$, $b_1 = 0$, $a_2 = 10$, $b_2 = 0$, $a_3 = -24$ and $b_3 = 0$.

9. Writing the function as

$$\frac{30}{s^2 + 25} = 6\frac{5}{s^2 + 5^2},$$

we see that $\mathcal{L}^{-1}(Y(s)) = 6\sin 5t$.

11. Using partial fractions, we write

$$\frac{2}{s^2 + 3s - 4} = \frac{2}{5}\left[\frac{1}{s - 1} - \frac{1}{s + 4}\right].$$

Thus we see that $\mathcal{L}^{-1}(Y(s)) = \frac{2}{5}(e^t - e^{-4t})$.

13. Completing the square in the denominator, we have

$$\frac{5s + 25}{s^2 + 10s + 74} = \frac{5(s + 5)}{(s + 5)^2 + 49}.$$

Thus we see that $\mathcal{L}^{-1}(Y(s)) = 5e^{-5t}\cos 7t$.

15. Completing the square in the denominator, we have

$$\frac{2s + 1}{s^2 - 2s + 2} = \frac{2s + 1}{(s - 1)^2 + 1} = \frac{2(s - 1)}{(s - 1)^2 + 1} + \frac{3}{(s - 1)^2 + 1}.$$

Therefore, we see that $\mathcal{L}^{-1}(Y(s)) = 2e^t\cos t + 3e^t\sin t$.

17. Completing the square in the denominator, we have

$$\frac{1-2s}{s^2+4s+5} = \frac{5-2(s+2)}{(s+2)^2+1}.$$

Therefore, we see that $\mathcal{L}^{-1}(Y(s)) = 5e^{-2t}\sin t - 2e^{-2t}\cos t$.

19. First we write

$$\frac{3s+2}{(s-2)(s+2)(s+1)} = \frac{a}{s-2} + \frac{b}{s+2} + \frac{c}{s+1}.$$

Then a, b, c must satisfy the equation

$$a(s+2)(s+1) + b(s-2)(s+1) + c(s-2)(s+2) = 3s+2.$$

We can rewrite this equation as

$$(a+b+c)s^2 + (3a-b)s + (2a-2b-4c) = 3s+2.$$

Equating like coefficients, we have $a+b+c = 0$, $3a-b = 3$ and $2a-2b-4c = 2$. Solving this system of equations, we have $a = 2/3$, $b = -1$ and $c = 1/3$. Therefore, we can write

$$\frac{3s+2}{(s-2)(s+2)(s+1)} = \frac{2/3}{s-2} - \frac{1}{s+2} + \frac{1/3}{s+1},$$

which implies

$$F(s) = \frac{2}{s-2} - \frac{3}{s+2} + \frac{1}{s+1}.$$

By the linearity of $\mathcal{L}^{-1}$, we have

$$\mathcal{L}^{-1}(F(s)) = 2\mathcal{L}^{-1}\left(\frac{1}{s-2}\right) - 3\mathcal{L}^{-1}\left(\frac{1}{s+2}\right) + \mathcal{L}^{-1}\left(\frac{1}{s+1}\right) = 2e^{2t} - 3e^{-2t} + e^{-t}.$$

21. First we write

$$\frac{s^2-7s+28}{(s^2-8s+25)(s+3)} = \frac{as+b}{s^2-8s+25} + \frac{c}{s+3}.$$

Then a, b, c must satisfy the equation

$$(as+b)(s+3) + c(s^2-8s+25) = s^2 - 7s + 28.$$

We can rewrite this equation as

$$(a+c)s^2 + (3a+b-8c)s + (3b+25c) = s^2 - 7s + 28.$$

Equating like coefficients, we have $a+c = 1$, $3a+b-8c = -7$, and $3b+25c = 28$. Solving this system of equations, we have $a = 0$, $b = 1$ and $c = 1$. Therefore, we can write

$$\frac{s^2-7s+28}{(s^2-8s+25)(s+3)} = \frac{1}{s^2-8s+25} + \frac{1}{s+3},$$

which implies

$$F(s) = \frac{12}{s^2 - 8s + 25} + \frac{12}{s + 3}.$$

By the linearity of $\mathcal{L}^{-1}$, we have

$$\mathcal{L}^{-1}(F(s)) = \mathcal{L}^{-1}\left(\frac{12}{s^2 - 8s + 25}\right) + \mathcal{L}^{-1}\left(\frac{12}{s+3}\right) = \mathcal{L}^{-1}\left(\frac{12}{(s-4)^2 + 3^2}\right) + \mathcal{L}^{-1}\left(\frac{12}{s+3}\right)$$

$$= 4\mathcal{L}^{-1}\left(\frac{3}{(s-4)^2 + 3^2}\right) + 12\mathcal{L}^{-1}\left(\frac{1}{s+3}\right) = 4e^{4t}\sin 3t + 12e^{-3t}.$$

23. First we write

$$\frac{s^3 - 2s^2 - 6s - 6}{(s^2 + 2s + 2)s^2} = \frac{as + b}{s^2 + 2s + 2} + \frac{c}{s} + \frac{d}{s^2}.$$

Then a, b, c, d must satisfy the equation

$$(as + b)s^2 + cs(s^2 + 2s + 2) + d(s^2 + 2s + 2) = s^3 - 2s^2 - 6s - 6.$$

This equation can be rewritten as

$$(a + c)s^3 + (b + 2c + d)s^2 + (2c + 2d)s + 2d = s^3 - 2s^2 - 6s - 6.$$

Equating like coefficients, we have $a + c = 1$, $b + 2c + d = -2$, $2c + 2d = -6$ and $2d = -6$. Solving this system of equations, we have $a = 1$, $b = 1$, $c = 0$, and $d = -3$. Therefore,

$$F(s) = \frac{s + 1}{s^2 + 2s + 2} - \frac{3}{s^2} = \frac{s + 1}{(s + 1)^2 + 1} - \frac{3}{s^2}.$$

By the linearity of $\mathcal{L}^{-1}$, we have

$$\mathcal{L}^{-1}(F(s)) = \mathcal{L}^{-1}\left(\frac{s + 1}{(s + 1)^2 + 1}\right) - 3\mathcal{L}^{-1}\left(\frac{1}{s^2}\right) = e^{-t}\cos t - 3t.$$

25. Using a computer algebra system, we obtain that

$$\mathcal{L}^{-1}(Y(s)) = \left(1 - \frac{7}{3}t\right)e^{-t}\cos 3t - \left(\frac{8}{9} + \frac{4}{3}t\right)e^{-t}\sin 3t.$$

27. Using a computer algebra system, we obtain that

$$\mathcal{L}^{-1}(Y(s)) = \left(\frac{1}{250} + \frac{13}{25}t - \frac{1}{40}t^2\right)e^t\cos t + \left(-\frac{261}{500} + \frac{3}{200}t + \frac{1}{5}t^2\right)e^t\sin t$$

$$- \left(\frac{1}{250} + \frac{1}{100}t\right)e^{-2t}.$$

5.4 Solving Differential Equations with Laplace Transforms

1. Applying the Laplace transform to the differential equation, we get $\mathcal{L}(y'') - 4\mathcal{L}(y') - 12\mathcal{L}(y) = 0$, which is $[s^2\mathcal{L}(y) - sy(0) - y'(0)] - 4[s\mathcal{L}(y) - y(0)] - 12\mathcal{L}(y) = 0$. Using the initial conditions, this gives $[s^2\mathcal{L}(y) - 8s] - 4[s\mathcal{L}(y) - 8] - 12\mathcal{L}(y) = 0$, which turns into $[s^2 - 4s - 12]\mathcal{L}(y) = 8s - 32$, thus (using partial fractions)

$$\mathcal{L}(y) = \frac{8s - 32}{s^2 - 4s - 12} = \frac{2}{s - 6} + \frac{6}{s + 2},$$

which implies that

$$y(t) = 2e^{6t} + 6e^{-2t}.$$

3. Applying the Laplace transform to the differential equation, we get $\mathcal{L}(y'') - 8\mathcal{L}(y') + 25\mathcal{L}(y) = 0$, which is $[s^2\mathcal{L}(y) - sy(0) - y'(0)] - 8[s\mathcal{L}(y) - y(0)] + 25\mathcal{L}(y) = 0$. Using the initial conditions, this gives $[s^2\mathcal{L}(y) - 3] - 8s\mathcal{L}(y) + 25\mathcal{L}(y) = 0$, which turns into $[s^2 - 8s + 25]\mathcal{L}(y) = 3$, and then

$$\mathcal{L}(y) = \frac{3}{s^2 - 8s + 25}.$$

Completing the square in the denominator, we have

$$\mathcal{L}(y) = \frac{3}{(s - 4)^2 + 9},$$

which implies that

$$y(t) = e^{4t}\sin 3t.$$

5. Applying the Laplace transform to the differential equation, we get $\mathcal{L}(y'') - 2\mathcal{L}(y') + 4\mathcal{L}(y) = 0$, which is $[s^2\mathcal{L}(y) - sy(0) - y'(0)] - 2[s\mathcal{L}(y) - y(0)] + 4\mathcal{L}(y) = 0$. Using the initial conditions, this gives $[s^2\mathcal{L}(y) - 2s] - 2[s\mathcal{L}(y) - 2] + 4\mathcal{L}(y) = 0$, which turns into $[s^2 - 2s + 4]\mathcal{L}(y) = 2s - 4$, and then

$$\mathcal{L}(y) = \frac{2s - 4}{s^2 - 2s + 4}.$$

Completing the square in the denominator, we have

$$\mathcal{L}(y) = \frac{2s - 4}{(s - 1)^2 + 3} = \frac{2(s - 1)}{(s - 1)^2 + 3} - \frac{2}{(s - 1)^2 + 3},$$

which implies that

$$y(t) = 2e^t\cos(\sqrt{3}\,t) - \frac{2\sqrt{3}}{3}e^t\sin(\sqrt{3}\,t).$$

7. Applying the Laplace transform to the differential equation, we get $\mathcal{L}(y'') + \omega^2\mathcal{L}(y) = \mathcal{L}(\cos 2t)$, which is

$$[s^2\mathcal{L}(y) - sy(0) - y'(0)] + \omega^2\mathcal{L}(y) = \frac{s}{s^2 + 4}.$$

Using the initial conditions, this gives

$$[s^2 \mathcal{L}(y) - s] + \omega^2 \mathcal{L}(y) = \frac{s}{s^2 + 4},$$

which turns into

$$[s^2 + \omega^2]\mathcal{L}(y) = s + \frac{s}{s^2 + 4},$$

and then

$$\mathcal{L}(y) = \frac{s}{s^2 + \omega^2} + \frac{s}{(s^2 + \omega^2)(s^2 + 4)}.$$

Using partial fractions in the second term, we rewrite $\mathcal{L}(y)$ as

$$\mathcal{L}(y) = \frac{s}{s^2 + \omega^2} + \frac{1}{4 - \omega^2}\left[\frac{s}{s^2 + \omega^2} - \frac{s}{s^2 + 4}\right],$$

which implies

$$y = \cos \omega t + \frac{1}{4 - \omega^2}[\cos \omega t - \cos 2t] = \frac{1}{4 - \omega^2}[(5 - \omega^2)\cos \omega t - \cos 2t].$$

9.   Applying the Laplace transform to the differential equation, we get $\mathcal{L}(y'') - 2\mathcal{L}(y') + 2\mathcal{L}(y) = \mathcal{L}(e^{-t})$, which is

$$[s^2 \mathcal{L}(y) - sy(0) - y'(0)] - 2[s\mathcal{L}(y) - y(0)] + 2\mathcal{L}(y) = \frac{1}{s+1}.$$

Using the initial conditions, this gives

$$[s^2 \mathcal{L}(y) - 1] - 2[s\mathcal{L}(y)] + 2\mathcal{L}(y) = \frac{1}{s+1},$$

which turns into

$$[s^2 - 2s + 2]\mathcal{L}(y) = 1 + \frac{1}{s+1},$$

and then

$$\mathcal{L}(y) = \frac{1}{s^2 - 2s + 2} + \frac{1}{(s^2 - 2s + 2)(s + 1)}.$$

Using partial fractions in the second term, we get

$$\frac{1}{(s^2 - 2s + 2)(s + 1)} = \frac{1}{5}\frac{1}{s+1} + \frac{1}{5}\frac{3 - s}{s^2 - 2s + 2}.$$

Thus

$$\mathcal{L}(y) = \frac{1}{5}\frac{1}{s+1} + \frac{1}{5}\frac{8 - s}{s^2 - 2s + 2}.$$

Completing the square in the denominator for the last term, we have

$$\frac{8 - s}{s^2 - 2s + 2} = -\frac{(s - 1) - 7}{(s - 1)^2 + 1}.$$

Therefore,

$$\mathcal{L}(y) = \frac{1}{5}\frac{1}{s+1} - \frac{1}{5}\frac{(s-1)-7}{(s-1)^2+1},$$

which implies

$$y = \frac{1}{5}e^{-t} - \frac{1}{5}e^t \cos t + \frac{7}{5}e^t \sin t.$$

11. Applying the Laplace transform to the differential equation, we get $\mathcal{L}(y^{(4)}) - 4\mathcal{L}(y''') + 6\mathcal{L}(y'') - 4\mathcal{L}(y') + \mathcal{L}(y) = 0$, which is $[s^4\mathcal{L}(y) - s^3y(0) - s^2y'(0) - sy''(0) - y'''(0)] - 4[s^3\mathcal{L}(y) - s^2y(0) - sy'(0) - y''(0)] + 6[s^2\mathcal{L}(y) - sy(0) - y'(0)] - 4[s\mathcal{L}(y) - y(0)] + \mathcal{L}(y) = 0$. Using the initial conditions, this gives $[s^4\mathcal{L}(y) - s^2 - 1] - 4[s^3\mathcal{L}(y) - s] + 6[s^2\mathcal{L}(y) - 1] - 4[s\mathcal{L}(y)] + \mathcal{L}(y) = 0$, which turns into $[s^4 - 4s^3 + 6s^2 - 4s + 1]\mathcal{L}(y) - s^2 - 1 + 4s - 6 = 0$, and then

$$\mathcal{L}(y) = \frac{s^2 - 4s + 7}{s^4 - 4s^3 + 6s^2 - 4s + 1} = \frac{s^2 - 4s + 7}{(s-1)^4}.$$

Using partial fractions, we write

$$\mathcal{L}(y) = \frac{s^2 - 4s + 7}{(s-1)^4} = \frac{a}{s-1} + \frac{b}{(s-1)^2} + \frac{c}{(s-1)^3} + \frac{d}{(s-1)^4}.$$

Then a, b, c, d must satisfy the equation

$$a(s-1)^3 + b(s-1)^2 + c(s-1) + d = s^2 - 4s + 7.$$

This equation can be rewritten as

$$as^3 + (-3a+b)s^2 + (3a-2b+c)s + (-a+b-c+d) = s^2 - 4s + 7.$$

Equating like coefficients, we have $a = 0$, $-3a+b = 1$, $3a-2b+c = -4$ and $-a+b-c+d = 7$. Solving this system of equations, we have $a = 0$, $b = 1$, $c = -2$ and $d = 4$. Thus

$$\mathcal{L}(y) = \frac{1}{(s-1)^2} - \frac{2}{(s-1)^3} + \frac{4}{(s-1)^4},$$

which implies that

$$y = \mathcal{L}^{-1}\left(\frac{1}{(s-1)^2}\right) - \mathcal{L}^{-1}\left(\frac{2}{(s-1)^3}\right) + \frac{2}{3}\mathcal{L}^{-1}\left(\frac{6}{(s-1)^4}\right) = te^t - t^2e^t + \frac{2}{3}t^3e^t.$$

13. Applying the Laplace transform to the differential equation, we get $\mathcal{L}(y^{(4)}) - 9\mathcal{L}(y) = 0$, which is $[s^4\mathcal{L}(y) - s^3y(0) - s^2y'(0) - sy''(0) - y'''(0)] - 9\mathcal{L}(y) = 0$. Using the initial conditions, this gives $[s^4\mathcal{L}(y) - s^3 + 3s] - 9\mathcal{L}(y) = 0$, which turns into $(s^4 - 9)\mathcal{L}(y) = s^3 - 3s$, and then

$$\mathcal{L}(y) = \frac{s^3 - 3s}{s^4 - 9} = \frac{s}{s^2 + 3}.$$

Thus

$$y = \mathcal{L}^{-1}\left(\frac{s}{s^2 + 3}\right) = \cos(\sqrt{3}\,t).$$

15. The system can be written as

$$
\begin{aligned}
y_1' &= 5y_1 - 2y_2 \\
y_2' &= 6y_1 - 2y_2
\end{aligned}
$$

with the initial conditions $y_1(0) = 1$ and $y_2(0) = 0$. Applying the Laplace transform to each equation, we have

$$
\begin{aligned}
\mathcal{L}(y_1') &= 5\mathcal{L}(y_1) - 2\mathcal{L}(y_2) \\
\mathcal{L}(y_2') &= 6\mathcal{L}(y_1) - 2\mathcal{L}(y_2),
\end{aligned}
$$

which implies

$$
\begin{aligned}
s\mathcal{L}(y_1) - y_1(0) &= 5\mathcal{L}(y_1) - 2\mathcal{L}(y_2) \\
s\mathcal{L}(y_2) - y_2(0) &= 6\mathcal{L}(y_1) - 2\mathcal{L}(y_2).
\end{aligned}
$$

Letting $Y_1 = \mathcal{L}(y_1)$, $Y_2 = \mathcal{L}(y_2)$ and plugging in the initial conditions, we have

$$
\begin{aligned}
sY_1 - 1 &= 5Y_1 - 2Y_2 \\
sY_2 - 0 &= 6Y_1 - 2Y_2.
\end{aligned}
$$

These equations can be written in matrix form as

$$
\begin{pmatrix} s-5 & 2 \\ -6 & s+2 \end{pmatrix} \begin{pmatrix} Y_1 \\ Y_2 \end{pmatrix} = \begin{pmatrix} 1 \\ 0 \end{pmatrix}.
$$

The solution of this system is given by

$$
\begin{pmatrix} Y_1 \\ Y_2 \end{pmatrix} = \frac{1}{(s-5)(s+2)+12} \begin{pmatrix} s+2 & -2 \\ 6 & s-5 \end{pmatrix} \begin{pmatrix} 1 \\ 0 \end{pmatrix} = \frac{1}{s^2 - 3s + 2} \begin{pmatrix} s+2 \\ 6 \end{pmatrix}.
$$

We first consider $Y_1(s)$. Using partial fractions, we write

$$
Y_1(s) = \frac{s+2}{s^2 - 3s + 2} = \frac{s+2}{(s-2)(s-1)} = \frac{a}{s-2} + \frac{b}{s-1}.
$$

Then a, b must satisfy the equation $a(s-1) + b(s-2) = s+2$. We can simplify this equation, rewriting it as $(a+b)s + (-a - 2b) = s + 2$. Equating like coefficients, we have $a + b = 1$ and $-a - 2b = 2$. The solution of this system is $a = 4$ and $b = -3$. Thus

$$
Y_1(s) = \frac{4}{s-2} - \frac{3}{s-1},
$$

and then

$$
y_1(t) = 4\mathcal{L}^{-1}\left(\frac{1}{s-2}\right) - 3\mathcal{L}^{-1}\left(\frac{1}{s-1}\right) = 4e^{2t} - 3e^t.
$$

Next, we consider $Y_2(s)$. Using partial fractions, we write

$$
Y_2(s) = \frac{6}{s^2 - 3s + 2} = \frac{6}{(s-2)(s-1)} = \frac{6}{s-2} - \frac{6}{s-1},
$$

and then

$$y_2(t) = 6\mathcal{L}^{-1}\left(\frac{1}{s-2}\right) - 6\mathcal{L}^{-1}\left(\frac{1}{s-1}\right) = 6e^{2t} - 6e^t.$$

17. The system can be written as

$$y_1' = 6y_2, \qquad y_2' = -6y_1$$

with the initial conditions $y_1(0) = 5$ and $y_2(0) = 4$. Applying the Laplace transform to each equation, we have

$$\mathcal{L}(y_1') = 6\mathcal{L}(y_2), \qquad \mathcal{L}(y_2') = -6\mathcal{L}(y_1),$$

which implies

$$s\mathcal{L}(y_1) - y_1(0) = 6\mathcal{L}(y_2), \qquad s\mathcal{L}(y_2) - y_2(0) = -6\mathcal{L}(y_1).$$

Letting $Y_1 = \mathcal{L}(y_1)$, $Y_2 = \mathcal{L}(y_2)$ and plugging in the initial conditions, we have

$$sY_1 - 5 = 6Y_2, \qquad sY_2 - 4 = -6Y_1.$$

These equations can be written in matrix form as

$$\begin{pmatrix} s & -6 \\ 6 & s \end{pmatrix} \begin{pmatrix} Y_1 \\ Y_2 \end{pmatrix} = \begin{pmatrix} 5 \\ 4 \end{pmatrix}.$$

The solution of this system is given by

$$\begin{pmatrix} Y_1 \\ Y_2 \end{pmatrix} = \frac{1}{s^2 + 36} \begin{pmatrix} s & 6 \\ -6 & s \end{pmatrix} \begin{pmatrix} 5 \\ 4 \end{pmatrix} = \frac{1}{s^2 + 36} \begin{pmatrix} 5s + 24 \\ 4s - 30 \end{pmatrix}.$$

Now $Y_1(s) = \dfrac{5s + 24}{s^2 + 36}$, thus

$$y_1(t) = 5\mathcal{L}^{-1}\left(\frac{s}{s^2 + 36}\right) + 4\mathcal{L}^{-1}\left(\frac{6}{s^2 + 36}\right) = 5\cos 6t + 4\sin 6t.$$

Next, $Y_2(s) = \dfrac{4s - 30}{s^2 + 36}$, thus

$$y_2(t) = 4\mathcal{L}^{-1}\left(\frac{s}{s^2 + 36}\right) - 5\mathcal{L}^{-1}\left(\frac{6}{s^2 + 26}\right) = 4\cos 6t - 5\sin 6t.$$

19. The system can be written as

$$\begin{aligned} y_1' &= 2y_1 - 64y_2 \\ y_2' &= y_1 - 14y_2 \end{aligned}$$

with the initial conditions $y_1(0) = 0$ and $y_2(0) = 1$. Applying the Laplace transform to each equation, we have

$$\begin{aligned} \mathcal{L}(y_1') &= 2\mathcal{L}(y_1) - 64\mathcal{L}(y_2) \\ \mathcal{L}(y_2') &= \mathcal{L}(y_1) - 14\mathcal{L}(y_2), \end{aligned}$$

which implies

$$s\mathcal{L}(y_1) - y_1(0) = 2\mathcal{L}(y_1) - 64\mathcal{L}(y_2)$$
$$s\mathcal{L}(y_2) - y_2(0) = \mathcal{L}(y_1) - 14\mathcal{L}(y_2).$$

Letting $Y_1 = \mathcal{L}(y_1)$, $Y_2 = \mathcal{L}(y_2)$ and plugging in the initial conditions, we have

$$sY_1 - 0 = 2Y_1 - 64Y_2$$
$$sY_2 - 1 = Y_1 - 14Y_2.$$

These equations can be written in matrix form as

$$\begin{pmatrix} s-2 & 64 \\ -1 & s+14 \end{pmatrix} \begin{pmatrix} Y_1 \\ Y_2 \end{pmatrix} = \begin{pmatrix} 0 \\ 1 \end{pmatrix}.$$

The solution of this system is given by

$$\begin{pmatrix} Y_1 \\ Y_2 \end{pmatrix} = \frac{1}{(s-2)(s+14)+64} \begin{pmatrix} s+14 & -64 \\ 1 & s-2 \end{pmatrix} \begin{pmatrix} 0 \\ 1 \end{pmatrix} = \frac{1}{s^2+12s+36} \begin{pmatrix} -64 \\ s-2 \end{pmatrix}.$$

Now $Y_1(s) = -\dfrac{64}{(s+6)^2}$, thus

$$y_1(t) = -64\mathcal{L}^{-1}\left(\frac{1}{(s+6)^2}\right) = -64te^{-6t}.$$

Next, $Y_2(s) = \dfrac{s-2}{(s+6)^2} = \dfrac{s+6-8}{(s+6)^2} = \dfrac{1}{s+6} - \dfrac{8}{(s+6)^2}$, thus

$$y_2(t) = \mathcal{L}^{-1}\left(\frac{1}{s+6}\right) - 8\mathcal{L}^{-1}\left(\frac{1}{(s+6)^2}\right) = e^{-6t} - 8te^{-6t}.$$

21. Applying the Laplace transform, we have

$$\mathcal{L}(y_1') = 5\mathcal{L}(y_1) - \mathcal{L}(y_2) + \mathcal{L}(e^{-t})$$
$$\mathcal{L}(y_2') = \mathcal{L}(y_1) + 3\mathcal{L}(y_2) + \mathcal{L}(2e^t)$$

Letting $Y_1 = \mathcal{L}(y_1)$ and $Y_2 = \mathcal{L}(y_2)$ and using the property of the transform of a derivative, we have

$$sY_1 - y_1(0) = 5Y_1 - Y_2 + \frac{1}{s+1}$$
$$sY_2 - y_2(0) = Y_1 + 3Y_2 + \frac{2}{s-1}.$$

Plugging in the initial conditions, we obtain

$$sY_1 + 3 = 5Y_1 - Y_2 + \frac{1}{s+1}$$
$$sY_2 - 2 = Y_1 + 3Y_2 + \frac{2}{s-1}.$$

This system can be written in matrix form as

$$\begin{pmatrix} s-5 & 1 \\ -1 & s-3 \end{pmatrix} \begin{pmatrix} Y_1 \\ Y_2 \end{pmatrix} = \begin{pmatrix} \frac{-3s-2}{s+1} \\ \frac{2s}{s-1} \end{pmatrix}.$$

The solution of this system is given by

$$\begin{pmatrix} Y_1 \\ Y_2 \end{pmatrix} = \frac{1}{(s-4)^2} \begin{pmatrix} \frac{-3s^3+8s^2-3s-6}{s^2-1} \\ \frac{2s^3-11s^2-9s+2}{s^2-1} \end{pmatrix}.$$

Taking the inverse Laplace transform, we obtain

$$y_1(t) = \left(-\frac{82t}{15} - \frac{589}{225} \right) e^{4t} - \frac{4}{25}e^{-t} - \frac{2}{9}e^t$$

and

$$y_2(t) = \left(-\frac{82t}{15} + \frac{641}{225} \right) e^{4t} + \frac{1}{25}e^{-t} - \frac{8}{9}e^t.$$

23. Applying the Laplace transform, we have

$$\begin{aligned} \mathcal{L}(y_1') &= -2\mathcal{L}(y_1) + \mathcal{L}(y_2) \\ \mathcal{L}(y_2') &= \mathcal{L}(y_1) - 2\mathcal{L}(y_2) + \mathcal{L}(\sin t). \end{aligned}$$

Letting $Y_1 = \mathcal{L}(y_1)$ and $Y_2 = \mathcal{L}(y_2)$ and using the property of the transform of a derivative, we get

$$\begin{aligned} sY_1 - y_1(0) &= -2Y_1 + Y_2 \\ sY_2 - y_2(0) &= Y_1 - 2Y_2 + \frac{1}{s^2+1}. \end{aligned}$$

Plugging in the initial conditions, we obtain

$$\begin{aligned} sY_1 &= -2Y_1 + Y_2 \\ sY_2 &= Y_1 - 2Y_2 + \frac{1}{s^2+1}. \end{aligned}$$

This system can be written in matrix form as

$$\begin{pmatrix} s+2 & -1 \\ -1 & s+2 \end{pmatrix} \begin{pmatrix} Y_1 \\ Y_2 \end{pmatrix} = \begin{pmatrix} 0 \\ \frac{1}{s^2+1} \end{pmatrix}.$$

The solution of this system is given by

$$\begin{pmatrix} Y_1 \\ Y_2 \end{pmatrix} = \frac{1}{s^2+4s+3} \begin{pmatrix} \frac{1}{s^2+1} \\ \frac{s+2}{s^2+1} \end{pmatrix}.$$

Taking the inverse Laplace transform, we get

$$y_1(t) = \frac{1}{4}e^{-t} - \frac{1}{20}e^{-3t} - \frac{1}{5}\cos t + \frac{1}{10}\sin t$$

and

$$y_2(t) = \frac{1}{20}e^{-3t} - \frac{3}{10}\cos t + \frac{2}{5}\sin t + \frac{1}{4}e^{-t}.$$

25. The system cannot have a solution, because the initial conditions $x'(0) = 1$, $y'(0) = 2$ do not satisfy the second equation at $t = 0$.

5.5 Discontinuous Functions and Periodic Functions

1.

3.

5.

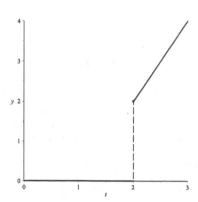

7. Using the Heaviside function, we can write $f(t) = (t - 9)^5 u_9(t)$. The Laplace transform has the property that $\mathcal{L}(u_c(t)f(t - c)) = e^{-cs}\mathcal{L}(f(t))$. Therefore,

$$\mathcal{L}(u_9(t)(t - 9)^5) = e^{-9s}\mathcal{L}(t^5) = \frac{120e^{-9s}}{s^6}.$$

9. Using the Heaviside function, we can write $f(t) = (t - \pi)[u_\pi(t) - u_{2\pi}(t)]$. The Laplace transform has the property that $\mathcal{L}(u_c(t)f(t - c)) = e^{-cs}\mathcal{L}(f(t))$. Therefore,

$$\mathcal{L}((t - \pi)[u_\pi(t) - u_{2\pi}(t)]) = \mathcal{L}(u_\pi(t)(t - \pi)) - \mathcal{L}(u_{2\pi}(t)(t - 2\pi)) - \pi\mathcal{L}(u_{2\pi}(t))$$

$$= e^{-\pi s}\mathcal{L}(t) - e^{-2\pi s}\mathcal{L}(t) - \pi e^{-2\pi s}\mathcal{L}(1) = \frac{e^{-\pi s}}{s^2} - \frac{e^{-2\pi s}}{s^2} - \frac{\pi e^{-2\pi s}}{s}.$$

11. The Laplace transform has the property that $\mathcal{L}(u_c(t)f(t - c)) = e^{-cs}\mathcal{L}(f(t))$. Therefore,

$$\mathcal{L}((t - 4)u_3(t) - (t - 3)u_4(t)) = \mathcal{L}((t - 3)u_3) - \mathcal{L}(u_3) - \mathcal{L}((t - 4)u_4) - \mathcal{L}(u_4)$$

$$= e^{-3s}\mathcal{L}(t) - e^{-3s}\mathcal{L}(1) - e^{-4s}\mathcal{L}(t) - e^{-4s}\mathcal{L}(1) = e^{-3s}\left[\frac{1}{s^2} - \frac{1}{s}\right] - e^{-4s}\left[\frac{1}{s^2} + \frac{1}{s}\right].$$

13. Using the fact that $\mathcal{L}^{-1}(e^{-cs}G(s)) = u_c(t)g(t - c)$, we see that

$$\mathcal{L}^{-1}\left(\frac{5!e^{-s}}{(s - 5)^6}\right) = u_1(t)g(t - 1),$$

where

$$g(t) = \mathcal{L}^{-1}\left(\frac{5!}{(s - 5)^6}\right) = t^5e^{5t}.$$

Therefore,

$$\mathcal{L}^{-1}(F(s)) = u_1(t)(t - 1)^5e^{5(t-1)}.$$

15. Using the fact that $\mathcal{L}^{-1}(e^{-cs}G(s)) = u_c(t)g(t - c)$, we see that

$$\mathcal{L}^{-1}\left(\frac{e^{-2s}2(s - 1)}{s^2 - 2s + 2}\right) = u_2(t)g(t - 2),$$

where

$$g(t) = \mathcal{L}^{-1}\left(\frac{2(s - 1)}{s^2 - 2s + 2}\right).$$

Completing the square in the denominator, we write

$$\frac{2(s - 1)}{s^2 - 2s + 2} = \frac{2(s - 1)}{(s - 1)^2 + 1}.$$

Therefore, $g(t) = 2e^t \cos t$, and then

$$\mathcal{L}^{-1}(F(s)) = 2u_2(t)e^{t-2}\cos(t - 2).$$

17. Using the fact that $\mathcal{L}^{-1}(e^{-cs}G(s)) = u_c(t)g(t - c)$, we see that

$$\mathcal{L}^{-1}\left(\frac{(s - 2)e^{-s}}{s^2 - 4s + 3}\right) = u_1(t)g(t - 1),$$

where

$$g(t) = \mathcal{L}^{-1}\left(\frac{s - 2}{s^2 - 4s + 3}\right).$$

Using partial fractions, we write

$$\frac{s-2}{s^2 - 4s + 3} = \frac{1}{2}\left[\frac{1}{s-1} + \frac{1}{s-3}\right].$$

Therefore,

$$g(t) = \frac{1}{2}[e^t + e^{3t}],$$

and then

$$\mathcal{L}^{-1}(F(s)) = \frac{1}{2}[e^{t-1} + e^{3(t-1)}]u_1(t).$$

19. By the definition of the Laplace transform (for $s > 0$),

$$\mathcal{L}(f(t)) = \int_0^\infty f(t)e^{-st}\,dt = \int_0^1 e^{-st}\,dt = \frac{1 - e^{-s}}{s}.$$

21. Using the definition of the Laplace transform, we have (for $s > 0$)

$$\mathcal{L}(f(t)) = \int_0^\infty f(t)e^{-st}\,dt = \int_0^1 e^{-st}\,dt + \int_2^3 e^{-st}\,dt + \int_4^5 e^{-st}\,dt + \ldots + \int_{2n}^{2n+1} e^{-st}\,dt$$

$$= \frac{1}{s}\left[1 - e^{-s} + e^{-2s} - e^{-3s} + \ldots + e^{-2ns} - e^{-(2n+1)s}\right] = \frac{1 - e^{-(2n+2)s}}{s(1 + e^{-s})}$$

23. Using Theorem 5.5.3, we know that

$$\mathcal{L}(f(t)) = \frac{F_T(s)}{1 - e^{-sT}},$$

where T is the period and

$$F_T(s) = \int_0^T e^{-st} f(t)\,dt.$$

Here, $T = 1$. Therefore,

$$F_T(s) = \int_0^1 e^{-st} f(t)\,dt = \int_0^1 e^{-st}t\,dt = \frac{1 - se^{-s} - e^{-s}}{s^2}.$$

Thus (for $s > 0$)

$$\mathcal{L}(f(t)) = \frac{1 - se^{-s} - e^{-s}}{s^2(1 - e^{-s})}.$$

25.(a) For $s > 0$:

$$\mathcal{L}(f(t)) = \mathcal{L}(1) - \mathcal{L}(u_1(t)) = \frac{1}{s} - \frac{e^{-s}}{s}.$$

(b)

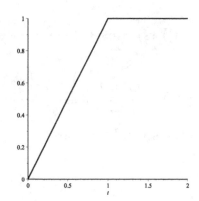

Let $F(s) = \mathcal{L}[1 - u_1(t)]$. Then

$$\mathcal{L}\left[\int_0^t [1 - u_1(\xi)]\,d\xi\right] = \frac{F(s)}{s} = \frac{1 - e^{-s}}{s^2}.$$

(c)

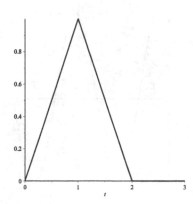

Let $G(s) = \mathcal{L}(g(t))$. Then

$$\mathcal{L}(h(t)) = G(s) - e^{-s}G(s) = \frac{1 - e^{-s}}{s^2} - e^{-s}\frac{1 - e^{-s}}{s^2} = \frac{(1 - e^{-s})^2}{s^2}.$$

5.6   Differential Equations with Discontinuous Forcing Functions

1. Let $Y(s) = \mathcal{L}(y)$. Applying the Laplace transform to the equation, we have $\mathcal{L}(y'')+\mathcal{L}(y) = \mathcal{L}(f(t))$, which is $[s^2Y(s) - sy(0) - y'(0)] + Y(s) = \mathcal{L}(f(t))$. Applying the initial conditions, we get $s^2Y(s) - 5s - 3 + Y(s) = \mathcal{L}(f(t))$. The forcing function $f(t)$ can be written as $f(t) = u_0(t) - u_{\pi/2}(t)$. Therefore,

$$\mathcal{L}(f(t)) = \mathcal{L}(u_0(t)) - \mathcal{L}(u_{\pi/2}(t)) = \frac{1 - e^{-\pi s/2}}{s}.$$

Thus the equation for Y becomes

$$[s^2 + 1]Y(s) = 5s + 3 + \frac{1 - e^{-\pi s/2}}{s},$$

and then

$$Y(s) = \frac{5s + 3}{s^2 + 1} + \frac{1 - e^{-\pi s/2}}{s(s^2 + 1)}.$$

Using partial fractions, we write the second term as

$$(1 - e^{-\pi s/2})\left[\frac{1}{s} - \frac{s}{s^2 + 1}\right].$$

We obtain that

$$Y(s) = \frac{4s + 3}{s^2 + 1} + \frac{1}{s} - \frac{e^{-\pi s/2}}{s} + \frac{se^{-\pi s/2}}{s^2 + 1}.$$

Then, using the fact that $\mathcal{L}^{-1}(e^{-cs}G(s)) = u_c(t)g(t - c)$, we conclude that

$$y(t) = 4\cos t + 3\sin t + 1 - u_{\pi/2}(t) + u_{\pi/2}(t)\cos(t - \pi/2).$$

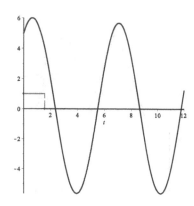

3. Let $Y(s) = \mathcal{L}(y)$. Applying the Laplace transform to the equation, we have $\mathcal{L}(y'') + 4\mathcal{L}(y) = \mathcal{L}(\sin(t) - u_{2\pi}(t)\sin(t - 2\pi))$, which is

$$[s^2Y(s) - sy(0) - y'(0)] + 4Y(s) = \frac{1 - e^{-2\pi s}}{s^2 + 1}.$$

Applying the initial conditions, we get

$$s^2 Y(s) + 4Y(s) = \frac{1 - e^{-2\pi s}}{s^2 + 1}.$$

Therefore, the equation for Y becomes

$$[s^2 + 4]Y(s) = \frac{1 - e^{-2\pi s}}{s^2 + 1},$$

thus

$$Y(s) = \frac{1 - e^{-2\pi s}}{(s^2 + 4)(s^2 + 1)}.$$

Using partial fractions, we can write

$$Y(s) = \frac{1}{3}(1 - e^{-2\pi s}) \left[\frac{1}{s^2 + 1} - \frac{1}{s^2 + 4} \right].$$

Therefore, we conclude that

$$y(t) = \frac{1}{3} \left[\sin t - \frac{1}{2} \sin 2t - u_{2\pi}(t) \sin(t - 2\pi) + \frac{1}{2} u_{2\pi}(t) \sin(2(t - 2\pi)) \right].$$

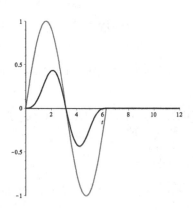

5. First, we write the nonhomogeneous term as $f(t) = u_0(t) - u_{10}(t)$. Then, let $Y(s) = \mathcal{L}(y)$. Applying the Laplace transform, we have $\mathcal{L}(y'') + 3\mathcal{L}(y') + 2\mathcal{L}(y) = \mathcal{L}(u_0(t) - u_{10}(t))$, which is

$$[s^2 Y(s) - sy(0) - y'(0)] + 3[sY(s) - y(0)] + 2Y(s) = \frac{1 - e^{-10s}}{s}.$$

Applying the initial conditions, we get

$$s^2 Y(s) + 3sY(s) + 2Y(s) = \frac{1 - e^{-10s}}{s}.$$

Therefore, the equation for Y becomes

$$[s^2 + 3s + 2]Y(s) = \frac{1 - e^{-10s}}{s},$$

thus

$$Y(s) = \frac{1 - e^{-10s}}{s(s^2 + 3s + 2)}.$$

Using partial fractions, we can write

$$Y(s) = \frac{1}{2}(1 - e^{-10s})\left[\frac{1}{s} + \frac{1}{s+2} - \frac{2}{s+1}\right].$$

Therefore, we conclude that

$$y(t) = \frac{1}{2}[1 + e^{-2t} - 2e^{-t}] - \frac{1}{2}u_{10}(t)[1 + e^{-2(t-10)} - 2e^{-(t-10)}].$$

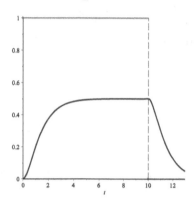

The solution increases to a temporary steady value of $y = 1/2$. After the forcing ceases, the response decays exponentially to $y = 0$.

7. Let $Y(s) = \mathcal{L}(y)$. Applying the Laplace transform to the equation, we have $\mathcal{L}(y'') + \mathcal{L}(y) = \mathcal{L}(u_{3\pi}(t))$, which is

$$[s^2 Y(s) - sy(0) - y'(0)] + Y(s) = \frac{e^{-3\pi s}}{s}.$$

Applying the initial conditions, we get

$$s^2 Y(s) - s + Y(s) = \frac{e^{-3\pi s}}{s}.$$

Therefore, the equation for Y becomes

$$[s^2 + 1]Y(s) = s + \frac{e^{-3\pi s}}{s},$$

thus

$$Y(s) = \frac{s}{s^2 + 1} + \frac{e^{-3\pi s}}{s(s^2 + 1)}.$$

Using partial fractions, we can write

$$Y(s) = \frac{s}{s^2 + 1} + e^{-3\pi s}\left[\frac{1}{s} - \frac{s}{s^2 + 1}\right].$$

Therefore, we conclude that

$$y(t) = \cos t + u_{3\pi}(t)[1 - \cos(t - 3\pi)] = \cos t + u_{3\pi}(t)[1 + \cos t].$$

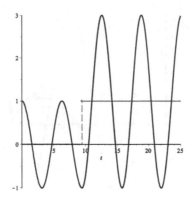

Due to the initial conditions, the solution temporarily oscillated about $y = 0$. After the forcing is applied, the response is a steady oscillation about $y_m = 1$.

9. First, we can write the nonhomogeneous term as $g(t) = t/2 - u_6(t)(t-6)/2$. Then, let $Y(s) = \mathcal{L}(y)$. Applying the Laplace transform to the equation, we have $\mathcal{L}(y'') + \mathcal{L}(y) = \frac{1}{2}\mathcal{L}(t - u_6(t)(t-6))$, which is

$$[s^2 Y(s) - sy(0) - y'(0)] + Y(s) = \frac{1}{2}\left[\frac{1 - e^{-6s}}{s^2}\right].$$

Applying the initial conditions, we have

$$s^2 Y(s) - 6s - 8 + Y(s) = \frac{1}{2}\frac{1 - e^{-6s}}{s^2}.$$

Therefore, the equation for Y becomes

$$[s^2 + 1]Y(s) = 6s + 8 + \frac{1}{2}\frac{1 - e^{-6s}}{s^2},$$

thus

$$Y(s) = \frac{6s + 8}{s^2 + 1} + \frac{1}{2}\frac{1 - e^{-6s}}{s^2(s^2 + 1)}.$$

Using partial fractions, we can write

$$Y(s) = 6\frac{s}{s^2 + 1} + 8\frac{1}{s^2 + 1} + \frac{1}{2}(1 - e^{-6s})\left[\frac{1}{s^2} - \frac{1}{s^2 + 1}\right].$$

Therefore, we conclude that

$$y(t) = 6\cos t + 8\sin t + \frac{1}{2}\left[t - \sin t - u_6(t)(t-6) + u_6(t)\sin(t-6)\right].$$

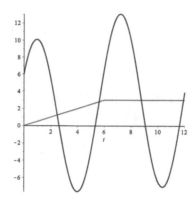

The solution increases in response to the ramp input, and, thereafter oscillated about a mean value of $y_m = 3$.

11. Let $Y(s) = \mathcal{L}(y)$. Applying the Laplace transform to the equation, we have $\mathcal{L}(y'') + 4\mathcal{L}(y) = \mathcal{L}(u_\pi(t) - u_{3\pi}(t))$, which is

$$[s^2 Y(s) - sy(0) - y'(0)] + 4Y(s) = \frac{e^{-\pi s} - e^{-3\pi s}}{s}.$$

Applying the initial conditions, we get

$$s^2 Y(s) - 3s - 7 + 4Y(s) = \frac{e^{-\pi s} - e^{-3\pi s}}{s}.$$

Therefore, the equation for Y becomes

$$[s^2 + 4]Y(s) = 3s + 7 + \frac{e^{-\pi s} - e^{-3\pi s}}{s},$$

thus

$$Y(s) = \frac{3s + 7}{s^2 + 4} + \frac{e^{-\pi s} - e^{-3\pi s}}{s(s^2 + 4)}.$$

Using partial fractions, we can write

$$Y(s) = 3\frac{s}{s^2 + 4} + \frac{7}{2}\frac{2}{s^2 + 4} + (e^{-\pi s} - e^{-3\pi s})\frac{1}{4}\left[\frac{1}{s} - \frac{s}{s^2 + 4}\right].$$

Therefore, we conclude that

$$y(t) = 3\cos 2t + \frac{7}{2}\sin 2t + \frac{1}{4}[1 - \cos(2(t - \pi))]u_\pi(t) - \frac{1}{4}[1 - \cos(2(t - 3\pi))]u_{3\pi}(t).$$

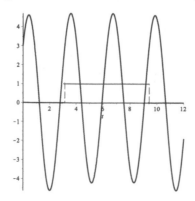

Since there is no damping term, the solution responds immediately to the forcing term. There is a temporary oscillation about $y = 1/4$.

13. Let $Y(s) = \mathcal{L}(y)$. Applying the Laplace transform to the equation, we have $\mathcal{L}(y'''') + 5\mathcal{L}(y'') + 4\mathcal{L}(y) = \mathcal{L}(1 - u_\pi(t))$, which is

$$[s^4Y(s) - s^3y(0) - s^2y'(0) - sy''(0) - y'''(0)] + 5[s^2Y(s) - sy(0) - y'(0)] + 4Y(s) = \frac{1 - e^{-\pi s}}{s}.$$

Applying the initial conditions, we get

$$s^4Y(s) + 5s^2Y(s) + 4Y(s) = \frac{1 - e^{-\pi s}}{s}.$$

Therefore, the equation for Y becomes

$$[s^4 + 5s^2 + 4]Y(s) = \frac{1 - e^{-\pi s}}{s},$$

thus

$$Y(s) = \frac{1 - e^{-\pi s}}{s(s^4 + 5s^2 + 4)}.$$

Using partial fractions, we can write

$$Y(s) = (1 - e^{-\pi s})\frac{1}{12}\left[\frac{3}{s} + \frac{s}{s^2 + 4} - \frac{4s}{s^2 + 1}\right].$$

Therefore, we conclude that

$$y(t) = \frac{1}{12}[3 + \cos 2t - 4\cos t] - u_\pi(t)\frac{1}{12}[3 + \cos(2(t - \pi)) - 4\cos(t - \pi)].$$

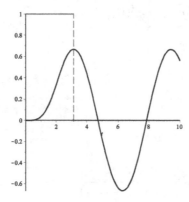

After an initial transient, the solution oscillates about $y_m = 0$.

15. The specified function is defined by

$$f(t) = \begin{cases} 0 & 0 \le t < t_0 \\ \frac{h}{k}(t - t_0) & t_0 \le t < t_0 + k \\ -\frac{h}{k}(t - t_0 - 2k) & t_0 + k \le t < t_0 + 2kk \\ h & t_0 + 2k \le t. \end{cases}$$

Using u_c, this function can be written as

$$f(t) = \frac{h}{k}(t - t_0)u_{t_0}(t) - \frac{2h}{k}(t - t_0 - k)u_{t_0+k}(t) + \frac{h}{k}(t - t_0 - 2k)u_{t_0+2k}(t).$$

17. In this problem, we consider the initial value problem

$$y'' + 4y = \frac{1}{k}[(t - 5)u_5(t) - (t - 5 - k)u_{5+k}(t)].$$

(a) The plot shows $k = 5$.

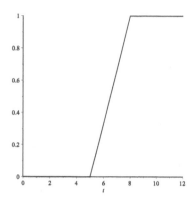

The problem is identical to Example 1 when $k = 5$.

(b) Let $Y(s) = \mathcal{L}(y)$. Applying the Laplace transform to the equation, we have $\mathcal{L}(y'') + 4\mathcal{L}(y) = \mathcal{L}(f(t))$, which is

$$[s^2 Y(s) - sy(0) - y'(0)] + 4Y(s) = \frac{e^{-5s}}{ks^2} - \frac{e^{-(5+k)s}}{ks^2}.$$

Applying the initial conditions, we have

$$s^2 Y(s) + 4Y(s) = \frac{e^{-5s}}{ks^2} - \frac{e^{-(5+k)s}}{ks^2}.$$

Therefore, the equation for Y becomes

$$Y(s) = \frac{e^{-5s}}{ks^2(s^2 + 4)} - \frac{e^{-(5+k)s}}{ks^2(s^2 + 4)}.$$

Using partial fractions, we can write

$$Y(s) = (e^{-5s} - e^{-(5+k)s})\frac{1}{4k}\left[\frac{1}{s^2} - \frac{1}{s^2 + 4}\right].$$

Thus we conclude that

$$y(t) = \frac{u_5(t)}{4k}\left[(t - 5) - \frac{1}{2}\sin(2(t - 5))\right] - \frac{u_{5+k}(t)}{4k}\left[(t - 5 - k) + \frac{1}{2}\sin(2(t - 5 - k))\right].$$

(c) We note that for $t > 5 + k$, the solution is given by

$$y(t) = \frac{1}{4} - \frac{1}{8k}\sin(2t - 10) + \frac{1}{8k}\sin(2t - 10 - 2k) = \frac{1}{4} - \frac{\sin k}{4k}\cos(2t - 10 - k).$$

Therefore, for $t > 5 + k$, the solution oscillates about $y_m = 1/4$ with an amplitude of $A = |\sin k|/4k$. The figures show the cases $k = 3$, $k = 4$, and $k = 5$, respectively.

19.(a) Let $Y(s) = \mathcal{L}(y)$. Applying the Laplace transform to the equation, we have

$$[s^2 Y(s) - sy(0) - y'(0)] + Y(s) = \mathcal{L}\left(1 + 2\sum_{k=1}^{n}(-1)^k u_{k\pi}(t)\right) = \frac{1 + 2\displaystyle\sum_{k=1}^{n}(-1)^k e^{-k\pi s}}{s}.$$

Applying the initial conditions, we have

$$[s^2 + 1]Y(s) = \frac{1 + 2\displaystyle\sum_{k=1}^{n}(-1)^k e^{-k\pi s}}{s}.$$

Therefore,

$$Y(s) = \frac{1 + 2\displaystyle\sum_{k=1}^{n}(-1)^k e^{-k\pi s}}{s(s^2 + 1)}.$$

Using partial fractions, we see that

$$Y(s) = \left(1 + 2\sum_{k=1}^{n}(-1)^k e^{-k\pi s}\right)\left[\frac{1}{s} - \frac{s}{s^2 + 1}\right].$$

Thus we conclude that

$$\begin{aligned}
y(t) &= 1 + 2\sum_{k=1}^{n}(-1)^k u_{k\pi}(t) - \cos(t) - 2\sum_{k=1}^{n}(-1)^k u_{k\pi}(t)\cos(t - k\pi) \\
&= 1 - \cos(t) + 2\sum_{k=1}^{n}(-1)^k[1 - \cos(t - k\pi)]u_{k\pi}(t).
\end{aligned}$$

(b)

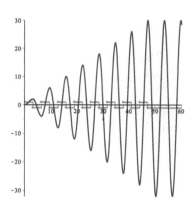

(c) As $n \to \infty$, the amplitude of the solution increases. Below is a graph for the case $n = 30$.

21.(a)

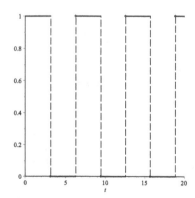

(b) Let $Y(s) = \mathcal{L}(y)$. Applying the Laplace transform to the equation, we have

$$[s^2Y(s) - sy(0) - y'(0)] + Y(s) = \frac{1 + \displaystyle\sum_{k=1}^{n}(-1)^k e^{-k\pi s}}{s}.$$

Applying the initial conditions, we get

$$[s^2 + 1]Y(s) = \frac{1 + \sum_{k=1}^{n}(-1)^k e^{-k\pi s}}{s}.$$

Therefore,

$$Y(s) = \frac{1 + \sum_{k=1}^{n}(-1)^k e^{-k\pi s}}{s(s^2 + 1)}.$$

Using partial fractions, we see that

$$Y(s) = \left(1 + \sum_{k=1}^{n}(-1)^k e^{-k\pi s}\right)\left[\frac{1}{s} - \frac{s}{s^2 + 1}\right].$$

Thus we conclude that

$$
\begin{aligned}
y(t) &= 1 + \sum_{k=1}^{n}(-1)^k u_{k\pi}(t) - \cos(t) - \sum_{k=1}^{n}(-1)^k u_{k\pi}(t)\cos(t - k\pi) \\
&= 1 - \cos(t) + \sum_{k=1}^{n}(-1)^k[1 - \cos(t - k\pi)]u_{k\pi}(t).
\end{aligned}
$$

(c)

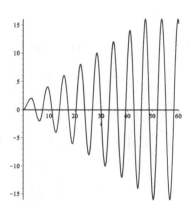

There is no damping there. Each interval of forcing adds energy to the system. Therefore, the amplitude will continue to increase until increase. For $n = 15$, $g(t) = 0$ when $t > 15\pi$. Therefore, the oscillation will eventually become steady.

(d) If n is even, the forcing term $g(t) = 1$ for large values of t, while for n odd, the forcing term $g(t) = 0$ for large values of t. Therefore, for n even, the solution will eventually oscillate about the line $y = 1$, while for n odd, the solution will eventually oscillate about the line $y = 0$. As n increases, the amount of forcing increases, causing the amplitude of the oscillations to increase until they reach a steady state after $t > n\pi$.

23.(a) Let $Y(s) = \mathcal{L}(y)$. Applying the Laplace transform to the equation, we have

$$[s^2 Y(s) - sy(0) - y'(0)] + Y(s) = \frac{1 + 2\sum_{k=1}^{n}(-1)^k e^{(-11k/4)s}}{s}.$$

Applying the initial conditions, we get

$$[s^2 + 1]Y(s) = \frac{1 + 2\sum_{k=1}^{n}(-1)^k e^{(-11k/4)s}}{s}.$$

Therefore,

$$Y(s) = \frac{1 + 2\sum_{k=1}^{n}(-1)^k e^{(-11k/4)s}}{s(s^2 + 1)}.$$

Using partial fractions, we see that

$$Y(s) = \left(1 + 2\sum_{k=1}^{n}(-1)^k e^{(-11k/4)s}\right)\left[\frac{1}{s} - \frac{s}{s^2 + 1}\right].$$

Thus we conclude that

$$
\begin{aligned}
y(t) &= 1 + 2\sum_{k=1}^{n}(-1)^k u_{11k/4}(t) - \cos(t) - 2\sum_{k=1}^{n}(-1)^k u_{11k/4}(t)\cos(t - 11k/4) \\
&= 1 - \cos(t) + 2\sum_{k=1}^{n}(-1)^k[1 - \cos(t - 11k/4)]u_{11k/4}(t).
\end{aligned}
$$

(b)

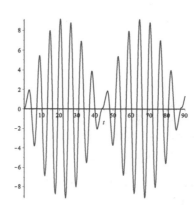

(c) Based on the plot, the "slow period" appears to be approximately 88 and the "fast period" appears to be about 6.

(d) The natural frequency of the system is $\omega_0 = 1$. The forcing function is initially periodic with period $T = 11/2$. Therefore, the corresponding forcing frequency is $\omega = 1.1424$. Therefore, the "slow frequency" is

$$\omega_s = \frac{|\omega - \omega_0|}{2} = 0.0712$$

and the "fast frequency' is

$$\omega_f = \frac{|\omega + \omega_0|}{2} = 1.0712.$$

Based on these values, the slow period is predicted as 88.247 and the fast period is predicted to be 5.8656.

5.7 Impulse Functions

1. Let $Y(s) = \mathcal{L}(y)$ and apply the Laplace transform to the differential equation. We obtain

$$[s^2 Y(s) - sy(0) - y'(0)] + 2[sY(s) - y(0)] + 2Y(s) = e^{-\pi s}.$$

Applying the initial conditions, we get

$$[s^2 Y(s) - 1] + 2sY(s) + 2Y(s) = e^{-\pi s},$$

which can be rewritten as

$$[s^2 + 2s + 2]Y(s) - 1 = e^{-\pi s}.$$

Therefore,

$$Y(s) = \frac{1 + e^{-\pi s}}{s^2 + 2s + 2} = \frac{1}{(s+1)^2 + 1} + \frac{e^{-\pi s}}{(s+1)^2 + 1}.$$

This implies that

$$y(t) = e^{-t} \sin t + u_\pi(t) e^{-(t-\pi)} \sin(t - \pi).$$

3. Let $Y(s) = \mathcal{L}(y)$ and apply the Laplace transform to the differential equation. We obtain

$$[s^2 Y(s) - sy(0) - y'(0)] + 3[sY(s) - y(0)] + 2Y(s) = e^{-5s} + \frac{e^{-10s}}{s}.$$

Applying the initial conditions, we get

$$[s^2 Y(s) - 1/2] + 3sY(s) + 2Y(s) = e^{-5s} + \frac{e^{-10s}}{s},$$

which can be rewritten as

$$[s^2 + 3s + 2]Y(s) - 1/2 = e^{-5s} + \frac{e^{-10s}}{s}.$$

Therefore,

$$Y(s) = \frac{1/2 + e^{-5s}}{s^2 + 3s + 2} + \frac{e^{-10s}}{s(s^2 + 3s + 2)} = \frac{1}{2} + \left[\frac{e^{-5s}}{s+1} - \frac{e^{-5s}}{s+2}\right] + \left[\frac{e^{-10s}}{2(s+2)} - \frac{e^{-10s}}{s+1} + \frac{e^{-10s}}{2s}\right].$$

This implies that

$$y(t) = \frac{1}{2}[e^{-t} - e^{-2t}] + u_5(t)[e^{-(t-5)} - e^{-2(t-5)}] + u_{10}(t)\left[\frac{1}{2}e^{-2(t-10)} - e^{-(t-10)} + \frac{1}{2}\right].$$

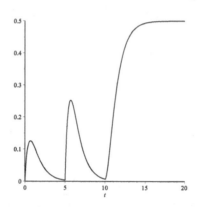

5. Let $Y(s) = \mathcal{L}(y)$ and apply the Laplace transform to the differential equation. We obtain

$$[s^2 Y(s) - sy(0) - y'(0)] + 2[sY(s) - y(0)] + 3Y(s) = \frac{1}{s^2 + 1} + e^{-3\pi s}.$$

Applying the initial conditions, we get

$$[s^2 + 2s + 3]Y(s) = \frac{1}{s^2 + 1} + e^{-3\pi s}.$$

Therefore,

$$Y(s) = \frac{1}{(s^2 + 1)(s^2 + 2s + 3)} + \frac{e^{-3\pi s}}{s^2 + 2s + 3} = \frac{1}{4}\left[\frac{-s + 1}{s^2 + 1} + \frac{s + 1}{(s+1)^2 + 2}\right] + \frac{e^{-3\pi s}}{(s+1)^2 + 2}.$$

This implies that

$$y(t) = \frac{1}{4}\left[-\cos(t) + \sin(t) + e^{-t}\cos(\sqrt{2}t)\right] + \frac{\sqrt{2}}{2}u_{3\pi}(t)e^{-(t-3\pi)}\sin(\sqrt{2}(t - 3\pi)).$$

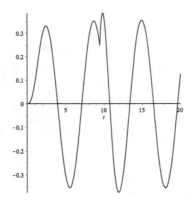

7. Let $Y(s) = \mathcal{L}(y)$ and apply the Laplace transform to the differential equation. We obtain

$$[s^2 Y(s) - sy(0) - y'(0)] + Y(s) = e^{-2\pi s}.$$

Applying the initial conditions, we get

$$[s^2 Y(s) - 1] + Y(s) = e^{-2\pi s},$$

which can be rewritten as

$$[s^2 + 1]Y(s) - 1 = e^{-2\pi s}.$$

Therefore,

$$Y(s) = \frac{1}{s^2 + 1} + \frac{e^{-2\pi s}}{s^2 + 1}.$$

This implies that

$$y(t) = \sin t + u_{2\pi}(t)\sin(t - 2\pi) = [1 + u_{2\pi}(t)]\sin t.$$

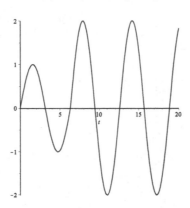

9. Let $Y(s) = \mathcal{L}(y)$ and apply the Laplace transform to the differential equation. We obtain

$$[s^2 Y(s) - sy(0) - y'(0)] + Y(s) = \frac{e^{-(\pi/2)s}}{s} + 3e^{-(3\pi/2)s} - \frac{e^{-2\pi s}}{s}.$$

Applying the initial conditions, we get

$$[s^2 + 1]Y(s) = \frac{e^{-(\pi/2)s}}{s} + 3e^{-(3\pi/2)s} - \frac{e^{-2\pi s}}{s}.$$

Therefore,

$$Y(s) = \frac{e^{-(\pi/2)s}}{s(s^2+1)} + \frac{3e^{-(3\pi/2)s}}{s^2+1} - \frac{e^{-2\pi s}}{s(s^2+1)} = \left[\frac{e^{-(\pi/2)s}}{s} - \frac{se^{-(\pi/2)s}}{s^2+1}\right] + \frac{3e^{-(3\pi/2)s}}{s^2+1} - \left[\frac{e^{-2\pi s}}{s} - \frac{se^{-2\pi s}}{s^2+1}\right].$$

This implies that

$$y(t) = u_{\pi/2}(t)[1 - \cos(t - \pi/2)] + 3u_{3\pi/2}(t)\sin(t - 3\pi/2) - u_{2\pi}(t)[1 - \cos(t - 2\pi)].$$

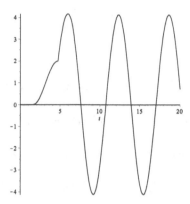

11. Let $Y(s) = \mathcal{L}(y)$ and apply the Laplace transform to the differential equation. We obtain

$$[s^2Y(s) - sy(0) - y'(0)] + 2[sY(s) - y(0)] + 2Y(s) = \frac{s}{s^2+1} + e^{-(\pi/2)s}.$$

Applying the initial conditions, we get

$$[s^2 + 2s + 2]Y(s) = \frac{s}{s^2+1} + e^{-(\pi/2)s}.$$

Therefore,

$$\begin{aligned}
Y(s) &= \frac{s}{(s^2+1)(s^2+2s+2)} + \frac{e^{-(\pi/2)s}}{s^2+2s+2}\\
&= \frac{1}{5}\left[\frac{s}{s^2+1} + \frac{2}{s^2+1} - \frac{s+1}{(s+1)^2+1} - \frac{3}{(s+1)^2+1}\right] + \frac{e^{-(\pi/2)s}}{(s+1)^2+1}.
\end{aligned}$$

This implies that

$$y(t) = \frac{1}{5}\left[\cos t + 2\sin t - e^{-t}\cos t - 3e^{-t}\sin t\right] + u_{\pi/2}(t)e^{-(t-\pi/2)}\sin(t - \pi/2).$$

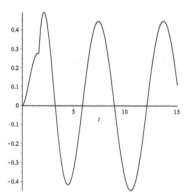

13.(a) The equation will be of the form

$$2y'' + y' + 2y = \delta(t - 5) + k\delta(t - t_0).$$

We need to find k and t_0 so that the system will rest again after exactly one cycle. Applying the Laplace transform and using the initial conditions, this equation becomes

$$[2s^2 + s + 2]Y(s) = e^{-5s} + ke^{-t_0s}.$$

Therefore,

$$Y(s) = \frac{e^{-5s} + ke^{-t_0s}}{2s^2 + s + 2} = (e^{-5s} + ke^{-t_0s})\frac{1}{2}\left[\frac{1}{(s + 1/4)^2 + (15/16)}\right].$$

This implies that

$$y(t) = u_5(t)\frac{2}{\sqrt{15}}e^{-(t-5)/4}\sin\left(\frac{\sqrt{15}}{4}(t - 5)\right) + ku_{t_0}(t)\frac{2}{\sqrt{15}}e^{-(t-t_0)/4}\sin\left(\frac{\sqrt{15}}{4}(t - t_0)\right).$$

In order for this system to return to equilibrium after one cycle, we need the magnitude of these sinusoidal waves to be the same. Therefore, we need $e^{5/4} = -ke^{t_0/4}$, and we need the phase to differ by 2π. That is, we need

$$\frac{\sqrt{15}}{4}(t - 5) = \frac{\sqrt{15}}{4}(t - t_0) + 2\pi.$$

Solving this equation, we see that $t_0 = 5 + \dfrac{8\pi}{\sqrt{15}}$. Solving the equation above for k, we see that $k = -e^{-2\pi/\sqrt{15}}$.

(b) From the analysis above, we see that the solution will be zero after time t_0.

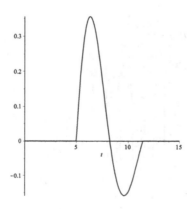

15.(a) The solution of this initial value problem is

$$y(t) = \frac{10k}{3\sqrt{11}}e^{-(t-1)/10}\sin\left(\frac{3\sqrt{11}}{10}(t - 1)\right)u_1(t).$$

This function is a multiple of the answer in problem 14(d), with $\gamma = 1/5$. Therefore, the peak value occurs at $t_1 \approx 2.4780$. The maximum value is calculated as $y(2.4780) \approx 0.8626k$. We find that the appropriate value of k is $k_1 = 6/0.8626 \approx 6.9557$.

(b) Based on problem 14(d) with $\gamma = 1/10$, the solution is

$$y(t) = \frac{20k}{\sqrt{399}} e^{-(t-1)/20} \sin\left(\frac{\sqrt{399}}{20}(t-1)\right) u_1(t).$$

Since this function is a multiple of the solution in problem 14(c), we have $t_1 \approx 2.5227$ with $y(t_1) \approx 0.9267k$. The solution reaches a maximum value of $y = 6$ for $k_1 = 6/0.9267 \approx 6.4746$.

(c) Similar to problem 14(d), for $0 < \gamma < 1$, the solution is given by

$$y(t) = \frac{2k}{\sqrt{4 - \gamma^2}} e^{-\gamma(t-1)/2} \sin\left(\sqrt{1 - \gamma^2/4}(t-1)\right) u_1(t).$$

As $\gamma \to 0$, we see that $y'(t) = 0$ at $t_1 \to 1 + \pi/2$. Setting $t_1 = 1 + \pi/2$ in $y(t)$ and letting $\gamma \to 0$, we find that $y_1 \to k$. Requiring that the peak value remains at $y = 6$, the limiting value of k is $k_1 = 6$. These conclusions agree with the case $\gamma = 0$ for which it is straightforward to show that $y(t) = k\sin(t-1)u_1(t)$.

17.(b) Applying the Laplace transform to the equation and using the initial conditions, we have

$$[s^2 + 1]Y(s) = \sum_{k=1}^{20} e^{-k\pi s}.$$

Therefore,

$$Y(s) = \frac{1}{s^2 + 1} \sum_{k=1}^{20} e^{-k\pi s},$$

which implies

$$y(t) = \sum_{k=1}^{20} \sin(t - k\pi)u_{k\pi}(t).$$

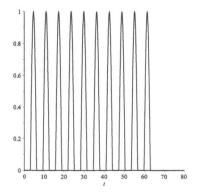

(c) After the sequence of impulses ends, the oscillator returns to equilibrium.

19.(b) Applying the Laplace transform to the equation and using the initial conditions, we have

$$[s^2 + 1]Y(s) = \sum_{k=1}^{20} e^{-(k\pi/2)s}.$$

Therefore,

$$Y(s) = \frac{1}{s^2 + 1} \sum_{k=1}^{20} e^{-(k\pi/2)s},$$

which implies

$$y(t) = \sum_{k=1}^{20} \sin(t - k\pi/2)u_{k\pi/2}(t).$$

(c) After the sequence of impulses ends, the oscillator returns to equilibrium.

21.(b) Applying the Laplace transform to the equation and using the initial conditions, we have

$$[s^2 + 1]Y(s) = \sum_{k=1}^{20} e^{-(2k-1)\pi s}.$$

Therefore,

$$Y(s) = \frac{1}{s^2 + 1} \sum_{k=1}^{20} e^{-(2k-1)\pi s},$$

which implies

$$y(t) = \sum_{k=1}^{20} \sin(t - (2k - 1)\pi)u_{(2k-1)\pi}(t).$$

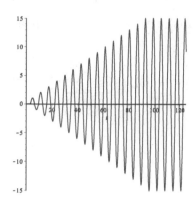

(c) After the sequence of impulses ends, the oscillator continues to oscillate at a constant amplitude.

23.(b) Applying the Laplace transform to the equation and using the initial conditions, we have

$$[s^2 + 0.1s + 1]Y(s) = \sum_{k=1}^{20} (-1)^{k+1} e^{-k\pi s}.$$

Therefore,

$$Y(s) = \frac{1}{s^2 + 0.1s + 1} \sum_{k=1}^{20} (-1)^{k+1} e^{-k\pi s} = \frac{1}{(s + 1/20)^2 + (399/400)} \sum_{k=1}^{20} (-1)^{k+1} e^{-k\pi s},$$

which implies

$$y(t) = \sum_{k=1}^{20} (-1)^{k+1} \frac{20}{\sqrt{399}} e^{-(t-k\pi)/20} \sin\left(\frac{\sqrt{399}}{20}(t - k\pi)\right) u_{k\pi}(t).$$

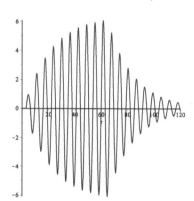

(c) After the sequence of impulses ends, the solution oscillates to zero due to the damping term.

25.(a) A fundamental set of solutions to the homogeneous problem is $y_1(t) = e^{-t}\cos t$ and $y_2(t) = e^{-t}\sin t$. Based on problem 22 from Section 4.8, a particular solution is given by

$$y_p(t) = \int_0^t \frac{y_1(s)y_2(t) - y_1(t)y_2(s)}{W(y_1, y_2)(s)} f(s)\, ds.$$

Using the functions y_1, y_2 above, we see that

$$y_p(t) = \int_0^t \frac{e^{-(s-t)}[\cos(s)\sin(t) - \sin(s)\cos(t)]}{e^{-2s}} f(s)\, ds = \int_0^t e^{-(t-s)}\sin(t-s)f(s)\, ds.$$

Given the fact that the initial conditions are zero, the solution of this initial value problem is

$$y_p(t) = \int_0^t e^{-(t-s)}\sin(t-s)f(s)\, ds.$$

(b) If $f(t) = \delta(t - \pi)$, then for $t < \pi$, $y(t) = 0$. On the other hand, for $t > \pi$,

$$y(t) = \int_0^t e^{-(t-s)}\sin(t-s)\delta(s-\pi)\, ds = e^{-(t-\pi)}\sin(t-\pi).$$

Therefore, $y(t) = u_\pi(t)e^{-(t-\pi)}\sin(t-\pi)$.

(c) Applying the Laplace transform to the ODE with $f(t) = \delta(t - \pi)$, we conclude that the Laplace transform satisfies

$$Y(s) = \frac{e^{-\pi s}}{s^2 + 2s + 2} = \frac{e^{-\pi s}}{(s+1)^2 + 1}.$$

Therefore,

$$y(t) = u_\pi(t)\sin(t - \pi).$$

5.8 Convolution Integrals and Their Applications

1.(a) We compute:

$$\begin{aligned}
f * (g_1 + g_2)(t) &= \int_0^t f(t-\tau)[g_1 + g_2](\tau)\, d\tau = \int_0^t f(t-\tau)g_1(\tau)\, d\tau + \int_0^t f(t-\tau)g_2(\tau)\, d\tau \\
&= f * g_1(t) + f * g_2(t).
\end{aligned}$$

(b) We compute:

$$\begin{aligned}
f * (g * h)(t) &= \int_0^t f(t-\tau)[g * h(\tau)]\, d\tau = \int_0^t f(t-\tau)\left[\int_0^\tau g(\tau-\eta)h(\eta)\, d\eta\right] d\tau \\
&= \int_0^t \int_0^\tau f(t-\tau)g(\tau-\eta)h(\eta)\, d\eta\, d\tau.
\end{aligned}$$

Interchanging the order of integration, we have

$$\begin{aligned}
\int_0^t \int_0^\tau f(t-\tau)g(\tau-\eta)h(\eta)\, d\eta\, d\tau &= \int_0^t \int_\eta^t f(t-\tau)g(\tau-\eta)h(\eta)\, d\tau\, d\eta \\
&= \int_0^t \left[\int_\eta^t f(t-\tau)g(\tau-\eta)\, d\tau\right] h(\eta)\, d\eta.
\end{aligned}$$

Now letting $\tau - \eta = u$, we obtain

$$\int_{\eta}^{t} f(t - \tau)g(\tau - \eta)\, d\tau = \int_{0}^{t-\eta} f(t - \eta - u)g(u)\, du = f * g(t - \eta).$$

Therefore,

$$\int_{0}^{t} f(t - \tau)[g * h(\tau)]\, d\tau = \int_{0}^{t} [f * g(t - \tau)]h(\tau)\, d\tau.$$

3. $\mathcal{L}[t^4] = 24/s^5$ and $\mathcal{L}[\cos 6t] = s/(s^2 + 36)$. Therefore,

$$\mathcal{L}\left[\int_{0}^{t} (t - \tau)^4 \cos 6\tau\, d\tau\right] = \left(\frac{24}{s^5}\right) \cdot \left(\frac{s}{s^2 + 36}\right) = \frac{24}{s^4(s^2 + 36)}.$$

5. $\mathcal{L}[t] = 1/s^2$ and $\mathcal{L}[e^t] = 1/(s - 1)$. Therefore,

$$\mathcal{L}\left[\int_{0}^{t} (t - \tau)e^{\tau}\, d\tau\right] = \left(\frac{1}{s^2}\right) \cdot \left(\frac{1}{s - 1}\right) = \frac{1}{s^2(s - 1)}.$$

7. $\mathcal{L}^{-1}[1/s^3] = t^2/2$ and $\mathcal{L}^{-1}[1/(s^2 + 1)] = \sin t$. Therefore,

$$\mathcal{L}^{-1}\left[\frac{1}{s^3(s^2 + 1)}\right] = \int_{0}^{t} \frac{1}{2}(t - \tau)^2 \sin \tau\, d\tau.$$

9. $\mathcal{L}^{-1}[1/(s + 3)^4] = \frac{1}{6}t^3 e^{-3t}$ and $\mathcal{L}^{-1}[1/(s^2 + 4)] = \frac{1}{2}\sin 2t$. Therefore,

$$\mathcal{L}^{-1}\left[\frac{1}{(s + 3)^4(s^2 + 4)}\right] = \int_{0}^{t} \frac{1}{12}(t - \tau)^3 e^{-3(t-\tau)} \sin 2\tau\, d\tau.$$

11. Consider

$$F(s) = \frac{1}{(s^2 + 1)^2} = \frac{1}{s^2 + 1} \cdot \frac{1}{s^2 + 1} = G(s) \cdot G(s);$$

then

$$\mathcal{L}^{-1}(F(s)) = \int_{0}^{t} g(t - \tau)g(\tau)\, d\tau,$$

where $g(t) = \mathcal{L}^{-1}(G(s)) = \sin t$. Therefore,

$$\mathcal{L}^{-1}(F(s)) = \int_{0}^{t} \sin(t - \tau) \sin \tau\, d\tau.$$

13.(a) We know that

$$f * g = \int_{0}^{t} (t - \tau)^m \tau^n\, d\tau.$$

Introducing a new variable u such that $\tau = t - tu$, we have $d\tau = -t\, du$. Therefore,

$$\int_{0}^{t} (t - \tau)^m \tau^n\, d\tau = -\int_{1}^{0} (tu)^m (t - tu)^n t\, du = t^{m+n+1} \int_{0}^{1} u^m (1 - u)^n\, du.$$

(b) From part (a),

$$\mathcal{L}[f * g] = \mathcal{L}\left[t^{m+n+1} \int_0^1 u^m (1-u)^n \, du\right] = \frac{(m+n+1)!}{s^{m+n+2}} \int_0^1 u^m (1-u)^n \, du.$$

By the Convolution Theorem, we know that $\mathcal{L}[f * g] = \mathcal{L}[f] \cdot \mathcal{L}[g]$. Therefore, calculating $\mathcal{L}[f] \cdot \mathcal{L}[g]$, we have

$$\mathcal{L}[f] \cdot \mathcal{L}[g] = \frac{m!}{s^{m+1}} \cdot \frac{n!}{s^{n+1}}.$$

Equating $\mathcal{L}[f * g]$ and $\mathcal{L}[f] \cdot \mathcal{L}[g]$, we obtain

$$\frac{(m+n+1)!}{s^{m+n+2}} \int_0^1 u^m (1-u)^n \, du = \frac{m!}{s^{m+1}} \cdot \frac{n!}{s^{n+1}}.$$

Therefore, we conclude that

$$\int_0^1 u^m (1-u)^n \, du = \frac{m!n!}{(m+n+1)!}.$$

(c) For the case when m and n are positive numbers, but not necessarily integers, we have

$$\mathcal{L}[f * g] = \mathcal{L}\left[t^{m+n+1} \int_0^1 u^m (1-u)^n \, du\right] = \frac{\Gamma(m+n+2)}{s^{m+n+2}} \int_0^1 u^m (1-u)^n \, du.$$

Further,

$$\mathcal{L}[f] \cdot \mathcal{L}[g] = \frac{\Gamma(m+1)}{s^{m+1}} \cdot \frac{\Gamma(n+1)}{s^{n+1}}.$$

Equating $\mathcal{L}[f * g]$ and $\mathcal{L}[f] \cdot \mathcal{L}[g]$, we obtain

$$\frac{\Gamma(m+n+2)}{s^{m+n+2}} \int_0^1 u^m (1-u)^n \, du = \frac{\Gamma(m+1)}{s^{m+1}} \cdot \frac{\Gamma(n+1)}{s^{n+1}}.$$

Therefore, we conclude that

$$\int_0^1 u^m (1-u)^n \, du = \frac{\Gamma(m+1)\Gamma(n+1)}{\Gamma(m+n+2)}.$$

15. Applying the Laplace transform to the equation, we have

$$[s^2 Y(s) - sy(0) - y'(0)] + 6[sY(s) - y(0)] + 25Y(s) = \frac{\alpha}{s^2 + \alpha^2}.$$

Applying the initial conditions, we get

$$[s^2 + 6s + 25]Y(s) = \frac{\alpha}{s^2 + \alpha^2}.$$

Therefore,

$$Y(s) = \frac{1}{s^2 + 6s + 25} \cdot \frac{\alpha}{s^2 + \alpha^2} = \frac{1}{(s+3)^2 + 16} \cdot \frac{\alpha}{s^2 + \alpha^2}.$$

Thus

$$y(t) = \frac{1}{4} \int_0^t e^{-3(t-\tau)} \sin 4(t - \tau) \sin \alpha\tau \, d\tau.$$

17. Applying the Laplace transform to the equation, we have

$$[s^2 Y(s) - sy(0) - y'(0)] + [sY(s) - y(0)] + \frac{5}{4}Y(s) = \frac{1}{s} - \frac{e^{-\pi s}}{s}.$$

Applying the initial conditions, we get

$$[s^2 + s + 5/4]Y(s) = s + \frac{1}{s} - \frac{e^{-\pi s}}{s}.$$

Therefore,

$$
\begin{aligned}
Y(s) &= \frac{s}{s^2 + s + 5/4} + \frac{1}{s^2 + s + 5/4} \cdot \frac{1 - e^{-\pi s}}{s} \\
&= \frac{s + 1/2}{(s + 1/2)^2 + 1} - \frac{1/2}{(s + 1/2)^2 + 1} + \frac{1}{(s + 1/2)^2 + 1} \cdot \frac{1 - e^{-\pi s}}{s}
\end{aligned}
$$

Thus

$$y(t) = e^{-t/2} \cos(t) - \frac{1}{2}e^{-t/2} \sin(t) + \int_0^t e^{-(t-\tau)/2} \sin(t - \tau)(1 - u_\pi(\tau)) \, d\tau.$$

19. Applying the Laplace transform to the equation, we have

$$[s^2 Y(s) - sy(0) - y'(0)] + 3[sY(s) - y(0)] + 2Y(s) = \frac{s}{s^2 + \alpha^2}.$$

Applying the initial conditions, we get

$$[s^2 + 3s + 2]Y(s) = s + 3 + \frac{s}{s^2 + \alpha^2}.$$

Therefore,

$$Y(s) = \frac{s + 3}{s^2 + 3s + 2} + \frac{s}{(s^2 + 3s + 2)(s^2 + \alpha^2)}.$$

Using partial fractions, we obtain

$$\frac{s + 3}{s^2 + 3s + 2} = \frac{2}{s + 1} - \frac{1}{s + 2}$$

and

$$\frac{1}{s^2 + 3s + 2} = \frac{1}{s + 1} - \frac{1}{s + 2}.$$

Thus we conclude that

$$y(t) = 2e^{-t} - e^{-2t} + \int_0^t \left(e^{-(t-\tau)} - e^{-2(t-\tau)} \right) \cos \alpha\tau \, d\tau.$$

21. Applying the Laplace transform to the equation, we have

$$[s^4 Y(s) - s^3 y(0) - s^2 y'(0) - s y''(0) - y'''(0)] + 17[s^2 Y(s) - sy(0) - y'(0)] + 16Y(s) = G(s).$$

Applying the initial conditions, we get

$$[s^4 + 17s^2 + 16]Y(s) = 2s^3 + 34s + G(s).$$

Therefore,

$$Y(s) = \frac{2s^3 + 34s}{s^4 + 17s^2 + 16} + \frac{G(s)}{s^4 + 17s^2 + 16}.$$

Using partial fractions, we write

$$\frac{2s^3 + 34s}{s^4 + 17s^2 + 16} = \frac{32s}{15(s^2 + 1)} - \frac{2s}{15(s^2 + 16)}$$

and

$$\frac{1}{s^4 + 17s^2 + 16} = \frac{1}{15(s^2 + 1)} - \frac{1}{15(s^2 + 16)}.$$

Therefore, we can conclude that

$$y(t) = \frac{32}{15}\cos t - \frac{2}{15}\cos 4t + \frac{1}{60}\int_0^t [4\sin(t - \tau) - \sin 4(t - \tau)]g(\tau)\,d\tau.$$

23. Taking the Laplace transform of the integral equation, we have

$$\mathcal{L}[\phi(t)] + \mathcal{L}\left[\int_0^t k(t - \xi)\phi(\xi)\,d\xi\right] = \mathcal{L}[f(t)],$$

which implies

$$\mathcal{L}[\phi(t)] + \mathcal{L}[k(t)] \cdot \mathcal{L}[\phi(t)] = \mathcal{L}[f(t)].$$

Therefore, $\mathcal{L}[\phi(t)](1 + \mathcal{L}[k(t)]) = \mathcal{L}[f(t)]$, and then

$$\Phi(s) = \mathcal{L}[\phi(t)] = \frac{\mathcal{L}[f(t)]}{1 + \mathcal{L}[k(t)]} = \frac{F(s)}{1 + K(s)}.$$

25.(a) Taking the Laplace transform of the equation, we have

$$\Phi(s) + \frac{1}{s^2}\Phi(s) = \frac{1}{s}.$$

Therefore,

$$\Phi(s) = \frac{s}{s^2 + 1},$$

which implies that $\phi(t) = \cos t$.

(b) Differentiating the equation once, we have

$$\phi'(t) + \int_0^t \phi(\xi)\,d\xi = 0.$$

Differentiating the equation again, we have

$$\phi''(t) + \phi(t) = 0.$$

Using equation (i), we see that $\phi(0) = 1$. Further, using the equation for the first derivative of (i), we see that $\phi'(0) = 0$.

(c) Letting $\Phi(s) = \mathcal{L}[\phi(t)]$ and taking the Laplace transform of the equation in part (ii), we have

$$[s^2\Phi(s) - s\phi(0) - \phi'(0)] + \Phi(s) = 0.$$

Applying the initial conditions, we get $[s^2 + 1]\Phi(s) = s$. Therefore,

$$\Phi(s) = \frac{s}{s^2 + 1},$$

which implies that $\phi(t) = \cos t$.

27.(a) Taking the Laplace transform of the equation, we have

$$\Phi(s) + \frac{2s}{s^2 + 1}\Phi(s) = \frac{1}{s + 1}.$$

Therefore,

$$\Phi(s) = \frac{s^2 + 1}{(s + 1)^3} = \frac{1}{s + 1} - \frac{2}{(s + 1)^2} + \frac{2}{(s + 1)^3},$$

which implies that $\phi(t) = e^{-t} - 2te^{-t} + t^2 e^{-t}$.

(b) Differentiating the equation once, we have

$$\phi'(t) + 2\phi(t) - 2\int_0^t \sin(t - \xi)\phi(\xi)\,d\xi = -e^{-t}.$$

Differentiating the equation again, we have

$$\phi''(t) + 2\phi'(t) - 2\int_0^t \cos(t - \xi)\phi(\xi)\,d\xi = e^{-t}.$$

Using the original equation, this gives us

$$\phi''(t) + 2\phi'(t) + \phi(t) = 2e^{-t}.$$

Using equation (i), we see that $\phi(0) = 1$. Further, using the equation for the first derivative of (i), we see that $\phi'(0) = -3$.

(c) Letting $\Phi(s) = \mathcal{L}[\phi(t)]$ and taking the Laplace transform of the equation in part (ii), we have

$$[s^2\Phi(s) - s\phi(0) - \phi'(0)] + 2[s\Phi(s) - \phi(0)] - 2\frac{s}{s^2 + 1}\Phi(s) = \frac{1}{s + 1}.$$

Applying the initial conditions, we get

$$\left[s^2 + 2s - \frac{2s}{s^2 + 1}\right]\Phi(s) = s - 1 + \frac{1}{s + 1}.$$

Therefore,

$$
\begin{aligned}
\Phi(s) &= \frac{(s-1)(s^2+1)}{s^4+2s^3+s^2} + \frac{s^2+1}{(s+1)(s^4+2s^3+s^2)} \\
&= -\frac{1}{s^2} - \frac{4}{(s+1)^2} + \frac{3}{s} - \frac{2}{s+1} + \frac{1}{s^2} + \frac{2}{(s+1)^2} - \frac{3}{s} + \frac{3}{s+1} + \frac{2}{(s+1)^3} \\
&= -\frac{2}{(s+1)^2} + \frac{1}{s+1} + \frac{2}{(s+1)^3},
\end{aligned}
$$

which implies $\phi(t) = -2te^{-t} + e^{-t} + t^2 e^{-t}$.

29.(a) Taking the Laplace transform of the equation, we have

$$
s\Phi(s) - 1 - \frac{1}{s^3}\Phi(s) = -\frac{1}{s^2}.
$$

Therefore,

$$
\Phi(s) = \frac{s}{s^2+1},
$$

which implies $\phi(t) = \cos t$.

(b) Differentiating the equation once, we have

$$
\phi''(t) - \int_0^t (t-\xi)\phi(\xi)\, d\xi = -1.
$$

Now

$$
-\int_0^t (t-\xi)\phi(\xi)\, d\xi = -t\int_0^t \phi(\xi)\, d\xi + \int_0^t \xi\phi(\xi)\, d\xi;
$$

using the product rule for differentiation and the fundamental theorem of calculus, after differentiating the equation again, we get

$$
\phi'''(t) - \int_0^t \phi(\xi)\, d\xi = 0.
$$

Differentiating again, we see that

$$
\phi''''(t) - \phi(t) = 0.
$$

Using the equations above for the derivatives of ϕ, we see that $\phi(0) = 1$, $\phi'(0) = 0$, $\phi''(0) = -1$ and $\phi'''(0) = 0$.

(c) Letting $\Phi(s) = \mathcal{L}[\phi(t)]$ and taking the Laplace transform of the equation in part (ii), we have

$$
[s^4\Phi(s) - s^3\phi(0) - s^2\phi'(0) - s\phi''(0) - \phi'''(0)] - \Phi(s) = 0.
$$

Applying the initial conditions, we get $[s^4 - 1]\Phi(s) = s^3 - s$. Therefore,

$$
\Phi(s) = \frac{s^3 - s}{s^4 - 1} = \frac{s(s^2-1)}{(s^2-1)(s^2+1)} = \frac{s}{s^2+1}.
$$

Therefore, $\phi(t) = \cos t$.

31.(a) We note that

$$\int_0^b \frac{f(y)}{\sqrt{b-y}}\,dy = \left(\frac{1}{\sqrt{y}} * f\right)(b).$$

Taking the Laplace transform of both sides of this equation, we see that

$$\frac{T_0}{s} = \frac{1}{\sqrt{2g}}F(s) \cdot \mathcal{L}\left[\frac{1}{\sqrt{y}}\right].$$

Then using the fact that

$$\mathcal{L}\left[\frac{1}{\sqrt{y}}\right] = \frac{\Gamma(1/2)}{s^{1/2}} = \sqrt{\frac{\pi}{s}},$$

we conclude that

$$\frac{T_0}{s} = \frac{1}{\sqrt{2g}}F(s) \cdot \sqrt{\pi s},$$

which implies

$$F(s) = \sqrt{\frac{2g}{\pi}} \cdot \frac{T_0}{\sqrt{s}}.$$

Taking the inverse transform, we see that

$$f(y) = \frac{T_0}{\pi}\sqrt{\frac{2g}{y}}.$$

(b) Combining equations (i) and (iv), we see that

$$\frac{2gT_0^2}{\pi^2 y} = 1 + \left(\frac{dx}{dy}\right)^2.$$

Solving this equation for dx/dy, we see that

$$\frac{dx}{dy} = \sqrt{\frac{2\alpha - y}{y}},$$

where $\alpha = gT_0^2/\pi^2$.

(c) Consider the change of variables $y = 2\alpha \sin^2(\theta/2)$. Using the chain rule, we have

$$\frac{dy}{dx} = 2\alpha \sin(\theta/2)\cos(\theta/2) \cdot \frac{d\theta}{dx}$$

and

$$\frac{dx}{dy} = \frac{1}{2\alpha \sin(\theta/2)\cos(\theta/2)} \cdot \frac{dx}{d\theta}.$$

It follows that

$$\frac{dx}{d\theta} = 2\alpha \sin(\theta/2)\cos(\theta/2)\sqrt{\frac{\cos^2(\theta/2)}{\sin^2(\theta/2)}} = 2\alpha \cos^2(\theta/2) = \alpha + \alpha\cos(\theta).$$

Now integrating this equation for $dx/d\theta$, we have

$$x(\theta) = \alpha\theta + \alpha\sin(\theta) + C.$$

Since the curve passes through the origin, we have $x(0) = 0 = y(0)$. Therefore, $C = 0$, and $x(\theta) = \alpha\theta + \alpha\sin(\theta)$. We also have $y(\theta) = 2\alpha\sin^2(\theta/2) = \alpha - \alpha\cos(\theta)$.

5.9 Linear Systems and Feedback Control

1. Let Y_1 represent the output before the junction point. Let Y_2 represent the output after the junction point. Let U_1 represent the output of controller G_1 and let U_2 represent the output of controller G_2. Then we have the following system of equations

$$
\begin{aligned}
Y_1 &= H_1[F - U_1] \\
U_1 &= G_1 Y_1 \\
Y_2 &= Y_1 + U_2 \\
U_2 &= G_2[F - U_1].
\end{aligned}
$$

Solving the equation for Y_1, we get $Y_1 = H_1 F - H_1 G_1 Y_1$, thus

$$Y_1 = \frac{H_1 F}{1 + H_1 G_1}.$$

Then $Y_2 = Y_1 + U_2$ implies that

$$Y_2 = \frac{H_1 F}{1 + H_1 G_1} + G_2\left[F - \frac{G_1 H_1 F}{1 + H_1 G_1}\right] = \frac{H_1 F + G_2 F}{1 + H_1 G_1}.$$

3. Consider the function $h(t) = t^k e^{\alpha t}$. We will prove that

$$\int_0^\infty |h(t)|\, dt \le \frac{k!}{|\alpha|^{k+1}}$$

by induction on k. First, for $k = 1$, we have

$$\int_0^\infty t e^{\alpha t}\, dt = \lim_{b\to\infty} \frac{t e^{\alpha t}}{\alpha}\Big|_0^b - \int_0^\infty \frac{e^{\alpha t}}{\alpha}\, dt = -\lim_{b\to\infty}\frac{e^{\alpha t}}{\alpha^2}\Big|_0^b = \frac{1}{\alpha^2}.$$

Now assume that

$$\int_0^\infty t^k e^{\alpha t}\, dt \le \frac{k!}{|\alpha|^{k+1}}$$

to prove that

$$\int_0^\infty t^{k+1} e^{\alpha t}\, dt \le \frac{(k+1)!}{|\alpha|^{k+2}}.$$

Integrating by parts, we see that

$$\int_0^\infty t^{k+1} e^{\alpha t}\, dt = \lim_{b\to\infty}\frac{t^{k+1}e^{\alpha t}}{\alpha}\Big|_0^b - \int_0^\infty \frac{e^{\alpha t}}{\alpha}(k+1)t^k\, dt = -\int_0^\infty \frac{e^{\alpha t}(k+1)t^k}{\alpha}\, dt.$$

From our inductive assumption,

$$\left| \int_0^\infty \frac{e^{\alpha t}(k+1)t^k}{\alpha}\, dt \right| \le \frac{k+1}{\alpha} \cdot \frac{k!}{|\alpha|^{k+1}} = \frac{(k+1)!}{|\alpha|^{k+2}}.$$

Further, $|t^k e^{\alpha t}\cos(\beta t)| \le t^k e^{\alpha t}$ and $|t^k e^{\alpha t}\sin(\beta t)| \le t^k e^{\alpha t}$. Therefore, the same bounds hold on integrals involving those functions.

5. The associated Routh table is given by

s^3	1	a_1	0
s^2	a_2	a_0	0
s	$\dfrac{a_1 a_2 - a_0}{a_2}$	0	
s^0	a_0		

In order for all the roots to have negative real parts, we need all entries in the first column to be positive. Therefore, we need (1) $a_2 > 0$, (2) $(a_1 a_2 - a_0)/a_2 > 0$ and (3) $a_0 > 0$. Therefore, all the roots will lie in the left half plane if and only if (1) $a_2 > 0$, (2) $a_1 a_2 > a_0$ and (3) $a_0 > 0$.

7. The associated Routh table is given by

s^3	1	9	0
s^2	5	5	0
s	8	0	
s^0	5		

There are no sign changes in the first column. Therefore, there are no roots with positive real parts. Using a computer, we see that the roots are -1 and $-2 \pm i$. As shown by the Routh table, none of the roots have positive real parts.

9. The associated Routh table is given by

s^4	1	-5	34	0
s^3	5	-35	0	
s^2	2	34	0	
s	-120	0		
s^0	34			

There are two sign changes in the first column. Therefore, there are two roots with positive real parts. Using a computer, we see that the roots are 1, 2 and $-4 \pm i$. As shown by the Routh table, exactly two of the roots have positive real parts.

11. The associated Routh table is given by

s^4	1	24	100	0
s^3	8	32	0	
s^2	20	100	0	
s	-8	0		
s^0	100			

There are two sign changes in the first column. Therefore, there are two roots with positive real parts. Using a computer, we see that the roots are approximately $0.141 \pm 2.141i$ and $-4.141 \pm 2.141i$. As shown by the Routh table, two of the roots have positive real parts.

13. The associated Routh table is given by

$$
\begin{array}{c|ccc}
s^3 & 1 & 6 & 0 \\
s^2 & 4 & 4+K & 0 \\
s & 5-K/4 & 0 & \\
s^0 & 4+K & &
\end{array}
$$

If we want all the poles to lie in the left half plane, we need all the entries in the first column to have the same sign. That is, we need $5-K/4 > 0$ and $4+K > 0$. We see that these two inequalities will be satisfied if and only if $-4 < K < 20$. Below we show a plot of the roots of the polynomial $s^3 + 4s^2 + 6s + 4 + K$ for $K = -12$, $K = 8$, and $K = 28$, respectively.

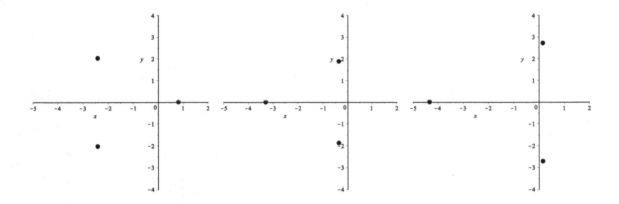

15. The associated Routh table is given by

$$
\begin{array}{c|cccc}
s^4 & 1 & 14 & 16+K & 0 \\
s^3 & 4 & 20 & 0 & \\
s^2 & 9 & 16+K & 0 & \\
s & 116/9 - 4K/9 & 0 & & \\
s^0 & 16+K & & &
\end{array}
$$

If we want all the poles to lie in the left half plane, we need all the entries in the first column to have the same sign. That is, we need $116 - 4K > 0$ and $16 + K > 0$. We see that these two inequalities will be satisfied if and only if $-16 < K < 29$. Below we show a plot of the roots of the polynomial $s^4 + 4s^3 + 14s^2 + 20s + 16 + K$ for $K = -38$, $K = 7$, and $K = 51$, respectively.

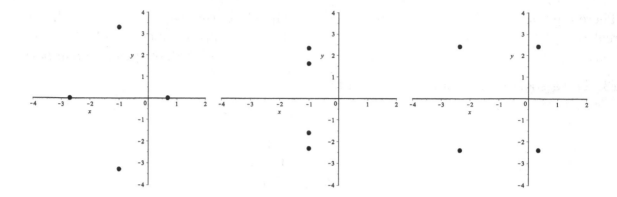

Chapter 6

Systems of First Order Linear Equations

6.1 Definitions and Examples

1.(a) We compute:

$$\mathbf{A} + 3\mathbf{B} = \begin{pmatrix} e^t & 2e^{-t} & e^{2t} \\ 5e^t & e^{-t} & -e^{2t} \\ -e^t & 3e^{-t} & 5e^{2t} \end{pmatrix} + \begin{pmatrix} 6e^t & 3e^{-t} & 12e^{2t} \\ -3e^t & 9e^{-t} & 3e^{2t} \\ 12e^t & -3e^{-t} & -3e^{2t} \end{pmatrix} = \begin{pmatrix} 7e^t & 5e^{-t} & 13e^{2t} \\ 2e^t & 10e^{-t} & 2e^{2t} \\ 11e^t & 0 & 2e^{2t} \end{pmatrix}.$$

(b) We compute:

$$\mathbf{AB} = \begin{pmatrix} -2 + 2e^{2t} + 4e^{3t} & 1 + 6e^{-2t} - e^t & 2e^t + 4e^{3t} - e^{4t} \\ -1 + 10e^{2t} - 4e^{3t} & 5 + 3e^{-2t} + e^t & e^t + 20e^{3t} + e^{4t} \\ -3 - 2e^{2t} + 20e^{3t} & -1 + 9e^{-2t} - 5e^t & 3e^t - 4e^{3t} - 5e^{4t} \end{pmatrix}.$$

(c) We differentiate elementwise:

$$\frac{d\mathbf{A}}{dt} = \begin{pmatrix} e^t & -2e^{-t} & 2e^{2t} \\ 5e^t & -e^{-t} & -2e^{2t} \\ -e^t & -3e^{-t} & 10e^{2t} \end{pmatrix}.$$

(d) We integrate elementwise:

$$\int_0^1 \mathbf{A}(t)\,dt = \begin{pmatrix} e-1 & 2-2/e & e^2/2 - 1/2 \\ 5e-5 & 1-1/e & 1/2 - e^2/2 \\ 1-e & 3-3/e & 5e^2/2 - 5/2 \end{pmatrix}.$$

3. First, we see that

$$\Psi' = \begin{pmatrix} e^t & -2e^{-2t} & 3e^{3t} \\ -4e^t & 2e^{-2t} & 6e^{3t} \\ -e^t & 2e^{-2t} & 3e^{3t} \end{pmatrix}.$$

At the same time,

$$\begin{pmatrix} 1 & -1 & 4 \\ 3 & 2 & -1 \\ 2 & 1 & -1 \end{pmatrix} \Psi = \begin{pmatrix} 1 & -1 & 4 \\ 3 & 2 & -1 \\ 2 & 1 & -1 \end{pmatrix} \begin{pmatrix} e^t & e^{-2t} & e^{3t} \\ -4e^t & -e^{-2t} & 2e^{3t} \\ -e^t & -e^{-2t} & e^{3t} \end{pmatrix}$$

$$= \begin{pmatrix} e^t & -2e^{-2t} & 3e^{3t} \\ -4e^t & 2e^{-2t} & 6e^{3t} \\ -e^t & 2e^{-2t} & 3e^{3t} \end{pmatrix}.$$

5. Let $x_1 = y$, $x_2 = y'$ and $x_3 = y''$. Then

$$x_1' = y' = x_2$$
$$x_2' = y'' = x_3$$
$$x_3' = y''' = -\frac{\sin t}{t} y'' - \frac{8}{t} y + \cos t = -\frac{\sin t}{t} x_3 - \frac{8}{t} x_1 + \cos t.$$

Therefore,

$$\begin{pmatrix} x_1 \\ x_2 \\ x_3 \end{pmatrix}' = \begin{pmatrix} 0 & 1 & 0 \\ 0 & 0 & 1 \\ -8/t & 0 & -\sin t/t \end{pmatrix} \begin{pmatrix} x_1 \\ x_2 \\ x_3 \end{pmatrix} + \begin{pmatrix} 0 \\ 0 \\ \cos t \end{pmatrix}.$$

7. Let $x_1 = y$, $x_2 = y'$, and $x_3 = y''$. Then

$$x_1' = y' = x_2$$
$$x_2' = y'' = x_3$$
$$x_3' = y''' = -ty'' - t^2 y' - t^2 y + \ln t = -tx_3 - t^2 x_2 - t^2 x_1 + \ln t.$$

Therefore,

$$\begin{pmatrix} x_1 \\ x_2 \\ x_3 \end{pmatrix}' = \begin{pmatrix} 0 & 1 & 0 \\ 0 & 0 & 1 \\ -t^2 & -t^2 & -t \end{pmatrix} \begin{pmatrix} x_1 \\ x_2 \\ x_3 \end{pmatrix} + \begin{pmatrix} 0 \\ 0 \\ \ln t \end{pmatrix}.$$

9. Let $y_1 = y$, $y_2 = y'$, $y_3 = y''$, $y_4 = y'''$, $y_5 = y^{(4)}$ and $y_6 = y^{(5)}$. Then

$$y_1' = y' = y_2$$
$$y_2' = y'' = y_3$$
$$y_3' = y''' = y_4$$
$$y_4' = y'''' = y_5$$
$$y_5' = y^{(5)} = y_6$$
$$y_6' = -\frac{x^2}{x^2 - 2} y'' - \frac{3}{x^2 - 2} y = -\frac{x^2}{x^2 - 2} y_3 - \frac{3}{x^2 - 2} y_1.$$

Therefore,

$$\begin{pmatrix} y_1 \\ y_2 \\ y_3 \\ y_4 \\ y_5 \\ y_6 \end{pmatrix}' = \begin{pmatrix} 0 & 1 & 0 & 0 & 0 & 0 \\ 0 & 0 & 1 & 0 & 0 & 0 \\ 0 & 0 & 0 & 1 & 0 & 0 \\ 0 & 0 & 0 & 0 & 1 & 0 \\ 0 & 0 & 0 & 0 & 0 & 1 \\ -\frac{3}{x^2-2} & 0 & -\frac{x^2}{x^2-2} & 0 & 0 & 0 \end{pmatrix} \begin{pmatrix} y_1 \\ y_2 \\ y_3 \\ y_4 \\ y_5 \\ y_6 \end{pmatrix}.$$

11. The system is given by

$$\begin{pmatrix} x_1 \\ x_2 \\ x_3 \end{pmatrix}' = \begin{pmatrix} -(k_{21} + k_{31} + k_{01}) & k_{12} & k_{13} \\ k_{21} & -(k_{02} + k_{12}) & 0 \\ k_{31} & 0 & -k_{13} \end{pmatrix} \begin{pmatrix} x_1 \\ x_2 \\ x_3 \end{pmatrix} + \begin{pmatrix} L(t) \\ 0 \\ 0 \end{pmatrix}.$$

Therefore,

$$\mathbf{K} = \begin{pmatrix} -(k_{21} + k_{31} + k_{01}) & k_{12} & k_{13} \\ k_{21} & -(k_{02} + k_{12}) & 0 \\ k_{31} & 0 & -k_{13} \end{pmatrix}$$

and

$$\mathbf{g}(t) = \begin{pmatrix} L(t) \\ 0 \\ 0 \end{pmatrix}.$$

13.(a) By equation (ii),

$$\frac{d}{dt}[x_1(t) + x_2(t)] = -L_{21}x_1 + L_{12}x_2 + L_{21}x_1 - L_{12}x_2 = 0.$$

(b) The system is given by

$$\begin{pmatrix} x_1 \\ x_2 \end{pmatrix}' = \begin{pmatrix} -L_{21} & L_{12} \\ L_{21} & -L_{12} \end{pmatrix} \begin{pmatrix} x_1 \\ x_2 \end{pmatrix},$$

thus

$$A - \lambda I = \begin{pmatrix} -L_{21} - \lambda & L_{12} \\ L_{21} & -L_{12} - \lambda \end{pmatrix},$$

which implies $\det(A - \lambda I) = \lambda^2 + (L_{12} + L_{21})\lambda$. Therefore, $\det(A - \lambda I) = 0$ implies that $\lambda(\lambda + L_{12} + L_{21}) = 0$. This gives $\lambda = 0$ or $\lambda = -(L_{12} + L_{21})$. Now $\lambda = 0$ implies

$$A - \lambda I = A = \begin{pmatrix} -L_{21} & L_{12} \\ L_{21} & -L_{12} \end{pmatrix}.$$

Thus

$$\mathbf{v}_1 = \begin{pmatrix} L_{12} \\ L_{21} \end{pmatrix}$$

is an eigenvector for $\lambda_1 = 0$, and

$$\mathbf{x}_1(t) = \begin{pmatrix} L_{12} \\ L_{21} \end{pmatrix}$$

is a solution of this system. Next, $\lambda = -(L_{12} + L_{21})$ implies

$$A - \lambda I = \begin{pmatrix} L_{12} & L_{12} \\ L_{21} & L_{21} \end{pmatrix}.$$

Thus

$$\mathbf{v}_2 = \begin{pmatrix} 1 \\ -1 \end{pmatrix}$$

is an eigenvector for $\lambda_2 = -(L_{12} + L_{21})$, and

$$\mathbf{x}_2(t) = e^{-(L_{12}+L_{21})t}\begin{pmatrix}1\\-1\end{pmatrix}$$

is a solution of the system. Therefore, the general solution is given by

$$\begin{pmatrix}x_1\\x_2\end{pmatrix} = c_1\begin{pmatrix}L_{12}\\L_{21}\end{pmatrix} + c_2 e^{-(L_{12}+L_{21})t}\begin{pmatrix}1\\-1\end{pmatrix}.$$

The initial condition $x_1(0) = \alpha$ and $x_2(0) = 0$ implies

$$\begin{pmatrix}x_1(0)\\x_2(0)\end{pmatrix} = c_1\begin{pmatrix}L_{12}\\L_{21}\end{pmatrix} + c_2\begin{pmatrix}1\\-1\end{pmatrix} = \begin{pmatrix}\alpha\\0\end{pmatrix}.$$

This equation gives

$$\begin{aligned}c_1 L_{12} + c_2 &= \alpha\\c_1 L_{21} - c_2 &= 0.\end{aligned}$$

The solution of this system is $c_1 = \alpha/(L_{12}+L_{21})$ and $c_2 = L_{21}\alpha/(L_{12}+L_{21})$. Thus we obtain that

$$\mathbf{x}(t) = \frac{\alpha}{L_{12}+L_{21}}\begin{pmatrix}L_{12}\\L_{21}\end{pmatrix} + \frac{L_{21}\alpha}{L_{12}+L_{21}}e^{-(L_{12}+L_{21})t}\begin{pmatrix}1\\-1\end{pmatrix}.$$

(c) We found

$$x_1(t) = \frac{\alpha}{L_{12}+L_{21}}L_{12} + \frac{L_{21}\alpha}{L_{12}+L_{21}}e^{-(L_{12}+L_{21})t}.$$

Thus

$$\overline{x}_1 = \lim_{t\to\infty}x_1(t) = \frac{\alpha L_{12}}{L_{12}+L_{21}}.$$

Also,

$$x_2(t) = \frac{\alpha}{L_{12}+L_{21}}L_{21} - \frac{L_{21}\alpha}{L_{12}+L_{21}}e^{-(L_{12}+L_{21})t}.$$

Therefore,

$$\overline{x}_2 = \lim_{t\to\infty}x_2(t) = \frac{\alpha L_{21}}{L_{12}+L_{21}}.$$

Since the decaying term $e^{-(L_{12}+L_{21})t}$ contains L_{12}, L_{21} in the exponent, the rate of decay depends on L_{12}, L_{21}.

(d) The plot shows the case $L_{12} = 1$, $L_{21} = 2$.

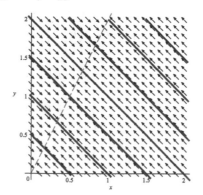

(e) The plot shows the case $L_{12} = 1$, $L_{21} = 2$, $\alpha = 1$.

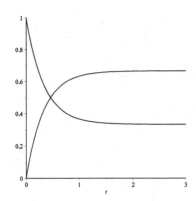

As the ratio L_{21}/L_{12} increases, the amount of tracer in compartment 1 decreases and the amount of tracer in compartment 2 increases. As $L_{21}/L_{12} \to \infty$, the amount of tracer in compartment 2 approaches α (the initial amount of tracer injected into the system) while the amount of tracer in compartment 1 approaches zero. Similarly, as $L_{21}/L_{12} \to 0$, the amount of tracer in compartment 1 approaches α while the amount of tracer in compartment 2 approaches zero.

15. If only the displacement of mass m_2 is observable, then we want

$$\mathbf{Cx} = \begin{pmatrix} 0 \\ x_2 \\ 0 \\ 0 \end{pmatrix}.$$

Therefore,

$$\mathbf{C} = \begin{pmatrix} 0 & 0 & 0 & 0 \\ 0 & 1 & 0 & 0 \\ 0 & 0 & 0 & 0 \\ 0 & 0 & 0 & 0 \end{pmatrix}.$$

6.2 Basic Theory of First Order Linear Systems

1. By Corollary 6.2.8, since the functions $p_1(t) = 5$, $p_4(t) = 4$ and $g(t) = t$ are continuous on $(-\infty, \infty)$, the solution is sure to exist for all $t \in (-\infty, \infty)$.

3. We will rewrite the equation and apply Corollary 6.2.8. We rewrite the equation as

$$y^{(4)} + \frac{e^t}{t(t-1)} y'' + \frac{7t^2}{t(t-1)} y = 0.$$

Since the coefficient functions $p_2(t) = \dfrac{e^t}{t(t-1)}$ and $p_4(t) = \dfrac{7t^2}{t(t-1)}$ are continuous for all $t \neq 0, 1$, the solution is sure to exist on the intervals $(-\infty, 0)$, $(0, 1)$ and $(1, \infty)$.

5. First, we rewrite the equation as

$$y^{(4)} + \frac{x+5}{x-1}y'' + \frac{\tan x}{x-1}y = 0.$$

The coefficient functions are $p_2(x) = \dfrac{x+5}{x-1}$ and $p_4(x) = \dfrac{\tan x}{x-1}$. These functions are continuous for all $x \neq 1, \pm\pi/2, \pm 3\pi/2, \ldots$. Therefore, the solution is sure to exist on the intervals $\ldots, (-3\pi/2, -\pi/2), (-\pi/2, 1), (1, \pi/2), (\pi/2, 3\pi/2), \ldots$

7. First, we let

$$\mathbf{X}(t) = \begin{pmatrix} e^t & e^{-t} & 2e^{4t} \\ 2e^t & -2e^{-t} & 2e^{4t} \\ -e^t & e^{-t} & -8e^{4t} \end{pmatrix}.$$

For the first method of computing $W[\mathbf{x}_1, \mathbf{x}_2, \mathbf{x}_3]$, we begin by computing $|\mathbf{X}(t)|$. We see that

$$|\mathbf{X}(t)| = e^t(16e^{3t} - 2e^{3t}) - e^{-t}(-16e^{5t} + 2e^{5t}) + 2e^{4t}(2-2) = 14e^{4t} + 14e^{4t} = 28e^{4t}.$$

Then, evaluating $|\mathbf{X}(0)|$, we have $|\mathbf{X}(0)| = 28$.
For the second method, we start by evaluating $\mathbf{X}(0)$. We see that

$$\mathbf{X}(0) = \begin{pmatrix} 1 & 1 & 2 \\ 2 & -2 & 2 \\ -1 & 1 & -8 \end{pmatrix}.$$

Then, $|\mathbf{X}(0)| = 28$.

9. First, $\mathbf{x}_i' = A\mathbf{x}_i$ for $i = 1, 2, 3$, thus the $\mathbf{x}_i$'s are solutions. Next, we calculate $W[\mathbf{x}_1, \mathbf{x}_2, \mathbf{x}_3](0)$. Now

$$\mathbf{X}(t) = \begin{pmatrix} e^{-t} & e^{-t} & 2e^{8t} \\ 0 & -4e^{-t} & e^{8t} \\ -e^{-t} & e^{-t} & 2e^{8t} \end{pmatrix}$$

implies

$$\mathbf{X}(0) = \begin{pmatrix} 1 & 1 & 2 \\ 0 & -4 & 1 \\ -1 & 1 & 2 \end{pmatrix}.$$

Therefore,

$$W[\mathbf{x}_1, \mathbf{x}_2, \mathbf{x}_3](0) = \det \mathbf{X}(0) = 1(-8-1) - 1(1+8) = -18.$$

Since $W[\mathbf{x}_1, \mathbf{x}_2, \mathbf{x}_3](0) \neq 0$, these functions are linearly independent and form a fundamental set of solutions.

11. We compute: $y_1 = 1$ gives $y_1''' + y_1' = 0$, $y_2 = \cos t$ gives $y_2''' + y_2' = \sin t - \sin t = 0$, and $y_3 = \sin t$ gives $y_3''' + y_3' = -\cos t + \cos t = 0$. Therefore, y_1, y_2, y_3 are all solutions of the differential equation. We now compute their Wronskian. We have

$$W[y_1, y_2, y_3] = \begin{vmatrix} 1 & \cos t & \sin t \\ 0 & -\sin t & \cos t \\ 0 & -\cos t & -\sin t \end{vmatrix},$$

which implies $W[y_1, y_2, y_3] = 1(\sin^2 t + \cos^2 t) = 1$.

13. We compute: $y_1 = e^{2t}$ gives $y_1''' + 4y_1'' - 4y_1' - 16y_1 = 8e^{2t} + 16e^{2t} - 8e^{2t} - 16e^{2t} = 0$, $y_2 = e^{-2t}$ gives $y_2''' + 4y_2'' - 4y_2' - 16y_2 = -8e^{-2t} + 16e^{-t} + 8e^{-2t} - 16e^{-2t} = 0$, and $y_3 = e^{-4t}$ gives $y_3''' + 4y_3'' - 4y_3' - 16y_3 = -64e^{-4t} + 64e^{-4t} + 16e^{-4t} - 16e^{-4t} = 0$. Therefore, y_1, y_2, y_3 are all solutions of the differential equation. We now compute their Wronskian. We have

$$W[y_1, y_2, y_3] = \begin{vmatrix} e^{2t} & e^{-2t} & e^{-4t} \\ 2e^{2t} & -2e^{-2t} & -4e^{-4t} \\ 4e^{2t} & 4e^{-2t} & 16e^{-4t} \end{vmatrix},$$

which implies

$$W[y_1, y_2, y_3] = e^{2t}(-32e^{-6t} + 16e^{-6t}) - e^{-2t}(32e^{-2t} + 16e^{-2t}) + e^{-4t}(8 + 8) = -48e^{-4t}.$$

15. We compute: $y_1 = 1$ gives $xy_1''' - y_1'' = 0$, $y_2 = x$ gives $xy_2''' - y_2'' = 0$, and $y_3 = x^3$ gives $xy_3''' - y_3'' = 6x - 6x = 0$. Therefore, y_1, y_2, y_3 are all solutions of the differential equation. We now compute their Wronskian. We have

$$W[y_1, y_2, y_3] = \begin{vmatrix} 1 & x & x^3 \\ 0 & 1 & 3x^2 \\ 0 & 0 & 6x \end{vmatrix},$$

which implies $W[y_1, y_2, y_3] = 6x$.

17. We compute:

$$\begin{aligned} L[c_1y_1 + c_2y_2] &= (c_1y_1 + c_2y_2)^{(n)} + p_1(t)(c_1y_1 + c_2y_2)^{(n-1)} + \ldots + p_n(t)(c_1y_1 + c_2y_2) \\ &= c_1y_1^{(n)} + c_2y_2^{(n)} + c_1p_1(t)y_1^{(n-1)} + c_2p_1(t)y_2^{(n-1)} + \ldots + c_1p_n(t)y_1 + c_2p_n(t)y_2 \\ &= c_1L[y_1] + c_2L[y_2]. \end{aligned}$$

6.3 Homogeneous Linear Systems with Constant Coefficients

1. We will write the system of equations in matrix form as $\mathbf{x}' = A\mathbf{x}$. Here, we have

$$\begin{pmatrix} x_1 \\ x_2 \\ x_3 \end{pmatrix}' = \begin{pmatrix} -4 & 1 & 0 \\ 1 & -5 & 1 \\ 0 & 1 & -4 \end{pmatrix} \begin{pmatrix} x_1 \\ x_2 \\ x_3 \end{pmatrix}.$$

To solve this system, we need to compute the eigenvalues and eigenvectors of A. We have

$$A - \lambda I = \begin{pmatrix} -4 - \lambda & 1 & 0 \\ 1 & -5 - \lambda & 1 \\ 0 & 1 & -4 - \lambda \end{pmatrix}.$$

Therefore, $\det(A - \lambda I) = -\lambda^3 - 13\lambda^2 - 54\lambda - 72 = -(\lambda + 3)(\lambda + 6)(\lambda + 4)$. Thus the eigenvalues are $\lambda = -3, -4, -6$. First, $\lambda = -3$ implies

$$A - \lambda I = \begin{pmatrix} -1 & 1 & 0 \\ 1 & -2 & 1 \\ 0 & 1 & -1 \end{pmatrix} \rightarrow \begin{pmatrix} 1 & -1 & 0 \\ 0 & 1 & -1 \\ 0 & 0 & 0 \end{pmatrix}$$

after elementary row operations. Therefore,

$$\mathbf{v}_1 = \begin{pmatrix} 1 \\ 1 \\ 1 \end{pmatrix}$$

is an eigenvector for $\lambda = -3$, and, consequently,

$$\mathbf{x}_1(t) = e^{-3t} \begin{pmatrix} 1 \\ 1 \\ 1 \end{pmatrix}$$

is a solution of the system. Next, $\lambda = -4$ implies

$$A - \lambda I = \begin{pmatrix} 0 & 1 & 0 \\ 1 & -1 & 1 \\ 0 & 1 & 0 \end{pmatrix} \rightarrow \begin{pmatrix} 1 & -1 & 1 \\ 0 & 1 & 0 \\ 0 & 0 & 0 \end{pmatrix}$$

after elementary row operations. Therefore,

$$\mathbf{v}_2 = \begin{pmatrix} 1 \\ 0 \\ -1 \end{pmatrix}$$

is an eigenvector for $\lambda = -4$, and, consequently,

$$\mathbf{x}_2(t) = e^{-4t} \begin{pmatrix} 1 \\ 0 \\ -1 \end{pmatrix}$$

is a solution of the system. Finally, $\lambda = -6$ implies

$$A - \lambda I = \begin{pmatrix} 2 & 1 & 0 \\ 1 & 1 & 1 \\ 0 & 1 & 2 \end{pmatrix} \rightarrow \begin{pmatrix} 1 & 1/2 & 0 \\ 0 & 1 & 2 \\ 0 & 0 & 0 \end{pmatrix}$$

after elementary row operations. Therefore,

$$\mathbf{v}_3 = \begin{pmatrix} 1 \\ -2 \\ 1 \end{pmatrix}$$

is an eigenvector for $\lambda = -6$, and, consequently,

$$\mathbf{x}_3(t) = e^{-6t} \begin{pmatrix} 1 \\ -2 \\ 1 \end{pmatrix}$$

is a solution of the system. Thus the general solution is

$$\mathbf{x}(t) = c_1 e^{-3t} \begin{pmatrix} 1 \\ 1 \\ 1 \end{pmatrix} + c_2 e^{-4t} \begin{pmatrix} 1 \\ 0 \\ -1 \end{pmatrix} + c_3 e^{-6t} \begin{pmatrix} 1 \\ -2 \\ 1 \end{pmatrix}.$$

3. We will write the system of equations in matrix form as $\mathbf{x}' = A\mathbf{x}$. Here, we have

$$\begin{pmatrix} x_1 \\ x_2 \\ x_3 \end{pmatrix}' = \begin{pmatrix} 2 & -4 & 2 \\ -4 & 2 & -2 \\ 2 & -2 & -1 \end{pmatrix} \begin{pmatrix} x_1 \\ x_2 \\ x_3 \end{pmatrix}.$$

To solve this system, we need to compute the eigenvalues and eigenvectors of A. We have

$$A - \lambda I = \begin{pmatrix} 2-\lambda & -4 & 2 \\ -4 & 2-\lambda & -2 \\ 2 & -2 & -1-\lambda \end{pmatrix}.$$

Therefore, $\det(A - \lambda I) = -\lambda^3 + 3\lambda^2 + 24\lambda + 28 = -(\lambda + 2)^2(\lambda - 7)$. Thus the eigenvalues are $\lambda = -2, 7$. First, $\lambda = -2$ implies

$$A - \lambda I = \begin{pmatrix} 4 & -4 & 2 \\ -4 & 4 & -2 \\ 2 & -2 & 1 \end{pmatrix} \rightarrow \begin{pmatrix} 2 & -2 & 1 \\ 0 & 0 & 0 \\ 0 & 0 & 0 \end{pmatrix}$$

after elementary row operations. Therefore,

$$\mathbf{v}_1 = \begin{pmatrix} 1 \\ 1 \\ 0 \end{pmatrix} \quad \text{and} \quad \mathbf{v}_2 = \begin{pmatrix} 1 \\ 0 \\ -2 \end{pmatrix}$$

are linearly independent eigenvectors for $\lambda = -2$, and, consequently,

$$\mathbf{x}_1(t) = e^{-2t} \begin{pmatrix} 1 \\ 1 \\ 0 \end{pmatrix} \quad \text{and} \quad \mathbf{x}_2(t) = e^{-2t} \begin{pmatrix} 1 \\ 0 \\ -2 \end{pmatrix}$$

are solutions of the system. Next, $\lambda = 7$ implies

$$A - \lambda I = \begin{pmatrix} -5 & -4 & 2 \\ -4 & -5 & -2 \\ 2 & -2 & -8 \end{pmatrix} \rightarrow \begin{pmatrix} 1 & -1 & -4 \\ 0 & 1 & 2 \\ 0 & 0 & 0 \end{pmatrix}$$

after elementary row operations. Therefore,

$$\mathbf{v}_3 = \begin{pmatrix} 2 \\ -2 \\ 1 \end{pmatrix}$$

is an eigenvector for $\lambda = 7$, and, consequently,

$$\mathbf{x}_3(t) = e^{7t} \begin{pmatrix} 2 \\ -2 \\ 1 \end{pmatrix}$$

is a solution of the system. Thus the general solution is

$$\mathbf{x}(t) = c_1 e^{-2t} \begin{pmatrix} 1 \\ 1 \\ 0 \end{pmatrix} + c_2 e^{-2t} \begin{pmatrix} 1 \\ 0 \\ -2 \end{pmatrix} + c_3 e^{7t} \begin{pmatrix} 2 \\ -2 \\ 1 \end{pmatrix}.$$

5. To solve this system, we need to compute the eigenvalues and eigenvectors of A. We have

$$A - \lambda I = \begin{pmatrix} 1 - \lambda & 1 & 6 \\ 1 & 6 - \lambda & 1 \\ 6 & 1 & 1 - \lambda \end{pmatrix}.$$

Therefore, $\det(A - \lambda I) = -\lambda^3 + 8\lambda^2 + 25\lambda - 200 = -(\lambda - 8)(\lambda - 5)(\lambda + 5)$. Therefore, the eigenvalues are $\lambda = -5, 5, 8$. First, $\lambda = -5$ implies

$$A - \lambda I = \begin{pmatrix} 6 & 1 & 6 \\ 1 & 11 & 1 \\ 6 & 1 & 6 \end{pmatrix} \rightarrow \begin{pmatrix} 1 & 0 & 1 \\ 0 & 1 & 0 \\ 0 & 0 & 0 \end{pmatrix}$$

after elementary row operations. Therefore,

$$\mathbf{v}_1 = \begin{pmatrix} 1 \\ 0 \\ -1 \end{pmatrix}$$

is an eigenvector for $\lambda = -5$, and, consequently,

$$\mathbf{x}_1(t) = e^{-5t} \begin{pmatrix} 1 \\ 0 \\ -1 \end{pmatrix}$$

is a solution of the system. Next, $\lambda = 5$ implies

$$A - \lambda I = \begin{pmatrix} -4 & 1 & 6 \\ 1 & 1 & 1 \\ 6 & 1 & -4 \end{pmatrix} \rightarrow \begin{pmatrix} 1 & 0 & -1 \\ 0 & 1 & 2 \\ 0 & 0 & 0 \end{pmatrix}$$

after elementary row operations. Therefore,

$$\mathbf{v}_2 = \begin{pmatrix} 1 \\ -2 \\ 1 \end{pmatrix}$$

is an eigenvector for $\lambda = 5$, and, consequently,

$$\mathbf{x}_2(t) = e^{5t} \begin{pmatrix} 1 \\ -2 \\ 1 \end{pmatrix}$$

is a solution of the system. Finally, $\lambda = 8$ implies

$$A - \lambda I = \begin{pmatrix} -7 & 1 & 6 \\ 1 & -2 & 1 \\ 6 & 1 & -7 \end{pmatrix} \rightarrow \begin{pmatrix} 1 & 0 & -1 \\ 0 & 1 & -1 \\ 0 & 0 & 0 \end{pmatrix}$$

after elementary row operations. Therefore,

$$\mathbf{v}_3 = \begin{pmatrix} 1 \\ 1 \\ 1 \end{pmatrix}$$

is an eigenvector for $\lambda = 8$, and, consequently,

$$\mathbf{x}_3(t) = e^{8t} \begin{pmatrix} 1 \\ 1 \\ 1 \end{pmatrix}$$

is a solution of the system. Thus, the general solution is

$$\mathbf{x}(t) = c_1 e^{-5t} \begin{pmatrix} 1 \\ 0 \\ -1 \end{pmatrix} + c_2 e^{5t} \begin{pmatrix} 1 \\ -2 \\ 1 \end{pmatrix} + c_3 e^{8t} \begin{pmatrix} 1 \\ 1 \\ 1 \end{pmatrix}.$$

7. To solve this system, we need to compute the eigenvalues and eigenvectors of A. We have

$$A - \lambda I = \begin{pmatrix} 1-\lambda & 1 & 1 \\ 2 & 1-\lambda & -1 \\ -8 & -5 & -3-\lambda \end{pmatrix}.$$

Therefore, $\det(A - \lambda I) = -\lambda^3 - \lambda^2 + 4\lambda + 4 = (2 - \lambda)(1 + \lambda)(2 + \lambda)$. Thus the eigenvalues are $\lambda = 2, -1, -2$. First, $\lambda = 2$ implies

$$A - \lambda I = \begin{pmatrix} -1 & 1 & 1 \\ 2 & -1 & -1 \\ -8 & -5 & -5 \end{pmatrix} \rightarrow \begin{pmatrix} 1 & 0 & 0 \\ 0 & 1 & 1 \\ 0 & 0 & 0 \end{pmatrix}$$

after elementary row operations. Therefore,

$$\mathbf{v}_1 = \begin{pmatrix} 0 \\ 1 \\ -1 \end{pmatrix}$$

is an eigenvector for $\lambda = 2$, and, consequently,

$$\mathbf{x}_1(t) = e^{2t} \begin{pmatrix} 0 \\ 1 \\ -1 \end{pmatrix}$$

is a solution of the system. Next, $\lambda = -1$ implies

$$A - \lambda I = \begin{pmatrix} 2 & 1 & 1 \\ 2 & 2 & -1 \\ -8 & -5 & -2 \end{pmatrix} \rightarrow \begin{pmatrix} 2 & 0 & 3 \\ 0 & 1 & -2 \\ 0 & 0 & 0 \end{pmatrix}$$

after elementary row operations. Therefore,

$$\mathbf{v}_2 = \begin{pmatrix} 3 \\ -4 \\ -2 \end{pmatrix}$$

is an eigenvector for $\lambda = -1$, and, consequently,

$$\mathbf{x}_2(t) = e^{-t} \begin{pmatrix} 3 \\ -4 \\ -2 \end{pmatrix}$$

is a solution of the system. Finally, $\lambda = -2$ implies

$$A - \lambda I = \begin{pmatrix} 3 & 1 & 1 \\ 2 & 3 & -1 \\ -8 & -5 & -1 \end{pmatrix} \rightarrow \begin{pmatrix} 7 & 0 & 4 \\ 0 & 7 & -5 \\ 0 & 0 & 0 \end{pmatrix}$$

after elementary row operations. Therefore,

$$\mathbf{v}_3 = \begin{pmatrix} 4 \\ -5 \\ -7 \end{pmatrix}$$

is an eigenvector for $\lambda = -2$, and, consequently,

$$\mathbf{x}_3(t) = e^{-2t} \begin{pmatrix} 4 \\ -5 \\ -7 \end{pmatrix}$$

is a solution of the system. Thus the general solution is

$$\mathbf{x}(t) = c_1 e^{2t} \begin{pmatrix} 0 \\ 1 \\ -1 \end{pmatrix} + c_2 e^{-t} \begin{pmatrix} 3 \\ -4 \\ -2 \end{pmatrix} + c_3 e^{-2t} \begin{pmatrix} 4 \\ -5 \\ -7 \end{pmatrix}.$$

9. We need to find the eigenvalues and eigenvectors.

$$A - \lambda I = \begin{pmatrix} 1-\lambda & 1 & 2 \\ 0 & 2-\lambda & 2 \\ -1 & 1 & 3-\lambda \end{pmatrix}$$

implies $\det(A - \lambda I) = -\lambda^3 + 6\lambda^2 - 11\lambda + 6 = -(\lambda - 1)(\lambda - 2)(\lambda - 3)$. Therefore, the eigenvalues are $\lambda = 1, 2, 3$. First, for $\lambda = 1$,

$$A - \lambda I = \begin{pmatrix} 0 & 1 & 2 \\ 0 & 1 & 2 \\ -1 & 1 & 2 \end{pmatrix} \rightarrow \begin{pmatrix} 1 & 0 & 0 \\ 0 & 1 & 2 \\ 0 & 0 & 0 \end{pmatrix}$$

after elementary row operations. Therefore,

$$\mathbf{v}_1 = \begin{pmatrix} 0 \\ -2 \\ 1 \end{pmatrix}$$

is an eigenvector for $\lambda = 1$ and

$$\mathbf{x}_1(t) = e^t \begin{pmatrix} 0 \\ -2 \\ 1 \end{pmatrix}$$

is a solution of the system. Second, for $\lambda = 2$,

$$A - \lambda I = \begin{pmatrix} -1 & 1 & 2 \\ 0 & 0 & 2 \\ -1 & 1 & 1 \end{pmatrix} \rightarrow \begin{pmatrix} 1 & -1 & 0 \\ 0 & 0 & 1 \\ 0 & 0 & 0 \end{pmatrix}$$

after elementary row operations. Therefore,

$$\mathbf{v}_2 = \begin{pmatrix} 1 \\ 1 \\ 0 \end{pmatrix}$$

is an eigenvector for $\lambda = 2$ and

$$\mathbf{x}_2(t) = e^{2t} \begin{pmatrix} 1 \\ 1 \\ 0 \end{pmatrix}$$

is a solution of the system. Last, for $\lambda = 3$,

$$A - \lambda I = \begin{pmatrix} -2 & 1 & 2 \\ 0 & -1 & 2 \\ -1 & 1 & 0 \end{pmatrix} \rightarrow \begin{pmatrix} 1 & 0 & -2 \\ 0 & 1 & -2 \\ 0 & 0 & 0 \end{pmatrix}$$

after elementary row operations. Therefore,

$$\mathbf{v}_3 = \begin{pmatrix} 2 \\ 2 \\ 1 \end{pmatrix}$$

is an eigenvector for $\lambda = 3$ and

$$\mathbf{x}_3(t) = e^{3t} \begin{pmatrix} 2 \\ 2 \\ 1 \end{pmatrix}$$

is a solution of the system. Thus the general solution is

$$\mathbf{x}(t) = c_1 e^t \begin{pmatrix} 0 \\ -2 \\ 1 \end{pmatrix} + c_2 e^{2t} \begin{pmatrix} 1 \\ 1 \\ 0 \end{pmatrix} + c_3 e^{3t} \begin{pmatrix} 2 \\ 2 \\ 1 \end{pmatrix}.$$

The initial condition implies

$$\mathbf{x}(0) = c_1 \begin{pmatrix} 0 \\ -2 \\ 1 \end{pmatrix} + c_2 \begin{pmatrix} 1 \\ 1 \\ 0 \end{pmatrix} + c_3 \begin{pmatrix} 2 \\ 2 \\ 1 \end{pmatrix} = \begin{pmatrix} 11 \\ 1 \\ 5 \end{pmatrix}.$$

The solution of this equation is $c_1 = 5$, $c_2 = 11$ and $c_3 = 0$. Therefore, the solution is

$$\mathbf{x}(t) = 5e^t \begin{pmatrix} 0 \\ -2 \\ 1 \end{pmatrix} + 11e^{2t} \begin{pmatrix} 1 \\ 1 \\ 0 \end{pmatrix}.$$

The solution $\mathbf{x}$ satisfies $|\mathbf{x}| \to \infty$ as $t \to \infty$.

11. We need to find the eigenvalues and eigenvectors.

$$A - \lambda I = \begin{pmatrix} -1-\lambda & 0 & 3 \\ 0 & -2-\lambda & 0 \\ 3 & 0 & -1-\lambda \end{pmatrix}$$

implies $\det(A - \lambda I) = -(\lambda+2)(\lambda^2+2\lambda-8) = -(\lambda+2)(\lambda+4)(\lambda-2)$. Thus the eigenvalues are $\lambda = -2, -4, 2$. First, for $\lambda = -2$,

$$A - \lambda I = \begin{pmatrix} 1 & 0 & 3 \\ 0 & 0 & 0 \\ 3 & 0 & 1 \end{pmatrix} \to \begin{pmatrix} 1 & 0 & 3 \\ 0 & 0 & 1 \\ 0 & 0 & 0 \end{pmatrix}$$

after elementary row operations. Therefore,

$$\mathbf{v}_1 = \begin{pmatrix} 0 \\ 1 \\ 0 \end{pmatrix}$$

is an eigenvector for $\lambda = -2$ and

$$\mathbf{x}_1(t) = e^{-2t} \begin{pmatrix} 0 \\ 1 \\ 0 \end{pmatrix}$$

is a solution of the system. Second, for $\lambda = -4$,

$$A - \lambda I = \begin{pmatrix} 3 & 0 & 3 \\ 0 & 2 & 0 \\ 3 & 0 & 3 \end{pmatrix} \to \begin{pmatrix} 1 & 0 & 1 \\ 0 & 1 & 0 \\ 0 & 0 & 0 \end{pmatrix}$$

after elementary row operations. Therefore,

$$\mathbf{v}_2 = \begin{pmatrix} 1 \\ 0 \\ -1 \end{pmatrix}$$

is an eigenvector for $\lambda = -4$ and

$$\mathbf{x}_2(t) = e^{-4t} \begin{pmatrix} 1 \\ 0 \\ -1 \end{pmatrix}$$

is a solution of the system. Last, for $\lambda = 2$,

$$A - \lambda I = \begin{pmatrix} -3 & 0 & 3 \\ 0 & -4 & 0 \\ 3 & 0 & -3 \end{pmatrix} \to \begin{pmatrix} 1 & 0 & -1 \\ 0 & 1 & 0 \\ 0 & 0 & 0 \end{pmatrix}$$

after elementary row operations. Therefore,

$$\mathbf{v}_3 = \begin{pmatrix} 1 \\ 0 \\ 1 \end{pmatrix}$$

is an eigenvector for $\lambda = 2$ and

$$\mathbf{x}_3(t) = e^{2t} \begin{pmatrix} 1 \\ 0 \\ 1 \end{pmatrix}$$

is a solution of the system. Thus the general solution is

$$\mathbf{x}(t) = c_1 e^{-2t} \begin{pmatrix} 0 \\ 1 \\ 0 \end{pmatrix} + c_2 e^{-4t} \begin{pmatrix} 1 \\ 0 \\ -1 \end{pmatrix} + c_3 e^{2t} \begin{pmatrix} 1 \\ 0 \\ 1 \end{pmatrix}.$$

The initial condition implies

$$\mathbf{x}(0) = c_1 \begin{pmatrix} 0 \\ 1 \\ 0 \end{pmatrix} + c_2 \begin{pmatrix} 1 \\ 0 \\ -1 \end{pmatrix} + c_3 \begin{pmatrix} 1 \\ 0 \\ 1 \end{pmatrix} = \begin{pmatrix} 2 \\ -1 \\ -2 \end{pmatrix}.$$

The solution of this equation is $c_1 = -1$, $c_2 = 2$ and $c_3 = 0$. Therefore, the solution is

$$\mathbf{x}(t) = e^{-2t} \begin{pmatrix} 0 \\ -1 \\ 0 \end{pmatrix} + 2e^{-4t} \begin{pmatrix} 1 \\ 0 \\ -1 \end{pmatrix}.$$

The solution tends to the origin approaching the eigenvector $(0, -1, 0)^T$ as $t \to \infty$.

13. The system is given by $\mathbf{x}' = kA\mathbf{x}$, where

$$\mathbf{x} = \begin{pmatrix} x_1 \\ x_2 \\ x_3 \\ x_4 \end{pmatrix}$$

and

$$A = \begin{pmatrix} -1 & 1 & 0 & 0 \\ 1 & -2 & 1 & 0 \\ 0 & 1 & -2 & 1 \\ 0 & 0 & 1 & -1 \end{pmatrix}.$$

To find the general solution, we look for the eigenvalues of A. We see that

$$A - \lambda I = \begin{pmatrix} -1-\lambda & 1 & 0 & 0 \\ 1 & -2-\lambda & 1 & 0 \\ 0 & 1 & -2-\lambda & 1 \\ 0 & 0 & 1 & -1-\lambda \end{pmatrix}.$$

Therefore, $\det(A - \lambda I) = \lambda^4 + 6\lambda^3 + 10\lambda^2 + 4\lambda = \lambda(\lambda+2)(\lambda^2+4\lambda+2)$. Thus the eigenvalues are $\lambda = 0, -2, -2 \pm \sqrt{2}$. First, $\lambda = 0$ implies

$$A - \lambda I = \begin{pmatrix} -1 & 1 & 0 & 0 \\ 1 & -2 & 1 & 0 \\ 0 & 1 & -2 & 1 \\ 0 & 0 & 1 & -1 \end{pmatrix} \to \begin{pmatrix} 1 & -1 & 0 & 0 \\ 0 & 1 & -1 & 0 \\ 0 & 0 & 1 & -1 \\ 0 & 0 & 0 & 0 \end{pmatrix}$$

after elementary row operations. Therefore,

$$\mathbf{v}_1 = \begin{pmatrix} 1 \\ 1 \\ 1 \\ 1 \end{pmatrix}$$

is an eigenvector for $\lambda = 0$ and

$$\mathbf{x}_1(t) = \begin{pmatrix} 1 \\ 1 \\ 1 \\ 1 \end{pmatrix}$$

is a solution of the system. Second, $\lambda = -2$ implies

$$A - \lambda I = \begin{pmatrix} 1 & 1 & 0 & 0 \\ 1 & 0 & 1 & 0 \\ 0 & 1 & 0 & 1 \\ 0 & 0 & 1 & 1 \end{pmatrix} \rightarrow \begin{pmatrix} 1 & 1 & 0 & 0 \\ 0 & 1 & -1 & 0 \\ 0 & 0 & 1 & 1 \\ 0 & 0 & 0 & 0 \end{pmatrix}.$$

Therefore,

$$\mathbf{v}_2 = \begin{pmatrix} 1 \\ -1 \\ -1 \\ 1 \end{pmatrix}$$

is an eigenvector for $\lambda = -2$ and

$$\mathbf{x}_2(t) = e^{-2t} \begin{pmatrix} 1 \\ -1 \\ -1 \\ 1 \end{pmatrix}$$

is a solution of the system. Third, $\lambda = -2 + \sqrt{2}$ implies

$$A - \lambda I = \begin{pmatrix} 1-\sqrt{2} & 1 & 0 & 0 \\ 1 & -\sqrt{2} & 1 & 0 \\ 0 & 1 & -\sqrt{2} & 1 \\ 0 & 0 & 1 & 1-\sqrt{2} \end{pmatrix} \rightarrow \begin{pmatrix} 1 & -\sqrt{2} & 1 & 0 \\ 0 & 1 & 1 & 0 \\ 0 & 0 & 1 & 1-\sqrt{2} \\ 0 & 0 & 0 & 0 \end{pmatrix}.$$

Therefore,

$$\mathbf{v}_3 = \begin{pmatrix} 1 \\ -1+\sqrt{2} \\ 1-\sqrt{2} \\ -1 \end{pmatrix}$$

is an eigenvector for $\lambda = -2 + \sqrt{2}$ and

$$\mathbf{x}_3(t) = e^{(-2+\sqrt{2})t} \begin{pmatrix} 1 \\ -1+\sqrt{2} \\ 1-\sqrt{2} \\ -1 \end{pmatrix}$$

is a solution of the system. Fourth, $\lambda = -2 - \sqrt{2}$ implies

$$A - \lambda I = \begin{pmatrix} 1 + \sqrt{2} & 1 & 0 & 0 \\ 1 & \sqrt{2} & 1 & 0 \\ 0 & 1 & \sqrt{2} & 1 \\ 0 & 0 & 1 & 1 + \sqrt{2} \end{pmatrix} \rightarrow \begin{pmatrix} 1 & \sqrt{2} & 1 & 0 \\ 0 & 1 & 1 & 0 \\ 0 & 0 & 1 & 1 + \sqrt{2} \\ 0 & 0 & 0 & 0 \end{pmatrix}.$$

Therefore,

$$\mathbf{v}_4 = \begin{pmatrix} 1 \\ -1 - \sqrt{2} \\ 1 + \sqrt{2} \\ -1 \end{pmatrix}$$

is an eigenvector for $\lambda = -2 - \sqrt{2}$ and

$$\mathbf{x}_4(t) = e^{(-2-\sqrt{2})t} \begin{pmatrix} 1 \\ -1 - \sqrt{2} \\ 1 + \sqrt{2} \\ -1 \end{pmatrix}$$

is a solution of the system. Thus the general solution is

$$\mathbf{x}(t) = c_1 \begin{pmatrix} 1 \\ 1 \\ 1 \\ 1 \end{pmatrix} + c_2 e^{-2t} \begin{pmatrix} 1 \\ -1 \\ -1 \\ 1 \end{pmatrix} + c_3 e^{(-2+\sqrt{2})t} \begin{pmatrix} 1 \\ -1 + \sqrt{2} \\ 1 - \sqrt{2} \\ -1 \end{pmatrix} + c_4 e^{(-2-\sqrt{2})t} \begin{pmatrix} 1 \\ -1 - \sqrt{2} \\ 1 + \sqrt{2} \\ -1 \end{pmatrix}$$

Below is a graph of the components of the eigenvectors of A.

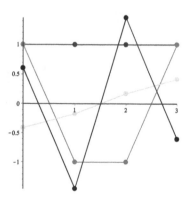

As $t \to \infty$, the solution approaches the equilibrium state $(1,1,1,1)^T$. The eigenvalue $\lambda_3 = -2 + \sqrt{2}$ controls the long term decay rate towards equilibrium.

15. We can see that

$$A - \lambda I = \begin{pmatrix} 2 - \lambda & -4 & -3 \\ 3 & -5 - \lambda & -3 \\ -2 & 2 & 1 - \lambda \end{pmatrix}.$$

Therefore, $\det(A-\lambda I) = -\lambda^3 - 2\lambda^2 + \lambda + 2 = -(\lambda+2)(\lambda+1)(\lambda-1)$. Thus the eigenvalues are given by $\lambda = -2, -1, 1$. The corresponding eigenvectors are given as follows. For $\lambda_1 = -2$,

$$A - \lambda I = \begin{pmatrix} 4 & -4 & -3 \\ 3 & -3 & -3 \\ -2 & 2 & 3 \end{pmatrix} \rightarrow \begin{pmatrix} 1 & -1 & -1 \\ 0 & 0 & 1 \\ 0 & 0 & 0 \end{pmatrix}$$

after elementary row operations. Therefore,

$$\mathbf{v}_1 = \begin{pmatrix} 1 \\ 1 \\ 0 \end{pmatrix}.$$

For $\lambda_2 = -1$,

$$A - \lambda I = \begin{pmatrix} 3 & -4 & -3 \\ 3 & -4 & -3 \\ -2 & 2 & 2 \end{pmatrix} \rightarrow \begin{pmatrix} 1 & -1 & -1 \\ 0 & 1 & 0 \\ 0 & 0 & 0 \end{pmatrix}$$

after elementary row operations. Therefore,

$$\mathbf{v}_2 = \begin{pmatrix} 1 \\ 0 \\ 1 \end{pmatrix}.$$

For $\lambda_3 = 1$,

$$A - \lambda I = \begin{pmatrix} 1 & -4 & -3 \\ 3 & -6 & -3 \\ -2 & 2 & 0 \end{pmatrix} \rightarrow \begin{pmatrix} 1 & -4 & -3 \\ 0 & 1 & 1 \\ 0 & 0 & 0 \end{pmatrix}$$

after elementary row operations. Therefore,

$$\mathbf{v}_3 = \begin{pmatrix} 1 \\ 1 \\ -1 \end{pmatrix}.$$

Thus the general solution is

$$\mathbf{x}(t) = c_1 e^{-2t} \begin{pmatrix} 1 \\ 1 \\ 0 \end{pmatrix} + c_2 e^{-t} \begin{pmatrix} 1 \\ 0 \\ 1 \end{pmatrix} + c_3 e^{t} \begin{pmatrix} 1 \\ 1 \\ -1 \end{pmatrix}.$$

If we want the solution to tend to $(0,0,0)^T$ as $t \to \infty$, we need $c_3 = 0$. That is, we need the initial condition $\mathbf{x}_0$ to satisfy

$$\mathbf{x}_0 = c_1 \begin{pmatrix} 1 \\ 1 \\ 0 \end{pmatrix} + c_2 \begin{pmatrix} 1 \\ 0 \\ 1 \end{pmatrix}.$$

Therefore, letting

$$\mathbf{u}_1 = \begin{pmatrix} 1 \\ 1 \\ 0 \end{pmatrix} \quad \text{and} \quad \mathbf{u}_2 = \begin{pmatrix} 1 \\ 0 \\ 1 \end{pmatrix},$$

and letting

$$S = \{\mathbf{u} : \mathbf{u} = a_1\mathbf{u}_1 + a_2\mathbf{u}_2, \ -\infty < a_1, a_2 < \infty\},$$

then for any $\mathbf{x}_0 \in S$, the solution $\mathbf{x}(t) \to (0,0,0)^T$ as $t \to \infty$.

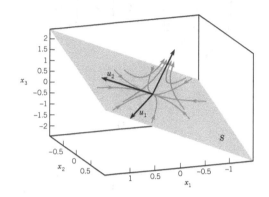

If $\mathbf{x}_0 \notin S$, then $\mathbf{x}(t)$ approaches the line determined by $\mathbf{v}_3 = (1, 1, -1)^T$ as $t \to \infty$.

17. The block diagram is given by

$$R_1 \xrightarrow{k_1} R_2 \xrightarrow{k_2} R_3$$

Writing the system in matrix form, we have

$$\begin{pmatrix} m_1 \\ m_2 \\ m_3 \end{pmatrix}' = \begin{pmatrix} -k_1 & 0 & 0 \\ k_1 & -k_2 & 0 \\ 0 & k_2 & 0 \end{pmatrix} \begin{pmatrix} m_1 \\ m_2 \\ m_3 \end{pmatrix}.$$

Then

$$A - \lambda I = \begin{pmatrix} -k_1 - \lambda & 0 & 0 \\ k_1 & -k_2 - \lambda & 0 \\ 0 & k_2 & -\lambda \end{pmatrix}$$

implies $\det(A - \lambda I) = -\lambda(\lambda + k_1)(\lambda + k_2)$. Therefore, $\lambda = 0, -k_1, -k_2$. We will first consider the case when $k_1 \neq k_2$. First, for $\lambda_1 = 0$,

$$A - \lambda I = \begin{pmatrix} -k_1 & 0 & 0 \\ k_1 & -k_2 & 0 \\ 0 & k_2 & 0 \end{pmatrix} \to \begin{pmatrix} 1 & 0 & 0 \\ 0 & 1 & 0 \\ 0 & 0 & 0 \end{pmatrix}$$

after elementary row operations. Therefore,

$$\mathbf{v}_1 = \begin{pmatrix} 0 \\ 0 \\ 1 \end{pmatrix}.$$

Next, for $\lambda_2 = -k_1$,

$$A - \lambda I = \begin{pmatrix} 0 & 0 & 0 \\ k_1 & k_1 - k_2 & 0 \\ 0 & k_2 & k_1 \end{pmatrix} \to \begin{pmatrix} k_1 & k_1 - k_2 & 0 \\ 0 & k_2 & k_1 \\ 0 & 0 & 0 \end{pmatrix}$$

after elementary row operations. Therefore,

$$\mathbf{v}_2 = \begin{pmatrix} k_2 - k_1 \\ k_1 \\ -k_2 \end{pmatrix}.$$

Third, for $\lambda_3 = -k_2$,

$$A - \lambda I = \begin{pmatrix} k_2 - k_1 & 0 & 0 \\ k_1 & 0 & 0 \\ 0 & k_2 & k_2 \end{pmatrix} \rightarrow \begin{pmatrix} 1 & 0 & 0 \\ 0 & 1 & 1 \\ 0 & 0 & 0 \end{pmatrix}$$

after elementary row operations. Therefore,

$$\mathbf{v}_3 = \begin{pmatrix} 0 \\ 1 \\ -1 \end{pmatrix}.$$

Therefore, the general solution is given by

$$\mathbf{x}(t) = c_1 \begin{pmatrix} 0 \\ 0 \\ 1 \end{pmatrix} + c_2 e^{-k_1 t} \begin{pmatrix} k_2 - k_1 \\ k_1 \\ -k_2 \end{pmatrix} + c_3 e^{-k_2 t} \begin{pmatrix} 0 \\ 1 \\ -1 \end{pmatrix}.$$

Using the initial conditions, we have

$$\mathbf{x}(0) = c_1 \begin{pmatrix} 0 \\ 0 \\ 1 \end{pmatrix} + c_2 \begin{pmatrix} k_2 - k_1 \\ k_1 \\ -k_2 \end{pmatrix} + c_3 \begin{pmatrix} 0 \\ 1 \\ -1 \end{pmatrix} = \begin{pmatrix} m_0 \\ 0 \\ 0 \end{pmatrix}.$$

Solving this system, we have $c_1 = m_0$, $c_2 = m_0/(k_2 - k - 1)$ and $c_3 = -m_0 k_1/(k_2 - k_1)$. Therefore, for $k_1 \neq k_2$, the solution is given by

$$\mathbf{x}(t) = m_0 \begin{pmatrix} 0 \\ 0 \\ 1 \end{pmatrix} + \frac{m_0}{k_2 - k_1} e^{-k_1 t} \begin{pmatrix} k_2 - k_1 \\ k_1 \\ -k_2 \end{pmatrix} - \frac{m_0 k_1}{k_2 - k_1} e^{-k_2 t} \begin{pmatrix} 0 \\ 1 \\ -1 \end{pmatrix}.$$

Now, let us consider the case when $k_1 = k_2$. As above, for $\lambda_1 = 0$, we have

$$\mathbf{v}_1 = \begin{pmatrix} 0 \\ 0 \\ 1 \end{pmatrix}$$

and, therefore,

$$\mathbf{x}_1(t) = \begin{pmatrix} 0 \\ 0 \\ 1 \end{pmatrix}$$

is one solution of the system. Now in this case, $\lambda_2 = -k_1$ is an eigenvalue with multiplicity two. As above, for $\lambda_2 = -k_1$,

$$A - \lambda I = \begin{pmatrix} 0 & 0 & 0 \\ k_1 & k_1 - k_2 & 0 \\ 0 & k_2 & k_1 \end{pmatrix} = \begin{pmatrix} 0 & 0 & 0 \\ k_1 & 0 & 0 \\ 0 & k_1 & k_1 \end{pmatrix} \rightarrow \begin{pmatrix} 1 & 0 & 0 \\ 0 & 1 & 1 \\ 0 & 0 & 0 \end{pmatrix}.$$

Therefore,

$$\mathbf{v}_2 = \begin{pmatrix} 0 \\ 1 \\ -1 \end{pmatrix}$$

is an associated eigenvector and

$$\mathbf{x}_2(t) = e^{-k_1 t} \begin{pmatrix} 0 \\ 1 \\ -1 \end{pmatrix}.$$

To find another solution associated with λ_2, we need to find $\mathbf{w}$ satisfying $(A - \lambda I)\mathbf{w} = \mathbf{v}_2$. That is, we need to find $\mathbf{w}$ satisfying

$$\begin{pmatrix} 0 & 0 & 0 \\ k_1 & 0 & 0 \\ 0 & k_1 & k_1 \end{pmatrix} \mathbf{w} = \begin{pmatrix} 0 \\ 1 \\ -1 \end{pmatrix}.$$

We see that

$$\mathbf{w} = \begin{pmatrix} 1/k_1 \\ 0 \\ -1/k_1 \end{pmatrix}$$

is such a vector. Therefore,

$$\mathbf{x}_3(t) = t e^{-k_1 t} \begin{pmatrix} 0 \\ 1 \\ -1 \end{pmatrix} + e^{-k_1 t} \begin{pmatrix} 1/k_1 \\ 0 \\ -1/k_1 \end{pmatrix}$$

is another solution of the system. Therefore, for $k_1 = k_2$, the general solution is given by

$$\mathbf{x}(t) = c_1 \begin{pmatrix} 0 \\ 0 \\ 1 \end{pmatrix} + c_2 e^{-k_1 t} \begin{pmatrix} 0 \\ 1 \\ -1 \end{pmatrix} + c_3 e^{-k_1 t} \begin{pmatrix} 1 \\ k_1 t \\ -k_1 t - 1 \end{pmatrix}.$$

Using our initial conditions, we have

$$\mathbf{x}(0) = c_1 \begin{pmatrix} 0 \\ 0 \\ 1 \end{pmatrix} + c_2 \begin{pmatrix} 0 \\ 1 \\ -1 \end{pmatrix} + c_3 \begin{pmatrix} 1 \\ 0 \\ -1 \end{pmatrix} = \begin{pmatrix} m_0 \\ 0 \\ 0 \end{pmatrix}.$$

The solution of this system is $c_1 = m_0$, $c_2 = 0$ and $c_3 = m_0$. Therefore, for $k_1 = k_2$, the solution of our equation is

$$\mathbf{x}(t) = m_0 \begin{pmatrix} 0 \\ 0 \\ 1 \end{pmatrix} + m_0 e^{-k_1 t} \begin{pmatrix} 1 \\ k_1 t \\ -k_1 t - 1 \end{pmatrix}.$$

19. The eigenvalues are given by $\lambda = -3, 3$. First, $\lambda_1 = -3$ has multiplicity 1 and an associated eigenvector

$$\mathbf{v}_1 = \begin{pmatrix} 1 \\ -2 \\ 0 \\ 1 \end{pmatrix}.$$

Second, $\lambda_2 = 3$ has multiplicity 3 with 3 associated eigenvectors,

$$\mathbf{v}_2 = \begin{pmatrix} -1 \\ 0 \\ 0 \\ 1 \end{pmatrix}, \quad \mathbf{v}_3 = \begin{pmatrix} 0 \\ 0 \\ 1 \\ 0 \end{pmatrix}, \quad \mathbf{v}_4 = \begin{pmatrix} 2 \\ 1 \\ 0 \\ 0 \end{pmatrix}.$$

Therefore, a fundamental set of solutions is given by

$$\left\{ e^{-3t} \begin{pmatrix} 1 \\ -2 \\ 0 \\ 1 \end{pmatrix}, e^{3t} \begin{pmatrix} -1 \\ 0 \\ 0 \\ 1 \end{pmatrix}, e^{3t} \begin{pmatrix} 0 \\ 0 \\ 1 \\ 0 \end{pmatrix}, e^{3t} \begin{pmatrix} 2 \\ 1 \\ 0 \\ 0 \end{pmatrix} \right\}.$$

21. The eigenvalues are given by $\lambda = -2, 2, -4, 4$. The associated eigenvectors are given by

$$\mathbf{v}_1 = \begin{pmatrix} -1 \\ 0 \\ 1 \\ 0 \end{pmatrix}, \mathbf{v}_2 = \begin{pmatrix} -1 \\ -1 \\ 1 \\ 1 \end{pmatrix}, \mathbf{v}_3 = \begin{pmatrix} 1 \\ 0 \\ 0 \\ 1 \end{pmatrix}, \mathbf{v}_4 = \begin{pmatrix} 0 \\ 1 \\ 0 \\ 1 \end{pmatrix},$$

respectively. Therefore, a fundamental set of solutions is given by

$$\left\{ e^{-2t} \begin{pmatrix} -1 \\ 0 \\ 1 \\ 0 \end{pmatrix}, e^{2t} \begin{pmatrix} -1 \\ -1 \\ 1 \\ 1 \end{pmatrix}, e^{-4t} \begin{pmatrix} 1 \\ 0 \\ 0 \\ 1 \end{pmatrix}, e^{4t} \begin{pmatrix} 0 \\ 1 \\ 0 \\ 1 \end{pmatrix} \right\}.$$

23. The eigenvalues are given by $\lambda = -2, -1, 1, 2$. First, the eigenvalues $\lambda = -2$ has multiplicity 2 and 2 associated eigenvectors

$$\mathbf{v}_1 = \begin{pmatrix} 0 \\ 1 \\ 0 \\ 1 \\ 0 \end{pmatrix}, \mathbf{v}_2 = \begin{pmatrix} 1 \\ 0 \\ 2 \\ 0 \\ 1 \end{pmatrix}.$$

All other eigenvalues have multiplicity 1. Eigenvectors for $\lambda = -1, 1, 2$ are given by

$$\mathbf{v}_3 = \begin{pmatrix} 1 \\ 0 \\ 0 \\ -1 \\ 1 \end{pmatrix}, \mathbf{v}_4 = \begin{pmatrix} 2 \\ 0 \\ 0 \\ 0 \\ 1 \end{pmatrix}, \mathbf{v}_5 = \begin{pmatrix} 1 \\ 1 \\ -1 \\ 1 \\ 0 \end{pmatrix},$$

respectively. Therefore, a fundamental set of solutions is given by

$$\left\{e^{-2t}\begin{pmatrix}0\\1\\0\\1\\0\end{pmatrix},\ e^{-2t}\begin{pmatrix}1\\0\\2\\0\\1\end{pmatrix},\ e^{-t}\begin{pmatrix}1\\0\\0\\-1\\1\end{pmatrix},\ e^{t}\begin{pmatrix}2\\0\\0\\0\\1\end{pmatrix},\ e^{2t}\begin{pmatrix}1\\1\\-1\\1\\0\end{pmatrix}\right\}.$$

6.4 Nondefective Matrices with Complex Eigenvalues

1. We find the eigenvalues first.

$$A-\lambda I=\begin{pmatrix}-2-\lambda & 2 & 1\\ -2 & 2-\lambda & 2\\ 2 & -3 & -3-\lambda\end{pmatrix}$$

implies $\det(A-\lambda I)=-\lambda^3-3\lambda^2-4\lambda-2=-(\lambda+1)(\lambda^2+2\lambda+2)$. Therefore, the eigenvalues are given by $\lambda=-1$ and $\lambda=-1\pm i$. First, for $\lambda=-1$, we have

$$A-\lambda I=\begin{pmatrix}-1 & 2 & 1\\ -2 & 3 & 2\\ 2 & -3 & -2\end{pmatrix}\rightarrow\begin{pmatrix}1 & -2 & -1\\ 0 & 1 & 0\\ 0 & 0 & 0\end{pmatrix}$$

after elementary row operations. Therefore,

$$\mathbf{v}_1=\begin{pmatrix}1\\0\\1\end{pmatrix}$$

is an associated eigenvector, and

$$\mathbf{x}_1(t)=e^{-t}\begin{pmatrix}1\\0\\1\end{pmatrix}$$

is a solution of the system. Next, for $\lambda=-1+i$,

$$A-\lambda I=\begin{pmatrix}-1-i & 2 & 1\\ -2 & 3-i & 2\\ 2 & -3 & -2-i\end{pmatrix}\rightarrow\begin{pmatrix}1 & 0 & 1/2-i/2\\ 0 & 1 & 1\\ 0 & 0 & 0\end{pmatrix}$$

after elementary row operations. Therefore,

$$\mathbf{v}_2=\begin{pmatrix}-1+i\\-2\\2\end{pmatrix}$$

is an associated eigenvector. Further,

$$\mathbf{u}(t) = e^{(-1+i)t}\left[\begin{pmatrix} -1 \\ -2 \\ 2 \end{pmatrix} + i\begin{pmatrix} 1 \\ 0 \\ 0 \end{pmatrix}\right]$$

is a solution of the system. We know that if $\mathbf{u}(t)$ is a solution, then $\mathrm{Re}(\mathbf{u})$ and $\mathrm{Im}(\mathbf{u})$ are also solutions. Consequently, we get the following two linearly independent solutions.

$$\mathbf{x}_2(t) = \mathrm{Re}(\mathbf{u}) = e^{-t}\begin{pmatrix} -\cos t - \sin t \\ -2\cos t \\ 2\cos t \end{pmatrix}$$

and

$$\mathbf{x}_3(t) = \mathrm{Im}(\mathbf{u}) = e^{-t}\begin{pmatrix} -\sin t + \cos t \\ -2\sin t \\ 2\sin t \end{pmatrix}.$$

We conclude that the general solution is given by

$$\mathbf{x}(t) = c_1 e^{-t}\begin{pmatrix} 1 \\ 0 \\ 1 \end{pmatrix} + c_2 e^{-t}\begin{pmatrix} -\cos t - \sin t \\ -2\cos t \\ 2\cos t \end{pmatrix} + c_3 e^{-t}\begin{pmatrix} -\sin t + \cos t \\ -2\sin t \\ 2\sin t \end{pmatrix}.$$

3. We find the eigenvalues first.

$$A - \lambda I = \begin{pmatrix} -\lambda & -2 & -1 \\ 1 & -1-\lambda & 1 \\ 1 & -2 & -2-\lambda \end{pmatrix}$$

implies $\det(A - \lambda I) = -\lambda^3 - 3\lambda^2 - 7\lambda - 5 = -(\lambda+1)(\lambda^2 + 2\lambda + 5)$. Therefore, the eigenvalues are given by $\lambda = -1$ and $\lambda = -1 \pm 2i$. First, for $\lambda = -1$, we have

$$A - \lambda I = \begin{pmatrix} 1 & -2 & -1 \\ 1 & 0 & 1 \\ 1 & -2 & -1 \end{pmatrix} \rightarrow \begin{pmatrix} 1 & -2 & -1 \\ 0 & 1 & 1 \\ 0 & 0 & 0 \end{pmatrix}$$

after elementary row operations. Therefore,

$$\mathbf{v}_1 = \begin{pmatrix} 1 \\ 1 \\ -1 \end{pmatrix}$$

is an associated eigenvector, and

$$\mathbf{x}_1(t) = e^{-t}\begin{pmatrix} 1 \\ 1 \\ -1 \end{pmatrix}$$

is a solution of the system. Next, for $\lambda = -1 + 2i$,

$$A - \lambda I = \begin{pmatrix} 1 - 2i & -2 & -1 \\ 1 & -2i & 1 \\ 1 & -2 & -1 - 2i \end{pmatrix} \rightarrow \begin{pmatrix} 1 & -2i & 1 \\ 0 & 1 & i \\ 0 & 0 & 0 \end{pmatrix}$$

after elementary row operations. Therefore,

$$\mathbf{v}_2 = \begin{pmatrix} 1 \\ -i \\ 1 \end{pmatrix}$$

is an associated eigenvector. Further,

$$\mathbf{u}(t) = e^{(-1+2i)t} \left[\begin{pmatrix} 1 \\ 0 \\ 1 \end{pmatrix} + i \begin{pmatrix} 0 \\ -1 \\ 0 \end{pmatrix} \right]$$

is a solution of the system. We know that if $\mathbf{u}(t)$ is a solution, then $\mathrm{Re}(\mathbf{u})$ and $\mathrm{Im}(\mathbf{u})$ are also solutions. Consequently, we get the following two linearly independent solutions.

$$\mathbf{x}_2(t) = \mathrm{Re}(\mathbf{u}) = e^{-t} \begin{pmatrix} \cos 2t \\ \sin 2t \\ \cos 2t \end{pmatrix}$$

and

$$\mathbf{x}_3(t) = \mathrm{Im}(\mathbf{u}) = e^{-t} \begin{pmatrix} \sin 2t \\ -\cos 2t \\ \sin 2t \end{pmatrix}.$$

We conclude that the general solution is given by

$$\mathbf{x}(t) = c_1 e^{-t} \begin{pmatrix} 1 \\ 1 \\ -1 \end{pmatrix} + c_2 e^{-t} \begin{pmatrix} \cos 2t \\ \sin 2t \\ \cos 2t \end{pmatrix} + c_3 e^{-t} \begin{pmatrix} \sin 2t \\ -\cos 2t \\ \sin 2t \end{pmatrix}.$$

5. We find the eigenvalues first.

$$A - \lambda I = \begin{pmatrix} -7 - \lambda & 6 & -6 \\ -9 & 5 - \lambda & -9 \\ 0 & -1 & -1 - \lambda \end{pmatrix}$$

implies $\det(A - \lambda I) = -\lambda^3 - 3\lambda^2 - 12\lambda - 10 = -(1 + \lambda)(\lambda^2 + 2\lambda + 10)$. Therefore, the eigenvalues are given by $\lambda = -1$ and $\lambda = -1 \pm 3i$. First, for $\lambda = -1$, we have

$$A - \lambda I = \begin{pmatrix} -6 & 6 & -6 \\ -9 & 6 & -9 \\ 0 & -1 & 0 \end{pmatrix} \rightarrow \begin{pmatrix} 1 & 0 & 1 \\ 0 & 1 & 0 \\ 0 & 0 & 0 \end{pmatrix}$$

after elementary row operations. Therefore,

$$\mathbf{v}_1 = \begin{pmatrix} 1 \\ 0 \\ -1 \end{pmatrix}$$

is an associated eigenvector, and

$$\mathbf{x}_1(t) = e^{-t} \begin{pmatrix} 1 \\ 0 \\ -1 \end{pmatrix}$$

is a solution of the system. Next, for $\lambda = -1 + 3i$,

$$A - \lambda I = \begin{pmatrix} -6 - 3i & 6 & -6 \\ -9 & 6 - 3i & -9 \\ 0 & -1 & -3i \end{pmatrix} \rightarrow \begin{pmatrix} 1 & 0 & 2 + 2i \\ 0 & 1 & 3i \\ 0 & 0 & 0 \end{pmatrix}$$

after elementary row operations. Therefore,

$$\mathbf{v}_2 = \begin{pmatrix} 2 + 2i \\ 3i \\ -1 \end{pmatrix}$$

is an associated eigenvector. Further,

$$\mathbf{u}(t) = e^{(-1+3i)t} \left[\begin{pmatrix} 2 \\ 0 \\ -1 \end{pmatrix} + i \begin{pmatrix} 2 \\ 3 \\ 0 \end{pmatrix} \right]$$

is a solution of the system. We know that if $\mathbf{u}(t)$ is a solution, then $\mathrm{Re}(\mathbf{u})$ and $\mathrm{Im}(\mathbf{u})$ are also solutions. Consequently, we get the following two linearly independent solutions.

$$\mathbf{x}_2(t) = \mathrm{Re}(\mathbf{u}) = e^{-t} \begin{pmatrix} 2\cos 3t - 2\sin 3t \\ -3\sin 3t \\ -\cos 3t \end{pmatrix}$$

and

$$\mathbf{x}_3(t) = \mathrm{Im}(\mathbf{u}) = e^{-t} \begin{pmatrix} 2\sin 3t + 2\cos 3t \\ 3\cos 3t \\ -\sin 3t \end{pmatrix}.$$

We conclude that the general solution is given by

$$\mathbf{x}(t) = c_1 e^{-t} \begin{pmatrix} 1 \\ 0 \\ -1 \end{pmatrix} + c_2 e^{-t} \begin{pmatrix} 2\cos 3t - 2\sin 3t \\ -3\sin 3t \\ -\cos 3t \end{pmatrix} + c_3 e^{-t} \begin{pmatrix} 2\sin 3t + 2\cos 3t \\ 3\cos 3t \\ -\sin 3t \end{pmatrix}.$$

7. We find the eigenvalues first.

$$A - \lambda I = \begin{pmatrix} 1 - \lambda & 1 & 1 \\ 2 & 1 - \lambda & -1 \\ -8 & -5 & -3 - \lambda \end{pmatrix}$$

implies $\det(A - \lambda I) = -\lambda^3 - \lambda^2 + 4\lambda + 4 = -(\lambda + 1)(\lambda^2 - 4)$. Therefore, the eigenvalues are given by $\lambda = -1, 2, -2$. First, for $\lambda = -1$, we have

$$A - \lambda I = \begin{pmatrix} 2 & 1 & 1 \\ 2 & 2 & -1 \\ -8 & -5 & -2 \end{pmatrix} \rightarrow \begin{pmatrix} 2 & 1 & 1 \\ 0 & 1 & -2 \\ 0 & 0 & 0 \end{pmatrix}$$

after elementary row operations. Therefore,

$$\mathbf{v}_1 = \begin{pmatrix} -3 \\ 4 \\ 2 \end{pmatrix}$$

is an associated eigenvector, and

$$\mathbf{x}_1(t) = e^{-t} \begin{pmatrix} -3 \\ 4 \\ 2 \end{pmatrix}$$

is a solution of the system. Next, for $\lambda = 2$,

$$A - \lambda I = \begin{pmatrix} -1 & 1 & 1 \\ 2 & -1 & -1 \\ -8 & -5 & -5 \end{pmatrix} \rightarrow \begin{pmatrix} 1 & -1 & -1 \\ 0 & 1 & 1 \\ 0 & 0 & 0 \end{pmatrix}$$

after elementary row operations. Therefore,

$$\mathbf{v}_2 = \begin{pmatrix} 0 \\ 1 \\ -1 \end{pmatrix}$$

is an associated eigenvector. Then, for $\lambda = -2$,

$$A - \lambda I = \begin{pmatrix} 3 & 1 & 1 \\ 2 & 3 & -1 \\ -8 & -5 & -1 \end{pmatrix} \rightarrow \begin{pmatrix} 1 & 1/3 & 1/3 \\ 0 & 1 & -5/7 \\ 0 & 0 & 0 \end{pmatrix}$$

after elementary row operations. Therefore,

$$\mathbf{v}_3 = \begin{pmatrix} -4 \\ 5 \\ 7 \end{pmatrix}$$

is an associated eigenvector. We conclude that the general solution is given by

$$\mathbf{x}(t) = c_1 e^{-t} \begin{pmatrix} -3 \\ 4 \\ 2 \end{pmatrix} + c_2 e^{2t} \begin{pmatrix} 0 \\ 1 \\ -1 \end{pmatrix} + c_3 e^{-2t} \begin{pmatrix} -4 \\ 5 \\ 7 \end{pmatrix}.$$

9.(a) We find the eigenvalues first.

$$A - \lambda I = \begin{pmatrix} 3/4 - \lambda & 29/4 & -11/2 \\ -3/4 & 3/4 - \lambda & -5/2 \\ 5/4 & 11/4 & -5/2 - \lambda \end{pmatrix}$$

implies $\det(A - \lambda I) = -\lambda^3 - \lambda^2 - 16\lambda - 16 = -(\lambda + 1)(\lambda^2 + 16)$. Therefore, the eigenvalues are given by $\lambda = -1, \pm 4i$. If $\lambda = -1$, then

$$A - \lambda I = \begin{pmatrix} 7/4 & 29/4 & -11/2 \\ -3/4 & 7/4 & -5/2 \\ 5/4 & 11/4 & -3/2 \end{pmatrix} \rightarrow \begin{pmatrix} 1 & 0 & 1 \\ 0 & 1 & -1 \\ 0 & 0 & 0 \end{pmatrix}.$$

Therefore,

$$\mathbf{v}_1 = \begin{pmatrix} 1 \\ -1 \\ -1 \end{pmatrix}$$

is an associated eigenvector. If $\lambda = 4i$, then

$$A - \lambda I = \begin{pmatrix} 3/4 - 4i & 29/4 & -11/2 \\ -3/4 & 3/4 - 4i & -5/2 \\ 5/4 & 11/4 & -5/2 - 4i \end{pmatrix} \rightarrow \begin{pmatrix} 1 & 0 & -2 - i \\ 0 & 1 & -i \\ 0 & 0 & 0 \end{pmatrix}.$$

Therefore,

$$\mathbf{v}_2 = \begin{pmatrix} 2 + i \\ i \\ 1 \end{pmatrix}$$

is an associated eigenvector, and

$$\mathbf{u}(t) = e^{4it} \left[\begin{pmatrix} 2 \\ 0 \\ 1 \end{pmatrix} + i \begin{pmatrix} 1 \\ 1 \\ 0 \end{pmatrix} \right]$$

is a solution of the system. We can write this solution as two linearly independent real-valued solutions by looking at the real and imaginary parts of $\mathbf{u}$. In particular, we see that

$$\mathbf{x}_1(t) = \text{Re}(\mathbf{u}) = \begin{pmatrix} 2 \\ 0 \\ 1 \end{pmatrix} \cos 4t - \begin{pmatrix} 1 \\ 1 \\ 0 \end{pmatrix} \sin 4t$$

and

$$\mathbf{x}_2(t) = \text{Im}(\mathbf{u}) = \begin{pmatrix} 2 \\ 0 \\ 1 \end{pmatrix} \sin 4t + \begin{pmatrix} 1 \\ 1 \\ 0 \end{pmatrix} \cos 4t$$

are two linearly independent solutions. We conclude that the general solution is given by

$$\mathbf{x}(t) = c_1 e^{-t} \begin{pmatrix} 1 \\ -1 \\ -1 \end{pmatrix} + c_2 \begin{pmatrix} 2 \cos 4t - \sin 4t \\ -\sin 4t \\ \cos 4t \end{pmatrix} + c_3 \begin{pmatrix} 2 \sin 4t + \cos 4t \\ \cos 4t \\ \sin 4t \end{pmatrix}.$$

If $c_1 = 0$, then the solution will lie on a closed curve in the plane spanned by

$$\mathbf{a} = \begin{pmatrix} 2 \\ 0 \\ 1 \end{pmatrix} \qquad \text{and} \qquad \mathbf{b} = \begin{pmatrix} 1 \\ 1 \\ 0 \end{pmatrix}.$$

(b)

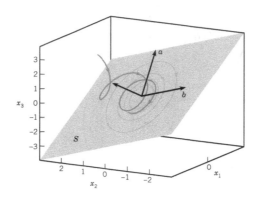

(c) For $\mathbf{x}_0 \notin S$, as $t \to \infty$, the solution will tend towards one of the closed curves in S, spiraling about the line spanned by the eigenvector $(1, -1, -1)^T$.

11.(a) Suppose that $c_1 \mathbf{a} + c_2 \mathbf{b} = 0$. Since $\mathbf{a}$ and $\mathbf{b}$ are the real and imaginary parts of the vector $\mathbf{v}_1$, $\mathbf{a} = (\mathbf{v}_1 + \overline{\mathbf{v}}_1)/2$ and $\mathbf{b} = (\mathbf{v}_1 - \overline{\mathbf{v}}_1)/2i$. Therefore,

$$c_1(\mathbf{v}_1 + \overline{\mathbf{v}}_1) - ic_2(\mathbf{v}_1 - \overline{\mathbf{v}}_1) = 0,$$

which leads to

$$(c_1 - ic_2)\mathbf{v}_1 + (c_1 + ic_2)\overline{\mathbf{v}}_1 = 0.$$

(b) Since $\mathbf{v}_1$ and $\overline{\mathbf{v}}_1$ are linearly independent, we must have

$$\begin{aligned} c_1 - ic_2 &= 0 \\ c_1 + ic_2 &= 0. \end{aligned}$$

It follows that $c_1 = c_2 = 0$.

(c) Consider the equation $c_1 \mathbf{x}_1(t_0) + c_2 \mathbf{x}_2(t_0) = 0$. Using equation (4), we can then write

$$c_1 e^{\mu t_0}(\mathbf{a} \cos \nu t_0 - \mathbf{b} \sin \nu t_0) + c_2 e^{\mu t_0}(\mathbf{a} \sin \nu t_0 + \mathbf{b} \cos \nu t_0) = 0.$$

Rearranging the terms and dividing by the exponential,

$$(c_1 + c_2) \cos(\nu t_0)\mathbf{a} + (c_2 - c_1) \sin(\nu t_0)\mathbf{b} = 0.$$

From part (b), since $\mathbf{a}$ and $\mathbf{b}$ are linearly independent, it follows that

$$(c_1 + c_2) \cos(\nu t_0) = (c_2 - c_1) \sin(\nu t_0) = 0.$$

Without loss of generality, we may assume that the trigonometric factors are nonzero. We then conclude that $c_1 + c_2 = 0$ and $c_2 - c_1 = 0$, which leads to $c_1 = c_2 = 0$. Therefore, $\mathbf{x}_1(t)$ and $\mathbf{x}_2(t)$ are linearly independent at the point t_0 and therefore at every point.

13.(a) We find the eigenvalues first.

$$A - \lambda I = \begin{pmatrix} -\lambda & 0 & 1 & 0 \\ 0 & -\lambda & 0 & 1 \\ -2 & 1 & -\lambda & 0 \\ 1 & -2 & 0 & -\lambda \end{pmatrix}$$

implies $\det(A - \lambda I) = \lambda^4 + 4\lambda^2 + 3 = (\lambda^2 + 3)(\lambda^2 + 1)$. Therefore, the eigenvalues of A are $\pm\sqrt{3}i$ and $\pm i$. First, for $\lambda = i$,

$$A - \lambda I = \begin{pmatrix} -i & 0 & 1 & 0 \\ 0 & -i & 0 & 1 \\ -2 & 1 & -i & 0 \\ 1 & -2 & 0 & -i \end{pmatrix} \rightarrow \begin{pmatrix} 1 & -2 & 0 & -i \\ 0 & 1 & 0 & i \\ 0 & 0 & 1 & -1 \\ 0 & 0 & 0 & 0 \end{pmatrix},$$

after elementary row operations. Therefore,

$$\mathbf{v}_1 = \begin{pmatrix} -i \\ -i \\ 1 \\ 1 \end{pmatrix}$$

is an associated eigenvector for $\lambda = i$. Since $A\mathbf{v}_1 = i\mathbf{v}_1$, then $A\bar{\mathbf{v}}_1 = -i\bar{\mathbf{v}}_1$. Therefore,

$$\mathbf{v}_2 = \begin{pmatrix} i \\ i \\ 1 \\ 1 \end{pmatrix}$$

is an eigenvector associated with $\lambda = -i$. Next, for $\lambda = \sqrt{3}i$,

$$A - \lambda I = \begin{pmatrix} -\sqrt{3}i & 0 & 1 & 0 \\ 0 & -\sqrt{3}i & 0 & 1 \\ -2 & 1 & -\sqrt{3}i & 0 \\ 1 & -2 & 0 & -\sqrt{3}i \end{pmatrix} \rightarrow \begin{pmatrix} 1 & -2 & 0 & -\sqrt{3}i \\ 0 & 1 & 0 & i/\sqrt{3} \\ 0 & 0 & 1 & 1 \\ 0 & 0 & 0 & 0 \end{pmatrix},$$

after elementary row operations. Therefore,

$$\mathbf{v}_3 = \begin{pmatrix} -i/\sqrt{3} \\ i/\sqrt{3} \\ 1 \\ -1 \end{pmatrix}$$

is an associated eigenvector for $\lambda = \sqrt{3}i$. Since $A\mathbf{v_3} = \sqrt{3}i\,\mathbf{v_3}$, then $A\bar{\mathbf{v}}_3 = -\sqrt{3}i\,\bar{\mathbf{v}}_3$. Therefore,

$$\mathbf{v}_4 = \begin{pmatrix} i/\sqrt{3} \\ -i/\sqrt{3} \\ 1 \\ -1 \end{pmatrix}$$

is an eigenvector associated with $\lambda = -\sqrt{3}i$.

(b) Let $\mathbf{u}_1(t)$ be the solution associated with $\lambda = i$. From part (a), we see that

$$\mathbf{u}_1(t) = e^{it} \begin{pmatrix} -i \\ -i \\ 1 \\ 1 \end{pmatrix}.$$

Taking the real and imaginary parts of $\mathbf{u}_1$, we have the following two linearly independent real-valued solutions,

$$\mathbf{x}_1(t) = \mathrm{Re}(\mathbf{u}_1) = \begin{pmatrix} 0 \\ 0 \\ 1 \\ 1 \end{pmatrix} \cos t - \begin{pmatrix} -1 \\ -1 \\ 0 \\ 0 \end{pmatrix} \sin t$$

and

$$\mathbf{x}_2(t) = \mathrm{Im}(\mathbf{u}_1) = \begin{pmatrix} 0 \\ 0 \\ 1 \\ 1 \end{pmatrix} \sin t + \begin{pmatrix} -1 \\ -1 \\ 0 \\ 0 \end{pmatrix} \cos t.$$

Then, let $\mathbf{u}_2(t)$ be the solution associated with $\lambda = \sqrt{3}i$. From part (a), we see that

$$\mathbf{u}_2(t) = e^{\sqrt{3}it} \begin{pmatrix} -i/\sqrt{3} \\ i/\sqrt{3} \\ 1 \\ -1 \end{pmatrix}.$$

Taking the real and imaginary parts of $\mathbf{u}_2$, we have two more solutions of our system,

$$\mathbf{x}_3(t) = \mathrm{Re}(\mathbf{u}_2) = \begin{pmatrix} 0 \\ 0 \\ 1 \\ -1 \end{pmatrix} \cos(\sqrt{3}t) - \begin{pmatrix} -1/\sqrt{3} \\ 1/\sqrt{3} \\ 0 \\ 0 \end{pmatrix} \sin(\sqrt{3}t)$$

and

$$\mathbf{x}_4(t) = \mathrm{Im}(\mathbf{u}_2) = \begin{pmatrix} 0 \\ 0 \\ 1 \\ -1 \end{pmatrix} \sin(\sqrt{3}t) + \begin{pmatrix} -1/\sqrt{3} \\ 1/\sqrt{3} \\ 0 \\ 0 \end{pmatrix} \cos(\sqrt{3}t).$$

Therefore, the general solution of the system is

$$\mathbf{x}(t) = c_1 \begin{pmatrix} \sin t \\ \sin t \\ \cos t \\ \cos t \end{pmatrix} + c_2 \begin{pmatrix} -\cos t \\ -\cos t \\ \sin t \\ \sin t \end{pmatrix} + c_3 \begin{pmatrix} \sin(\sqrt{3}t) \\ -\sin(\sqrt{3}t) \\ \sqrt{3}\cos(\sqrt{3}t) \\ -\sqrt{3}\cos(\sqrt{3}t) \end{pmatrix} + c_4 \begin{pmatrix} -\cos(\sqrt{3}t) \\ \cos(\sqrt{3}t) \\ \sqrt{3}\sin(\sqrt{3}t) \\ -\sqrt{3}\sin(\sqrt{3}t) \end{pmatrix}.$$

(c) Both masses will vibrate at the lowest frequency $\omega_1 = 1$ if we choose $c_3 = c_4 = 0$. In this case, the first mode of vibration occurs for any c_1 and c_2 not both zero and

$$x_1(t) = c_1 \sin t - c_2 \cos t = x_2(t).$$

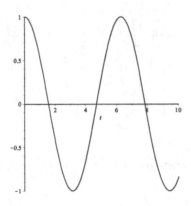

Both masses will vibrate at the second frequency $\omega_2 = \sqrt{3}$ if we choose $c_1 = c_2 = 0$. In this case, the first mode of vibration occurs for any c_3 and c_4 not both zero and

$$\begin{aligned} x_1(t) &= c_3 \sin(\sqrt{3}t) - c_4 \cos(\sqrt{3}t) \\ x_2(t) &= -c_3 \sin(\sqrt{3}t) + c_4 \cos(\sqrt{3}t). \end{aligned}$$

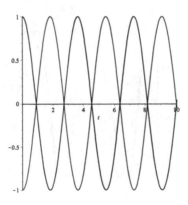

(d) Using the initial conditions and the general solution found in part (b), we have

$$\mathbf{x}(0) = c_1 \begin{pmatrix} 0 \\ 0 \\ 1 \\ 1 \end{pmatrix} + c_2 \begin{pmatrix} -1 \\ -1 \\ 0 \\ 0 \end{pmatrix} + c_3 \begin{pmatrix} 0 \\ 0 \\ \sqrt{3} \\ -\sqrt{3} \end{pmatrix} + c_4 \begin{pmatrix} -1 \\ 1 \\ 0 \\ 0 \end{pmatrix} = \begin{pmatrix} -1 \\ 3 \\ 0 \\ 0 \end{pmatrix}.$$

The solution of this system is $c_1 = 0$, $c_2 = -1$, $c_3 = 0$ and $c_4 = 2$. Therefore, the solution satisfying the given initial condition is

$$\mathbf{x}(t) = \begin{pmatrix} \cos t \\ \cos t \\ -\sin t \\ -\sin t \end{pmatrix} + 2 \begin{pmatrix} -\cos(\sqrt{3}t) \\ \cos(\sqrt{3}t) \\ \sqrt{3}\sin(\sqrt{3}t) \\ -\sqrt{3}\sin(\sqrt{3}t) \end{pmatrix}$$

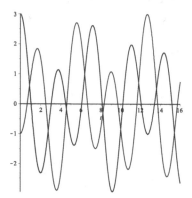

(e) In the case when the initial condition is $\mathbf{x}(0) = (1, 1, 0, 1)^T$, the solution is

$$\mathbf{x}(t) = \frac{1}{2} \begin{pmatrix} \sin t \\ \sin t \\ \cos t \\ \cos t \end{pmatrix} - \begin{pmatrix} -\cos t \\ -\cos t \\ \sin t \\ \sin t \end{pmatrix} - \frac{1}{2\sqrt{3}} \begin{pmatrix} \sin(\sqrt{3}t) \\ -\sin(\sqrt{3}t) \\ \sqrt{3}\cos(\sqrt{3}t) \\ -\sqrt{3}\cos(\sqrt{3}t) \end{pmatrix}.$$

Plots of x_1 and x_2 are shown below:

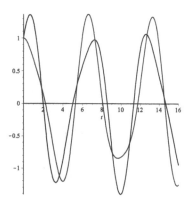

15. The eigenvalues are given by $-1, 1, -1 \pm 4i$. The corresponding eigenvectors are

$$\mathbf{v_1} = \begin{pmatrix} -1 \\ 0 \\ 0 \\ 1 \end{pmatrix}, \quad \mathbf{v_2} = \begin{pmatrix} 1 \\ 1 \\ 0 \\ 1 \end{pmatrix}, \quad \mathbf{v_3} = \begin{pmatrix} 1 \\ i \\ 1 \\ i \end{pmatrix}, \quad \mathbf{v_4} = \begin{pmatrix} 1 \\ -i \\ 1 \\ -i \end{pmatrix}.$$

Therefore, a fundamental set of solutions is given by

$$\left\{ e^{-t} \begin{pmatrix} -1 \\ 0 \\ 0 \\ 1 \end{pmatrix}, e^{t} \begin{pmatrix} 1 \\ 1 \\ 0 \\ 1 \end{pmatrix}, e^{(-1+4i)t} \begin{pmatrix} 1 \\ i \\ 1 \\ i \end{pmatrix}, e^{(-1-4i)t} \begin{pmatrix} 1 \\ -i \\ 1 \\ -i \end{pmatrix} \right\}.$$

In order to write the solutions as real-valued solutions, we look at the solution given from the eigenvalue $-1+4i$, and take the real and imaginary parts of that solution. In particular, for

$$\mathbf{u}(t) = e^{(-1+4i)t} \begin{pmatrix} 1 \\ i \\ 1 \\ i \end{pmatrix},$$

we have the two linearly independent real-valued solutions

$$\mathbf{x}_1(t) = \mathrm{Re}(\mathbf{u}(t)) = e^{-t} \left[\begin{pmatrix} 1 \\ 0 \\ 1 \\ 0 \end{pmatrix} \cos 4t - \begin{pmatrix} 0 \\ 1 \\ 0 \\ 1 \end{pmatrix} \sin 4t \right]$$

and

$$\mathbf{x}_2(t) = \mathrm{Im}(\mathbf{u}(t)) = e^{-t} \left[\begin{pmatrix} 1 \\ 0 \\ 1 \\ 0 \end{pmatrix} \sin 4t + \begin{pmatrix} 0 \\ 1 \\ 0 \\ 1 \end{pmatrix} \cos 4t \right].$$

Therefore, we conclude that a fundamental set of real-valued solutions is given by

$$\left\{ e^{-t} \begin{pmatrix} -1 \\ 0 \\ 0 \\ 1 \end{pmatrix}, e^{t} \begin{pmatrix} 1 \\ 1 \\ 0 \\ 1 \end{pmatrix}, e^{-t} \begin{pmatrix} \cos 4t \\ -\sin 4t \\ \cos 4t \\ -\sin 4t \end{pmatrix}, e^{-t} \begin{pmatrix} \sin 4t \\ \cos 4t \\ \sin 4t \\ \cos 4t \end{pmatrix} \right\}.$$

17. The eigenvalues are given by $\pm 2i, \pm 3i$. An eigenvector for $2i$ is given by

$$\mathbf{v}_1 = \begin{pmatrix} 1+i \\ 1-i \\ -1 \\ 1 \end{pmatrix}.$$

Therefore, one solution of the system is given by

$$\mathbf{u}_1(t) = e^{2it} \begin{pmatrix} 1+i \\ 1-i \\ -1 \\ 1 \end{pmatrix}.$$

Taking the real and imaginary parts of $\mathbf{u}_1$, we have the two linearly independent real-valued solutions

$$\mathbf{x}_1(t) = \begin{pmatrix} 1 \\ 1 \\ -1 \\ 1 \end{pmatrix} \cos 2t - \begin{pmatrix} 1 \\ -1 \\ 0 \\ 0 \end{pmatrix} \sin 2t$$

and

$$\mathbf{x}_2(t) = \begin{pmatrix} 1 \\ 1 \\ -1 \\ 1 \end{pmatrix} \sin 2t + \begin{pmatrix} 1 \\ -1 \\ 0 \\ 0 \end{pmatrix} \cos 2t.$$

Next, we look at an eigenvector for $3i$. An associated eigenvector is

$$\mathbf{v}_2 = \begin{pmatrix} 1 \\ 1+i \\ 1+i \\ 0 \end{pmatrix}.$$

Taking the real and imaginary parts, we have two more real-valued solutions,

$$\mathbf{x}_3(t) = \begin{pmatrix} 1 \\ 1 \\ 1 \\ 0 \end{pmatrix} \cos 3t - \begin{pmatrix} 0 \\ 1 \\ 1 \\ 0 \end{pmatrix} \sin 3t$$

and

$$\mathbf{x}_4(t) = \begin{pmatrix} 1 \\ 1 \\ 1 \\ 0 \end{pmatrix} \sin 3t + \begin{pmatrix} 0 \\ 1 \\ 1 \\ 0 \end{pmatrix} \cos 3t.$$

Therefore, a fundamental set of solutions is given by

$$\left\{ \begin{pmatrix} \cos 2t - \sin 2t \\ \cos 2t + \sin 2t \\ -\cos 2t \\ \cos 2t \end{pmatrix}, \begin{pmatrix} \sin 2t + \cos 2t \\ \sin 2t - \cos 2t \\ -\sin 2t \\ \sin 2t \end{pmatrix}, \begin{pmatrix} \cos 3t \\ \cos 3t - \sin 3t \\ \cos 3t - \sin 3t \\ 0 \end{pmatrix}, \begin{pmatrix} \sin 3t \\ \sin 3t + \cos 3t \\ \sin 3t + \cos 3t \\ 0 \end{pmatrix} \right\}.$$

6.5   Fundamental Matrices and the Exponential of a Matrix

1. We find the eigenvalues first.

$$A - \lambda I = \begin{pmatrix} 3 - \lambda & -2 \\ 2 & -2 - \lambda \end{pmatrix}$$

implies $\det(A - \lambda I) = \lambda^2 - \lambda - 2 = (\lambda + 1)(\lambda - 2)$. Thus the eigenvalues are $\lambda = 2, -1$. For $\lambda = 2$,

$$A - \lambda I = \begin{pmatrix} 1 & -2 \\ 2 & -4 \end{pmatrix} \rightarrow \begin{pmatrix} 1 & -2 \\ 0 & 0 \end{pmatrix},$$

and then

$$\mathbf{v}_1 = \begin{pmatrix} 2 \\ 1 \end{pmatrix}$$

is an eigenvector for $\lambda = 2$. Therefore,

$$\mathbf{x}_1(t) = \begin{pmatrix} 2 \\ 1 \end{pmatrix} e^{2t}$$

is a solution of the system. For $\lambda = -1$,

$$A - \lambda I = \begin{pmatrix} 4 & -2 \\ 2 & -1 \end{pmatrix} \rightarrow \begin{pmatrix} 2 & -1 \\ 0 & 0 \end{pmatrix},$$

and then

$$\mathbf{v}_2 = \begin{pmatrix} 1 \\ 2 \end{pmatrix}$$

is an eigenvector for $\lambda = -1$. Therefore,

$$\mathbf{x}_2(t) = \begin{pmatrix} 1 \\ 2 \end{pmatrix} e^{-t}$$

is a solution of the system. We conclude that

$$\mathbf{X}(t) = \begin{pmatrix} e^{-t} & 2e^{2t} \\ 2e^{-t} & e^{2t} \end{pmatrix}$$

is a fundamental matrix for this system. The fundamental matrix e^{At} is given by $e^{At} = \mathbf{X}(t)\mathbf{X}^{-1}(0)$. Thus

$$e^{At} = \frac{1}{3} \begin{pmatrix} e^{-t} & 2e^{2t} \\ 2e^{-t} & e^{2t} \end{pmatrix} \begin{pmatrix} -1 & 2 \\ 2 & -1 \end{pmatrix} = \frac{1}{3} \begin{pmatrix} -e^{-t} + 4e^{2t} & 2e^{-t} - 2e^{2t} \\ -2e^{-t} + 2e^{2t} & 4e^{-t} - e^{2t} \end{pmatrix}.$$

3. We find the eigenvalues first.

$$A - \lambda I = \begin{pmatrix} 3 - \lambda & -4 \\ 1 & -1 - \lambda \end{pmatrix}$$

implies $\det(A - \lambda I) = \lambda^2 - 2\lambda + 1 = (\lambda - 1)^2$. Therefore, $\lambda = 1$ is the only eigenvalue. For $\lambda = 1$,

$$A - \lambda I = \begin{pmatrix} 2 & -4 \\ 1 & -2 \end{pmatrix} \rightarrow \begin{pmatrix} 1 & -2 \\ 0 & 0 \end{pmatrix}.$$

Thus

$$\mathbf{v} = \begin{pmatrix} 2 \\ 1 \end{pmatrix}$$

is an eigenvector for $\lambda = 1$. Therefore,

$$\mathbf{x}_1(t) = \begin{pmatrix} 2 \\ 1 \end{pmatrix} e^t$$

is a solution of the system. To find another solution, we need to look for a vector $\mathbf{w}$ satisfying $(A - I)\mathbf{w} = \mathbf{v}$. That is, we need to find $\mathbf{w}$ such that

$$\begin{pmatrix} 2 & -4 \\ 1 & -2 \end{pmatrix} \mathbf{w} = \begin{pmatrix} 2 \\ 1 \end{pmatrix}.$$

The vector

$$\mathbf{w} = \begin{pmatrix} 1 \\ 0 \end{pmatrix}$$

satisfies this equation. Therefore,

$$\mathbf{x}_2(t) = \begin{pmatrix} 2 \\ 1 \end{pmatrix} te^t + \begin{pmatrix} 1 \\ 0 \end{pmatrix} e^t$$

is a second solution. We conclude that

$$\mathbf{X}(t) = \begin{pmatrix} 2e^t & 2te^t + e^t \\ e^t & te^t \end{pmatrix}$$

is a fundamental matrix for this system. The fundamental matrix e^{At} is given by $e^{At} = \mathbf{X}(t)\mathbf{X}^{-1}(0)$. Therefore,

$$e^{At} = \begin{pmatrix} 2e^t & 2te^t + e^t \\ e^t & te^t \end{pmatrix} \begin{pmatrix} 0 & 1 \\ 1 & -2 \end{pmatrix} = \begin{pmatrix} 2te^t + e^t & -4te^t \\ te^t & e^t - 2te^t \end{pmatrix}.$$

5. We find the eigenvalues first.

$$A - \lambda I = \begin{pmatrix} 1 - \lambda & -5/2 \\ 1/2 & -1 - \lambda \end{pmatrix}$$

implies $\det(A - \lambda I) = \lambda^2 + 1/4$. Therefore, the eigenvalues are $\lambda = \pm i/2$. For $\lambda = i/2$,

$$A - \lambda I = \begin{pmatrix} 1 - i/2 & -5/2 \\ 1/2 & -1 - i/2 \end{pmatrix} \rightarrow \begin{pmatrix} 1 & -2 - i \\ 0 & 0 \end{pmatrix}.$$

Thus

$$\mathbf{v}_1 = \begin{pmatrix} 2 + i \\ 1 \end{pmatrix}$$

is an eigenvector for $\lambda = i/2$. Therefore,

$$\mathbf{x}_1(t) = e^{it/2} \begin{pmatrix} 2 + i \\ 1 \end{pmatrix}$$

is a solution of the system. Taking the real and imaginary parts of $\mathbf{u}(t)$, we have the two linearly independent real-valued solutions

$$\mathbf{x}_1(t) \;=\; \begin{pmatrix} 2\cos(t/2) - \sin(t/2) \\ \cos(t/2) \end{pmatrix}$$

$$\mathbf{x}_2(t) \;=\; \begin{pmatrix} 2\sin(t/2) + \cos(t/2) \\ \sin(t/2) \end{pmatrix}.$$

We conclude that

$$\mathbf{X}(t) = \begin{pmatrix} 2\cos(t/2) - \sin(t/2) & 2\sin(t/2) + \cos(t/2) \\ \cos(t/2) & \sin(t/2) \end{pmatrix}$$

is a fundamental matrix for this system. The fundamental matrix e^{At} is given by $e^{At} = \mathbf{X}(t)\mathbf{X}^{-1}(0)$. Thus

$$
\begin{aligned}
e^{At} \;&=\; \begin{pmatrix} 2\cos(t/2) - \sin(t/2) & 2\sin(t/2) + \cos(t/2) \\ \cos(t/2) & \sin(t/2) \end{pmatrix} \begin{pmatrix} 0 & 1 \\ 1 & -2 \end{pmatrix} \\
&=\; \begin{pmatrix} 2\sin(t/2) + \cos(t/2) & -5\sin(t/2) \\ \sin(t/2) & \cos(t/2) - 2\sin(t/2) \end{pmatrix}.
\end{aligned}
$$

7. We find the eigenvalues first.

$$A - \lambda I = \begin{pmatrix} 5 - \lambda & -1 \\ 3 & 1 - \lambda \end{pmatrix}$$

implies $\det(A - \lambda I) = \lambda^2 - 6\lambda + 8 = (\lambda - 2)(\lambda - 4)$. Therefore, the eigenvalues are $\lambda = 2, 4$. For $\lambda = 2$,

$$A - \lambda I = \begin{pmatrix} 3 & -1 \\ 3 & -1 \end{pmatrix} \rightarrow \begin{pmatrix} 3 & -1 \\ 0 & 0 \end{pmatrix}.$$

Thus

$$\mathbf{v}_1 = \begin{pmatrix} 1 \\ 3 \end{pmatrix}$$

is an eigenvector for $\lambda = 2$. Therefore

$$\mathbf{x}_1(t) = \begin{pmatrix} 1 \\ 3 \end{pmatrix} e^{2t}$$

is a solution of the system. For $\lambda = 4$,

$$A - \lambda I = \begin{pmatrix} 1 & -1 \\ 3 & -3 \end{pmatrix} \rightarrow \begin{pmatrix} 1 & -1 \\ 0 & 0 \end{pmatrix}.$$

Thus

$$\mathbf{v}_2 = \begin{pmatrix} 1 \\ 1 \end{pmatrix}$$

is an eigenvector for $\lambda = 4$. Therefore

$$\mathbf{x}_2(t) = \begin{pmatrix} 1 \\ 1 \end{pmatrix} e^{4t}$$

is a solution of the system. We conclude that

$$\mathbf{X}(t) = \begin{pmatrix} e^{2t} & e^{4t} \\ 3e^{2t} & e^{4t} \end{pmatrix}$$

is a fundamental matrix for this system. The fundamental matrix e^{At} is given by $e^{At} = \mathbf{X}(t)\mathbf{X}^{-1}(0)$. Therefore

$$e^{At} = \frac{1}{2} \begin{pmatrix} e^{2t} & e^{4t} \\ 3e^{2t} & e^{4t} \end{pmatrix} \begin{pmatrix} -1 & 1 \\ 3 & -1 \end{pmatrix} = \frac{1}{2} \begin{pmatrix} -e^{2t} + 3e^{4t} & e^{2t} - e^{4t} \\ -3e^{2t} + 3e^{4t} & 3e^{2t} - e^{4t} \end{pmatrix}.$$

9. We find the eigenvalues first.

$$A - \lambda I = \begin{pmatrix} 2 - \lambda & -1 \\ 3 & -2 - \lambda \end{pmatrix}$$

implies $\det(A - \lambda I) = \lambda^2 - 1$. Therefore, the eigenvalues are $\lambda = 1, -1$. For $\lambda = 1$,

$$A - \lambda I = \begin{pmatrix} 1 & -1 \\ 3 & -3 \end{pmatrix} \rightarrow \begin{pmatrix} 1 & -1 \\ 0 & 0 \end{pmatrix}.$$

Therefore

$$\mathbf{v}_1 = \begin{pmatrix} 1 \\ 1 \end{pmatrix}$$

is an eigenvector for $\lambda = 1$. Thus

$$\mathbf{x}_1(t) = \begin{pmatrix} 1 \\ 1 \end{pmatrix} e^t$$

is a solution of the system. For $\lambda = -1$,

$$A - \lambda I = \begin{pmatrix} 3 & -1 \\ 3 & -1 \end{pmatrix} \rightarrow \begin{pmatrix} 3 & -1 \\ 0 & 0 \end{pmatrix}.$$

Therefore

$$\mathbf{v}_2 = \begin{pmatrix} 1 \\ 3 \end{pmatrix}$$

is an eigenvector for $\lambda = -1$. Thus

$$\mathbf{x}_2(t) = \begin{pmatrix} 1 \\ 3 \end{pmatrix} e^{-t}$$

is a solution of the system. We conclude that

$$\mathbf{X}(t) = \begin{pmatrix} e^t & e^{-t} \\ e^t & 3e^{-t} \end{pmatrix}$$

is a fundamental matrix for this system. The fundamental matrix e^{At} is given by $e^{At} = \mathbf{X}(t)\mathbf{X}^{-1}(0)$. Thus

$$e^{At} = \frac{1}{2}\begin{pmatrix} e^t & e^{-t} \\ e^t & 3e^{-t} \end{pmatrix}\begin{pmatrix} 3 & -1 \\ -1 & 1 \end{pmatrix} = \frac{1}{2}\begin{pmatrix} 3e^t - e^{-t} & -e^t + e^{-t} \\ 3e^t - 3e^{-t} & -e^t + 3e^{-t} \end{pmatrix}.$$

11. We find the eigenvalues first.

$$A - \lambda I = \begin{pmatrix} -3 - \lambda & 4 \\ -1 & -2 - \lambda \end{pmatrix}$$

implies $\det(A - \lambda I) = \lambda^2 + 5\lambda + 10$. Thus the eigenvalues are $\lambda = -5/2 \pm \sqrt{15}i/2$. For $\lambda = -5/2 + \sqrt{15}i/2$,

$$A - \lambda I = \begin{pmatrix} -1/2 - \sqrt{15}i/2 & 4 \\ -1 & 1/2 - \sqrt{15}i/2 \end{pmatrix} \rightarrow \begin{pmatrix} 2 & -1 + \sqrt{15}i \\ 0 & 0 \end{pmatrix}.$$

Therefore

$$\mathbf{v}_1 = \begin{pmatrix} 1 - \sqrt{15}i \\ 2 \end{pmatrix}$$

is an eigenvector for $\lambda = -5/2 + \sqrt{15}i/2$. Thus

$$\mathbf{u}(t) = e^{(-5+\sqrt{15}i)t/2}\begin{pmatrix} 1 - \sqrt{15}i \\ 2 \end{pmatrix}$$

is a solution of the system. Taking the real and imaginary parts of $\mathbf{u}(t)$, we have the two linearly independent real-valued solutions

$$\mathbf{x}_1(t) = e^{-5t/2}\begin{pmatrix} \cos(\sqrt{15}t/2) + \sqrt{15}\sin(\sqrt{15}t/2) \\ 2\cos(\sqrt{15}t/2) \end{pmatrix}$$

and

$$\mathbf{x}_2(t) = e^{-5t/2}\begin{pmatrix} \sin(\sqrt{15}t/2) - \sqrt{15}\cos(\sqrt{15}t/2) \\ 2\sin(\sqrt{15}t/2) \end{pmatrix}.$$

We conclude that

$$\mathbf{X}(t) = e^{-5t/2}\begin{pmatrix} \cos(\sqrt{15}t/2) + \sqrt{15}\sin(\sqrt{15}t/2) & \sin(\sqrt{15}t/2) - \sqrt{15}\cos(\sqrt{15}t/2) \\ 2\cos(\sqrt{15}t/2) & 2\sin(\sqrt{15}t/2) \end{pmatrix}$$

is a fundamental matrix for this system. The fundamental matrix e^{At} is given by $e^{At} = \mathbf{X}(t)\mathbf{X}^{-1}(0)$. Thus

$$e^{At} = e^{-5t/2}\begin{pmatrix} \cos(\sqrt{15}t/2) - (1/\sqrt{15})\sin(\sqrt{15}t/2) & (8/\sqrt{15})\sin(\sqrt{15}t/2) \\ -(2/\sqrt{15})\sin(\sqrt{15}t/2) & \cos(\sqrt{15}t/2) + (1/\sqrt{15})\sin(\sqrt{15}t/2) \end{pmatrix}.$$

13. The eigenvalues of this matrix are $\lambda = -1, -2, 2$. For $\lambda = -1$,

$$A - \lambda I = \begin{pmatrix} 2 & 1 & 1 \\ 2 & 2 & -1 \\ -8 & -5 & -2 \end{pmatrix} \rightarrow \begin{pmatrix} 2 & 0 & 3 \\ 0 & 1 & -2 \\ 0 & 0 & 0 \end{pmatrix}.$$

Therefore

$$\mathbf{v}_1 = \begin{pmatrix} 3 \\ -4 \\ -2 \end{pmatrix}$$

is an eigenvector for $\lambda = -1$. Thus

$$\mathbf{x}_1(t) = \begin{pmatrix} 3 \\ -4 \\ -2 \end{pmatrix} e^{-t}$$

is a solution of the system. For $\lambda = -2$,

$$A - \lambda I = \begin{pmatrix} 3 & 1 & 1 \\ 2 & 3 & -1 \\ -8 & -5 & -1 \end{pmatrix} \rightarrow \begin{pmatrix} 7 & 0 & 4 \\ 0 & 7 & -5 \\ 0 & 0 & 0 \end{pmatrix}.$$

Therefore

$$\mathbf{v}_2 = \begin{pmatrix} 4 \\ -5 \\ -7 \end{pmatrix}$$

is an eigenvector for $\lambda = -2$. Thus

$$\mathbf{x}_2(t) = \begin{pmatrix} 4 \\ -5 \\ -7 \end{pmatrix} e^{-2t}$$

is a solution of the system. For $\lambda = 2$,

$$A - \lambda I = \begin{pmatrix} -1 & 1 & 1 \\ 2 & -1 & -1 \\ -8 & -5 & -5 \end{pmatrix} \rightarrow \begin{pmatrix} 1 & 0 & 0 \\ 0 & 1 & 1 \\ 0 & 0 & 0 \end{pmatrix}.$$

Therefore

$$\mathbf{v}_3 = \begin{pmatrix} 0 \\ 1 \\ -1 \end{pmatrix}$$

is an eigenvector for $\lambda = 2$. Thus

$$\mathbf{x}_3(t) = \begin{pmatrix} 0 \\ 1 \\ -1 \end{pmatrix} e^{2t}$$

is a solution of the system. We conclude that

$$\mathbf{X}(t) = \begin{pmatrix} 3e^{-t} & 4e^{-2t} & 0 \\ -4e^{-t} & -5e^{-2t} & e^{2t} \\ -2e^{-t} & -7e^{-2t} & -e^{2t} \end{pmatrix}$$

is a fundamental matrix for this system. The fundamental matrix e^{At} is given by $e^{At} = \mathbf{X}(t)\mathbf{X}^{-1}(0)$. Therefore,

$$
e^{At} = \begin{pmatrix} 3e^{-t} & 4e^{-2t} & 0 \\ -4e^{-t} & -5e^{-2t} & e^{2t} \\ -2e^{-t} & -7e^{-2t} & -e^{2t} \end{pmatrix} \begin{pmatrix} 1 & 1/3 & 1/3 \\ -1/2 & -1/4 & -1/4 \\ 3/2 & 13/12 & 1/12 \end{pmatrix}
$$

$$
= \begin{pmatrix} -2e^{-2t} + 3e^{-t} & -e^{-2t} + e^{-t} & -e^{-2t} + e^{-t} \\ 5e^{-2t}/2 - 4e^{-t} + 3e^{2t}/2 & 5e^{-2t}/4 - 4e^{-t}/3 + 13e^{2t}/12 & 5e^{-2t}/4 - 4e^{-t}/3 + e^{2t}/12 \\ 7e^{-2t}/2 - 2e^{-t} - 3e^{2t}/2 & 7e^{-2t}/4 - 2e^{-t}/3 - 13e^{2t}/12 & 7e^{-2t}/4 - 2e^{-t}/3 - e^{2t}/12 \end{pmatrix}.
$$

15. We find that for this A matrix

$$
e^{At} = \frac{1}{3}e^{-3t} \begin{pmatrix} 3\cos 3t & -9\sin 3t \\ \sin 3t & 3\cos 3t \end{pmatrix}.
$$

Therefore, the solution of the initial value problem is given by

$$
\mathbf{x} = e^{At}\mathbf{x}_0 = \frac{1}{3}e^{-3t} \begin{pmatrix} 3\cos 3t & -9\sin 3t \\ \sin 3t & 3\cos 3t \end{pmatrix} \begin{pmatrix} 4 \\ 1 \end{pmatrix} = e^{-3t} \begin{pmatrix} 4\cos 3t - 3\sin 3t \\ \cos 3t + (4/3)\sin 3t \end{pmatrix}.
$$

17. First,

$$
sI_2 - A = \begin{pmatrix} s+4 & 1 \\ -1 & s+2 \end{pmatrix}
$$

implies

$$
(sI_2 - A)^{-1} = \frac{1}{(s+3)^2} \begin{pmatrix} s+2 & -1 \\ 1 & s+4 \end{pmatrix}.
$$

Therefore,

$$
e^{At} = \mathcal{L}^{-1}\left((sI_2 - A)^{-1}\right) = \begin{pmatrix} \mathcal{L}^{-1}\left\{\frac{s+2}{(s+3)^2}\right\} & \mathcal{L}^{-1}\left(-\frac{1}{(s+3)^2}\right) \\ \mathcal{L}^{-1}\left\{\frac{1}{(s+3)^2}\right\} & \mathcal{L}^{-1}\left(\frac{s+4}{(s+3)^2}\right) \end{pmatrix}.
$$

Using the facts that

$$
\frac{s+2}{(s+3)^2} = \frac{1}{s+3} - \frac{1}{(s+3)^2} \quad \text{and} \quad \frac{s+4}{(s+3)^2} = \frac{1}{s+3} + \frac{1}{(s+3)^2},
$$

combined with the table of Laplace transforms, we conclude that

$$
e^{At} = \begin{pmatrix} e^{-3t} - te^{-3t} & -te^{-3t} \\ te^{-3t} & e^{-3t} + te^{-3t} \end{pmatrix}.
$$

19. First,

$$
sI_2 - A = \begin{pmatrix} s+1 & 5 \\ -1 & s-3 \end{pmatrix}
$$

implies

$$
(sI_2 - A)^{-1} = \frac{1}{(s-1)^2 + 1} \begin{pmatrix} s-3 & -5 \\ 1 & s+1 \end{pmatrix}.
$$

Therefore,

$$e^{At} = \mathcal{L}^{-1}\left((sI_2 - A)^{-1}\right) = \begin{pmatrix} \mathcal{L}^{-1}\left(\frac{s-3}{(s-1)^2+1}\right) & \mathcal{L}^{-1}\left(-\frac{5}{(s-1)^2+1}\right) \\ \mathcal{L}^{-1}\left(\frac{1}{(s-1)^2+1}\right) & \mathcal{L}^{-1}\left(\frac{s+1}{(s-1)^2+1}\right) \end{pmatrix}.$$

Using the facts that

$$\frac{s-3}{(s-1)^2+1} = \frac{s-1}{(s-1)^2+1} - \frac{2}{(s-1)^2+1}$$

and

$$\frac{s+1}{(s-1)^2+1} = \frac{s-1}{(s-1)^2+1} + \frac{2}{(s-1)^2+1}$$

combined with the table of Laplace transforms, we conclude that

$$e^{At} = \begin{pmatrix} e^t\cos t - 2e^t\sin t & -5e^t\sin t \\ e^t\sin t & e^t\cos t + 2e^t\sin t \end{pmatrix}.$$

21.(a) Letting $x_1 = u$ and $x_2 = u'$, then $u'' = x_2'$. Therefore, in terms of the new variables, we have

$$x_2' + \omega^2 x_1 = 0$$

with the initial conditions $x_1(0) = u_0$ and $x_2(0) = v_0$. The equivalent first order system is

$$\begin{aligned} x_1' &= x_2 \\ x_2' &= -\omega^2 x_1, \end{aligned}$$

which can be expressed in the form

$$\begin{pmatrix} x_1 \\ x_2 \end{pmatrix}' = \begin{pmatrix} 0 & 1 \\ -\omega^2 & 0 \end{pmatrix}\begin{pmatrix} x_1 \\ x_2 \end{pmatrix}$$

with initial conditions

$$\begin{pmatrix} x_1(0) \\ x_2(0) \end{pmatrix} = \begin{pmatrix} u_0 \\ v_0 \end{pmatrix}.$$

(b) Setting

$$A = \begin{pmatrix} 0 & 1 \\ -\omega^2 & 0 \end{pmatrix},$$

we see that $A^2 = -\omega^2 I$, $A^3 = -\omega^2 A$, and $A^4 = \omega^4 I$. By induction, it follows that $A^{2k} = (-1)^k \omega^{2k} I$ and $A^{2k+1} = (-1)^k \omega^{2k} A$. Therefore,

$$\begin{aligned} e^{At} &= \sum_{k=0}^{\infty}\left[(-1)^k\frac{\omega^{2k}t^{2k}}{(2k)!}I + (-1)^k\frac{\omega^{2k}t^{2k+1}}{(2k+1)!}A\right] \\ &= \left[\sum_{k=0}^{\infty}(-1)^k\frac{\omega^{2k}t^{2k}}{(2k)!}\right]I + \frac{1}{\omega}\left[\sum_{k=0}^{\infty}(-1)^k\frac{\omega^{2k+1}t^{2k+1}}{(2k+1)!}\right]A \\ &= I\cos(\omega t) + A\frac{\sin(\omega t)}{\omega}. \end{aligned}$$

(c) Using the formula for $e^{\mathbf{A}t}$ found in part (b), we can conclude that the solution of the initial value problem is given by

$$\mathbf{x} = e^{\mathbf{A}t}\mathbf{x}_0 = \left[\cos(\omega t)\mathbf{I} + \frac{1}{\omega}\sin(\omega t)\mathbf{A}\right]\begin{pmatrix} u_0 \\ v_0 \end{pmatrix}$$

$$= \cos(\omega t)\begin{pmatrix} u_0 \\ v_0 \end{pmatrix} + \frac{1}{\omega}\sin(\omega t)\begin{pmatrix} v_0 \\ -\omega^2 u_0 \end{pmatrix}.$$

23.(a) If $T^{-1}AT = D$, then $TT^{-1}AT = TD$, therefore $AT = TD$, so $ATT^{-1} = TDT^{-1}$, and then $A = TDT^{-1}$. Therefore, $A^2 = (TDT^{-1})(TDT^{-1}) = TD^2T^{-1}$. Now, we will prove the general result by induction. Assume $A^n = TD^nT^{-1}$ to show that $A^{n+1} = TD^{n+1}T^{-1}$. We compute:

$$A^{n+1} = AA^n = ATD^nT^{-1} = (TDT^{-1})(TD^nT^{-1}) = TDD^nT^{-1} = TD^{n+1}T^{-1}.$$

(b)

$$\begin{aligned} e^{At} &= I_n + At + \frac{1}{2!}A^2t^2 + \frac{1}{3!}A^3t^3 + \cdots \\ &= TT^{-1} + TDT^{-t}t + \frac{1}{2!}TD^2T^{-1}t^2 + \frac{1}{3!}TD^3T^{-1}t^3 + \cdots \\ &= T\left[T^{-1} + DT^{-1}t + \frac{1}{2!}D^2T^{-1}t^2 + \frac{1}{3!}D^3T^{-1}t^3 + \cdots\right] \\ &= T\left[I_n + Dt + \frac{1}{2!}D^2t^2 + \frac{1}{3!}D^3t^3 + \cdots\right]T^{-1} = Te^{Dt}T^{-1}. \end{aligned}$$

6.6 Nonhomogeneous Linear Systems

1. By the product rule and the Fundamental Theorem of Calculus,

$$\begin{aligned} \mathbf{x}_p' &= \mathbf{X}'(t)\int_{t_1}^t \mathbf{X}^{-1}(s)\mathbf{g}(s)\,ds + \mathbf{X}(t)\mathbf{X}^{-1}(t)\mathbf{g}(t) = \mathbf{P}(t)\mathbf{X}(t)\int_{t_1}^t \mathbf{X}^{-1}(s)\mathbf{g}(s)\,ds + \mathbf{g}(t) \\ &= \mathbf{P}(t)\mathbf{x}_p + \mathbf{g}(t). \end{aligned}$$

3. The eigenvalues of

$$\begin{pmatrix} 1 & \sqrt{3} \\ \sqrt{3} & -1 \end{pmatrix}$$

are given by $\lambda_1 = 2$ and $\lambda_2 = -2$. Corresponding eigenvectors are given by

$$\mathbf{v}_1 = \begin{pmatrix} \sqrt{3} \\ 1 \end{pmatrix}, \qquad \mathbf{v}_2 = \begin{pmatrix} 1 \\ -\sqrt{3} \end{pmatrix}.$$

Therefore, two linearly independent solutions of the homogeneous equation are given by

$$\mathbf{x}_1(t) = \begin{pmatrix} \sqrt{3} \\ 1 \end{pmatrix}e^{2t}, \qquad \mathbf{x}_2(t) = \begin{pmatrix} 1 \\ -\sqrt{3} \end{pmatrix}e^{-2t}.$$

and

$$\mathbf{X}(t) = \begin{pmatrix} \sqrt{3}e^{2t} & e^{-2t} \\ e^{2t} & -\sqrt{3}e^{-2t} \end{pmatrix}$$

is a fundamental matrix. In order to calculate the general solution, we need to calculate $\int_{t_1}^{t} \mathbf{X}^{-1}(s)\mathbf{g}(s)\,ds$. We see that

$$\mathbf{X}^{-1}(s) = \frac{1}{4}\begin{pmatrix} \sqrt{3}e^{-2s} & e^{-2s} \\ e^{2s} & -\sqrt{3}e^{2s} \end{pmatrix}.$$

Therefore,

$$\int_{t_1}^{t} \mathbf{X}^{-1}(s)\mathbf{g}(s)\,ds = \frac{1}{4}\int_{t_1}^{t}\begin{pmatrix} \sqrt{3}e^{-2s} & e^{-2s} \\ e^{2s} & -\sqrt{3}e^{2s} \end{pmatrix}\begin{pmatrix} e^{s} \\ \sqrt{3}e^{-s} \end{pmatrix}\,ds$$

$$= \frac{1}{4}\int_{t_1}^{t}\begin{pmatrix} \sqrt{3}e^{-s} + \sqrt{3}e^{-3s} \\ e^{3s} - 3e^{s} \end{pmatrix}\,ds = \frac{1}{4}\begin{pmatrix} -\sqrt{3}e^{-t} - e^{-3t}/\sqrt{3} \\ e^{3t}/3 - 3e^{t} \end{pmatrix} + \mathbf{c}.$$

Then the general solution will be given by

$$\mathbf{x}(t) = \mathbf{X}(t)\mathbf{c} + \mathbf{X}(t)\int_{t_1}^{t}\mathbf{X}^{-1}(s)\mathbf{g}(s)\,ds$$

$$= \begin{pmatrix} \sqrt{3}e^{2t} & e^{-2t} \\ e^{2t} & -\sqrt{3}e^{-2t} \end{pmatrix}\mathbf{c} + \begin{pmatrix} \sqrt{3}e^{2t} & e^{-2t} \\ e^{2t} & -\sqrt{3}e^{-2t} \end{pmatrix}\left[\frac{1}{4}\begin{pmatrix} -\sqrt{3}e^{-t} - e^{-3t}/\sqrt{3} \\ e^{3t}/3 - 3e^{t} \end{pmatrix} + \mathbf{c}\right]$$

$$= c_1 e^{2t}\begin{pmatrix} \sqrt{3} \\ 1 \end{pmatrix} + c_2 e^{-2t}\begin{pmatrix} 1 \\ -\sqrt{3} \end{pmatrix} + \begin{pmatrix} -2e^{t}/3 - e^{-t} \\ -e^{t}/\sqrt{3} + 2e^{-t}/\sqrt{3} \end{pmatrix}.$$

5. The eigenvalues of

$$\begin{pmatrix} 1 & 1 \\ 4 & -2 \end{pmatrix}$$

are given by $\lambda_1 = -3$ and $\lambda_2 = 2$. Corresponding eigenvectors are given by

$$\mathbf{v}_1 = \begin{pmatrix} 1 \\ -4 \end{pmatrix}, \qquad \mathbf{v}_2 = \begin{pmatrix} 1 \\ 1 \end{pmatrix}.$$

Therefore, two linearly independent solutions of the homogeneous equation are given by

$$\mathbf{x}_1(t) = \begin{pmatrix} 1 \\ -4 \end{pmatrix}e^{-3t}, \qquad \mathbf{x}_2(t) = \begin{pmatrix} 1 \\ 1 \end{pmatrix}e^{2t}$$

and

$$\mathbf{X}(t) = \begin{pmatrix} e^{-3t} & e^{2t} \\ -4e^{-3t} & e^{2t} \end{pmatrix}$$

is a fundamental matrix for this equation. In order to calculate the general solution, we need to calculate $\int_{t_1}^{t}\mathbf{X}^{-1}(s)\mathbf{g}(s)\,ds$. We see that

$$\mathbf{X}^{-1}(s) = \frac{1}{5}\begin{pmatrix} e^{3s} & -e^{3s} \\ 4e^{-2s} & e^{-2s} \end{pmatrix}.$$

Therefore,

$$
\int_{t_1}^{t} \mathbf{X}^{-1}(s)\mathbf{g}(s)\,ds = \frac{1}{5}\int_{t_1}^{t}\begin{pmatrix} e^{3s} & -e^{3s} \\ 4e^{-2s} & e^{-2s} \end{pmatrix}\begin{pmatrix} e^{-2s} \\ -2e^{s} \end{pmatrix} ds
$$

$$
= \frac{1}{5}\int_{t_1}^{t}\begin{pmatrix} e^{s}+2e^{4s} \\ 4e^{-4s}-2e^{-s} \end{pmatrix} ds = \frac{1}{5}\begin{pmatrix} e^{t}+e^{4t}/2 \\ -e^{-4t}+2e^{-t} \end{pmatrix} + \mathbf{c}.
$$

Then the general solution will be given by

$$
\mathbf{x}(t) = \mathbf{X}(t)\mathbf{c} + \mathbf{X}(t)\int_{t_1}^{t}\mathbf{X}^{-1}(s)\mathbf{g}(s)\,ds
$$

$$
= \begin{pmatrix} e^{-3t} & e^{2t} \\ -4e^{-3t} & e^{2t} \end{pmatrix}\mathbf{c} + \frac{1}{5}\begin{pmatrix} e^{-3t} & e^{2t} \\ -4e^{-3t} & e^{2t} \end{pmatrix}\begin{pmatrix} e^{t}+e^{4t}/2 \\ -e^{-4t}+2e^{-t} \end{pmatrix}
$$

$$
= c_1 e^{-3t}\begin{pmatrix} 1 \\ -4 \end{pmatrix} + c_2 e^{2t}\begin{pmatrix} 1 \\ 1 \end{pmatrix} + \begin{pmatrix} e^{t}/2 \\ -e^{-2t} \end{pmatrix}.
$$

7. The eigenvalues of A are given by $\lambda_1 = -2$ and $\lambda_2 = -1$ with multiplicity 2. An eigenvector for λ_1 is given by

$$
\mathbf{v}_1 = \begin{pmatrix} 1 \\ -2 \\ 1 \end{pmatrix}.
$$

Two eigenvectors for λ_2 are given by

$$
\mathbf{v}_2 = \begin{pmatrix} -1 \\ 1 \\ 0 \end{pmatrix}, \qquad \mathbf{v}_3 = \begin{pmatrix} 1 \\ 0 \\ 1 \end{pmatrix}.
$$

Therefore, three solutions of the homogeneous equation are given by

$$
\mathbf{x}_1(t) = e^{-2t}\begin{pmatrix} 1 \\ -2 \\ 1 \end{pmatrix}, \quad \mathbf{x}_2(t) = e^{-t}\begin{pmatrix} -1 \\ 1 \\ 0 \end{pmatrix}, \quad \mathbf{x}_3(t) = e^{-t}\begin{pmatrix} 1 \\ 0 \\ 1 \end{pmatrix},
$$

and

$$
\mathbf{X}(t) = \begin{pmatrix} e^{-2t} & -e^{-t} & e^{-t} \\ -2e^{-2t} & e^{-t} & 0 \\ e^{-2t} & 0 & e^{-t} \end{pmatrix}
$$

is a fundamental matrix for this equation. Further,

$$
\mathbf{X}^{-1}(s) = \frac{1}{2}\begin{pmatrix} -e^{2s} & -e^{2s} & e^{2s} \\ -2e^{s} & 0 & 2e^{s} \\ e^{s} & e^{s} & e^{s} \end{pmatrix}.
$$

Therefore,

$$
\int_{t_1}^{t}\mathbf{X}^{-1}(s)\mathbf{g}(s)\,ds = \frac{1}{2}\int_{t_1}^{t}\begin{pmatrix} -e^{2s} & -e^{2s} & e^{2s} \\ -2e^{s} & 0 & 2e^{s} \\ e^{s} & e^{s} & e^{s} \end{pmatrix}\begin{pmatrix} 1 \\ s \\ 11e^{-3s} \end{pmatrix} ds
$$

$$
= \frac{1}{2}\int_{t_1}^{t}\begin{pmatrix} -e^{2s}-se^{2s}+11e^{-s} \\ -2e^{s}+22e^{-2s} \\ e^{s}+se^{s}+11e^{-2s} \end{pmatrix} ds = \frac{1}{2}\begin{pmatrix} -e^{2t}/4-te^{2t}/2-11e^{-t} \\ -2e^{t}-11e^{-2t} \\ te^{t}-11e^{-2t}/2 \end{pmatrix} + \mathbf{c}.
$$

Therefore, the general solution is given by

$$
\begin{aligned}
\mathbf{x}(t) \;=&\; \mathbf{X}(t)\mathbf{c} + \mathbf{X}(t)\int_{t_1}^{t}\mathbf{X}^{-1}(s)\mathbf{g}(s)\,ds \\[2mm]
=&\; \begin{pmatrix} e^{-2t} & -e^{-t} & e^{-t} \\ -2e^{-2t} & e^{-t} & 0 \\ e^{-2t} & 0 & e^{-t} \end{pmatrix}\mathbf{c} + \frac{1}{2}\begin{pmatrix} e^{-2t} & -e^{-t} & e^{-t} \\ -2e^{-2t} & e^{-t} & 0 \\ e^{-2t} & 0 & e^{-t} \end{pmatrix}\begin{pmatrix} -e^{2t}/4 - te^{2t}/2 - 11e^{-t} \\ -2e^{t} - 11e^{-2t} \\ te^{t} - 11e^{-2t}/2 \end{pmatrix} \\[2mm]
=&\; c_1 e^{-2t}\begin{pmatrix} 1 \\ -2 \\ 1 \end{pmatrix} + c_2 e^{-t}\begin{pmatrix} -1 \\ 1 \\ 0 \end{pmatrix} + c_3 e^{-t}\begin{pmatrix} 1 \\ 0 \\ 1 \end{pmatrix} + \begin{pmatrix} 7/8 + t/4 - 11e^{-3t}/4 \\ -3/4 + t/2 + 11e^{-3t}/2 \\ -1/8 + t/4 - 33e^{-3t}/4 \end{pmatrix}.
\end{aligned}
$$

9. The eigenvalues of A are given by $\lambda_1 = -2$, $\lambda_2 = -1$ and $\lambda_3 = 1$ with eigenvectors

$$
\mathbf{v}_1 = \begin{pmatrix} 1 \\ -2 \\ 1 \end{pmatrix}, \quad \mathbf{v}_2 = \begin{pmatrix} -1 \\ 0 \\ 1 \end{pmatrix}, \quad \mathbf{v}_3 = \begin{pmatrix} 1 \\ 1 \\ 1 \end{pmatrix}.
$$

Therefore, three solutions of the homogeneous equation are given by

$$
\mathbf{x}_1(t) = e^{-2t}\begin{pmatrix} 1 \\ -2 \\ 1 \end{pmatrix}, \quad \mathbf{x}_2(t) = e^{-t}\begin{pmatrix} -1 \\ 0 \\ 1 \end{pmatrix}, \quad \mathbf{x}_3(t) = e^{t}\begin{pmatrix} 1 \\ 1 \\ 1 \end{pmatrix},
$$

and

$$
\mathbf{X}(t) = \begin{pmatrix} e^{-2t} & -e^{-t} & e^{t} \\ -2e^{-2t} & 0 & e^{t} \\ e^{-2t} & e^{-t} & e^{t} \end{pmatrix}
$$

is a fundamental matrix for this equation. Further,

$$
\mathbf{X}^{-1}(s) = \frac{1}{6}\begin{pmatrix} e^{2s} & -2e^{2s} & e^{2s} \\ -3e^{s} & 0 & 3e^{s} \\ 2e^{-s} & 2e^{-s} & 2e^{-s} \end{pmatrix}.
$$

Therefore,

$$
\begin{aligned}
\int_{t_1}^{t}\mathbf{X}^{-1}(s)\mathbf{g}(s)\,ds \;=&\; \frac{1}{6}\int_{t_1}^{t}\begin{pmatrix} e^{2s} & -2e^{2s} & e^{2s} \\ -3e^{s} & 0 & 3e^{s} \\ 2e^{-s} & 2e^{-s} & 2e^{-s} \end{pmatrix}\begin{pmatrix} 0 \\ -\sin s \\ 0 \end{pmatrix}ds \\[2mm]
=&\; \frac{1}{6}\int_{t_1}^{t}\begin{pmatrix} 2e^{2s}\sin s \\ 0 \\ -2e^{-s}\sin s \end{pmatrix}ds = \frac{1}{6}\begin{pmatrix} -(2/5)e^{2t}\cos t + (4/5)e^{2t}\sin t \\ 0 \\ e^{-t}\cos t + e^{-t}\sin t \end{pmatrix} + \mathbf{c}.
\end{aligned}
$$

Therefore, the general solution is given by

$$
\mathbf{x}(t) = \mathbf{X}(t)\mathbf{c} + \mathbf{X}(t) \int_{t_1}^{t} \mathbf{X}^{-1}(s)\mathbf{g}(s)\, ds = \begin{pmatrix} e^{-2t} & -e^{-t} & e^{t} \\ -2e^{-2t} & 0 & e^{t} \\ e^{-2t} & e^{-t} & e^{t} \end{pmatrix} \mathbf{c}
$$

$$
+ \frac{1}{6} \begin{pmatrix} e^{-2t} & -e^{-t} & e^{t} \\ -2e^{-2t} & 0 & e^{t} \\ e^{-2t} & e^{-t} & e^{t} \end{pmatrix} \begin{pmatrix} -(2/5)e^{2t}\cos t + (4/5)e^{2t}\sin t \\ 0 \\ e^{-t}\cos t + e^{-t}\sin t \end{pmatrix}
$$

$$
= c_1 e^{-2t} \begin{pmatrix} 1 \\ -2 \\ 1 \end{pmatrix} + c_2 e^{-t} \begin{pmatrix} -1 \\ 0 \\ 1 \end{pmatrix} + c_3 e^{t} \begin{pmatrix} 1 \\ 1 \\ 1 \end{pmatrix} + \begin{pmatrix} (1/10)\cos t + (3/10)\sin t \\ -(1/10)\sin t + (3/10)\cos t \\ (1/10)\cos t + (3/10)\sin t \end{pmatrix}.
$$

11. Assuming another lattice point to the right of x_3 which is an absorbing boundary, the system of equations can be written as

$$
\begin{pmatrix} x_1 \\ x_2 \\ x_3 \end{pmatrix}' = k \begin{pmatrix} -2 & 1 & 0 \\ 1 & -2 & 1 \\ 0 & 1 & -2 \end{pmatrix} \begin{pmatrix} x_1 \\ x_2 \\ x_3 \end{pmatrix} + k \begin{pmatrix} f(t) \\ 0 \\ 0 \end{pmatrix}.
$$

For $k = 1$, $f(t) = 1$ with the given initial conditions, we show component plots of x_1, x_2, x_3 below.

As $t \to \infty$, we see that $\mathbf{x}(t) \to (3/4, 1/2, 1/4)^T$. For $k = 1$, $f(t) = 1 - \cos t$, we show component plots of x_1, x_2, x_3 below.

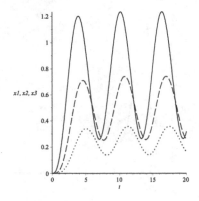

For $k = 1$, $f(t) = 1 - \cos 4t$, we show component plots of x_1, x_2, x_3 below.

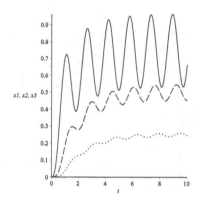

13. Using the values specified for m_1, m_2, k_1, k_2, the gain function is given by

$$\mathbf{G}(i\omega) = -(\mathbf{A} - i\omega\mathbf{I})^{-1}\mathbf{B}$$

where

$$\mathbf{A} = \begin{pmatrix} 0 & 0 & 1 & 0 \\ 0 & 0 & 0 & 1 \\ -2 & 3/2 & -\gamma/2 & 0 \\ 4/3 & -3 & 0 & -4\gamma/9 \end{pmatrix}$$

and

$$\mathbf{B} = \begin{pmatrix} 0 \\ 0 \\ e^{i\omega t} \\ 0 \end{pmatrix}.$$

We can compute $|G_1(i\omega)|$, $|G_2(i\omega)|$ by looking at the first and second components of the function $G(i\omega)$ above. For $\gamma = 1$, we have

$$|G_1(i\omega)| = \left| \frac{-18\omega^2 + 8i\omega + 54}{18\omega^4 - 17i\omega^3 - 94\omega^2 + 43i\omega + 72} \right|$$

$$|G_2(i\omega)| = \left| \frac{24}{18\omega^4 - 17i\omega^3 - 94\omega^2 + 43i\omega + 72} \right|$$

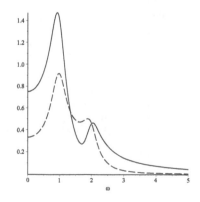

The gain functions corresponding to $\gamma = 0.5$ and $\gamma = 0.1$ are shown below:

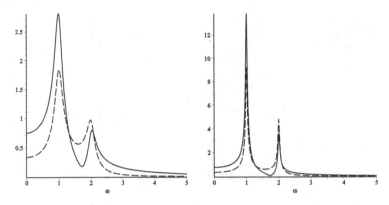

15. For

$$\mathbf{x}_p(t) = \cos t \left(\begin{array}{c} a_1 \\ a_2 \end{array} \right) + \sin t \left(\begin{array}{c} b_1 \\ b_2 \end{array} \right)$$

we can see that

$$\mathbf{x_p}' = -\sin t \left(\begin{array}{c} a_1 \\ a_2 \end{array} \right) + \cos t \left(\begin{array}{c} b_1 \\ b_2 \end{array} \right).$$

Therefore,

$$\left(\begin{array}{c} -a_1 \sin t + b_1 \cos t \\ -a_2 \sin t + b_2 \cos t \end{array} \right) = \left(\begin{array}{c} (-2a_1 + a_2)\cos t + (-2b_1 + b_2 + 1)\sin t \\ (a_1 - 2a_2)\cos t + (b_1 - 2b_2)\sin t \end{array} \right).$$

Equating like coefficients, we have $-2a_1 + a_2 - b_1 = 0$, $-2b_1 + b_2 + a_1 + 1 = 0$, $a_1 - 2a_2 - b_2 = 0$, $b_1 - 2b_2 + a_2 = 0$. Solving this system of equations, we have $a_1 = -3/10$, $a_2 = -1/5$, $b_1 = 2/5$, $b_2 = 1/10$. Therefore,

$$\mathbf{x_p}(t) = \cos t \left(\begin{array}{c} -3/10 \\ -1/5 \end{array} \right) + \sin t \left(\begin{array}{c} 2/5 \\ 1/10 \end{array} \right).$$

6.7   Defective Matrices

1. The only eigenvalue of A is $\lambda = 1$ with multiplicity 2. We see that

$$A - \lambda I = \left(\begin{array}{cc} 3 & -9 \\ 1 & -3 \end{array} \right) \quad \text{and} \quad (A - \lambda I)^2 = \left(\begin{array}{cc} 0 & 0 \\ 0 & 0 \end{array} \right).$$

Therefore, two linearly independent solutions of $(A - \lambda I)^2 \mathbf{v} = 0$ are

$$\mathbf{v}_1 = \left(\begin{array}{c} 1 \\ 0 \end{array} \right), \quad \text{and} \quad \mathbf{v}_2 = \left(\begin{array}{c} 0 \\ 1 \end{array} \right).$$

Then by Theorem 6.7.1, two linearly independent solutions of the system will be given by the following equation:

$$\mathbf{x}_k(t) = e^t [\mathbf{v}_k + t(A - I)\mathbf{v}_k]$$

for $k = 1, 2$. Using the values for $\mathbf{v}_k$ and $A - \lambda I$ above, we have

$$\mathbf{x}_1(t) = e^t\left[\begin{pmatrix} 1 \\ 0 \end{pmatrix} + t\begin{pmatrix} 3 \\ 1 \end{pmatrix}\right] \quad \text{and} \quad \mathbf{x}_2(t) = e^t\left[\begin{pmatrix} 0 \\ 1 \end{pmatrix} + t\begin{pmatrix} -9 \\ -3 \end{pmatrix}\right].$$

Therefore,

$$\mathbf{X}(t) = e^t\begin{pmatrix} 1 + 3t & -9t \\ t & 1 - 3t \end{pmatrix}$$

is a fundamental matrix for this system.

3. The only eigenvalue of A is $\lambda = 2$ with multiplicity 3. We see that

$$A - \lambda I = \begin{pmatrix} -1 & 1 & 1 \\ 2 & -1 & -1 \\ -3 & 2 & 2 \end{pmatrix}, (A - \lambda I)^2 = \begin{pmatrix} 0 & 0 & 0 \\ -1 & 1 & 1 \\ 1 & -1 & -1 \end{pmatrix}, (A - \lambda I)^3 = \begin{pmatrix} 0 & 0 & 0 \\ 0 & 0 & 0 \\ 0 & 0 & 0 \end{pmatrix}.$$

Therefore, three linearly independent solutions of $(A - \lambda I)^3\mathbf{v} = 0$ are

$$\mathbf{v}_1 = \begin{pmatrix} 1 \\ 0 \\ 0 \end{pmatrix}, \quad \mathbf{v}_2 = \begin{pmatrix} 0 \\ 1 \\ 0 \end{pmatrix}, \quad \mathbf{v}_3 = \begin{pmatrix} 0 \\ 0 \\ 1 \end{pmatrix}.$$

Then by Theorem 6.7.1, three linearly independent solutions of the system will be given by the following equation:

$$\mathbf{x}_k(t) = e^{2t}\left[\mathbf{v_k} + t(A - 2I)\mathbf{v_k} + \frac{t^2}{2}(A - 2I)^2\mathbf{v_k}\right]$$

for $k = 1, 2, 3$. Using the values for $\mathbf{v_k}$ and $A - \lambda I$ above, we have

$$\mathbf{x}_1(t) = e^{2t}\left[\begin{pmatrix} 1 \\ 0 \\ 0 \end{pmatrix} + t\begin{pmatrix} -1 \\ 2 \\ -3 \end{pmatrix} + \frac{t^2}{2}\begin{pmatrix} 0 \\ -1 \\ 1 \end{pmatrix}\right]$$

$$\mathbf{x}_2(t) = e^{2t}\left[\begin{pmatrix} 0 \\ 1 \\ 0 \end{pmatrix} + t\begin{pmatrix} 1 \\ -1 \\ 2 \end{pmatrix} + \frac{t^2}{2}\begin{pmatrix} 0 \\ 1 \\ -1 \end{pmatrix}\right]$$

$$\mathbf{x}_3(t) = e^{2t}\left[\begin{pmatrix} 0 \\ 0 \\ 1 \end{pmatrix} + t\begin{pmatrix} 1 \\ -1 \\ 2 \end{pmatrix} + \frac{t^2}{2}\begin{pmatrix} 0 \\ 1 \\ -1 \end{pmatrix}\right].$$

Therefore,

$$\mathbf{X}(t) = e^{2t}\begin{pmatrix} 1 - t & t & t \\ 2t - t^2/2 & 1 - t + t^2/2 & -t + t^2/2 \\ -3t + t^2/2 & 2t - t^2/2 & 1 + 2t - t^2/2 \end{pmatrix}$$

is a fundamental matrix for this system.

5. The eigenvalues of A are $\lambda = -1$ with multiplicity 1 and $\lambda = 2$ with multiplicity 2. For $\lambda = -1$,

$$A - \lambda I = \begin{pmatrix} -6 & 9 & -6 \\ -8 & 12 & -7 \\ -2 & 3 & 0 \end{pmatrix} \rightarrow \begin{pmatrix} 1 & -3/2 & 0 \\ 0 & 0 & 1 \\ 0 & 0 & 0 \end{pmatrix}.$$

Therefore,

$$\mathbf{v}_1 = \begin{pmatrix} 3/2 \\ 1 \\ 0 \end{pmatrix}$$

is an eigenvector associated with $\lambda = -1$ and

$$\mathbf{x}_1(t) = e^{-t} \begin{pmatrix} 3/2 \\ 1 \\ 0 \end{pmatrix}$$

is a solution of the system. For $\lambda = 2$,

$$A - \lambda I = \begin{pmatrix} -9 & 9 & -6 \\ -8 & 9 & -7 \\ -2 & 3 & -3 \end{pmatrix} \rightarrow \begin{pmatrix} 1 & -1 & 2/3 \\ 0 & 1 & -5/3 \\ 0 & 0 & 0 \end{pmatrix}.$$

As this eigenspace is only one-dimensional, we look at $(A - \lambda I)^2$. We see that for $\lambda = 2$,

$$(A - \lambda I)^2 = \begin{pmatrix} 21 & -18 & 9 \\ 14 & -12 & 6 \\ 0 & 0 & 0 \end{pmatrix}.$$

To find two solutions associated with the eigenvalue $\lambda = 2$, we look for two linearly independent solutions of $(A - \lambda I)^2 \mathbf{v} = 0$. We see that

$$(A - \lambda I)^2 \rightarrow \begin{pmatrix} 7 & -6 & 3 \\ 0 & 0 & 0 \\ 0 & 0 & 0 \end{pmatrix}$$

after elementary row operations. Therefore, any vector $\mathbf{v}$ lying in the plane $7x - 6y + 3z = 0$ will satisfy the equation $(A - \lambda I)^2 \mathbf{v} = 0$. We see that

$$\mathbf{v}_2 = \begin{pmatrix} 6 \\ 7 \\ 0 \end{pmatrix}, \qquad \mathbf{v}_3 = \begin{pmatrix} 3 \\ 0 \\ -7 \end{pmatrix}$$

are two linearly independent vectors satisfying this condition. Then, using Theorem 6.7.1, two solutions of the system will be given by the equation

$$\mathbf{x}_k(t) = e^{2t}[\mathbf{v}_k + t(A - 2I)\mathbf{v}_k]$$

for $k = 2, 3$. Using the values for $\mathbf{v}_k$ and $A - \lambda I$ above, we have

$$\mathbf{x}_2(t) = e^{2t}\left[\begin{pmatrix} 6 \\ 7 \\ 0 \end{pmatrix} + t\begin{pmatrix} 9 \\ 15 \\ 9 \end{pmatrix}\right] \quad \text{and} \quad \mathbf{x}_3(t) = e^{2t}\left[\begin{pmatrix} 3 \\ 0 \\ -7 \end{pmatrix} + t\begin{pmatrix} 15 \\ 25 \\ 15 \end{pmatrix}\right].$$

Therefore,

$$\mathbf{X}(t) = \begin{pmatrix} 3e^{-t}/2 & e^{2t}(6 + 9t) & e^{2t}(3 + 15t) \\ e^{-t} & e^{2t}(7 + 15t) & 25te^{2t} \\ 0 & 9te^{2t} & e^{2t}(-7 + 15t) \end{pmatrix}$$

is a fundamental matrix for this system.

7. The eigenvalues of A are $\lambda = -1 \pm i$ and $\lambda = -2$ with multiplicity 2. First, $\lambda = -1 + i$, has an eigenvector

$$\mathbf{v}_1 = \begin{pmatrix} -11 - 3i \\ -5 - 5i \\ -1 + 7i \\ 10 \end{pmatrix}.$$

Therefore,

$$\mathbf{u}(t) = e^{(-1+i)t}\begin{pmatrix} -11 - 3i \\ -5 - 5i \\ -1 + 7i \\ 10 \end{pmatrix}$$

is a solution of this system. Taking the real and imaginary parts of $\mathbf{u}$, we have the following two linearly independent real-valued solutions,

$$\mathbf{x}_1(t) = e^{-t}\left[\begin{pmatrix} -11 \\ -5 \\ -1 \\ 10 \end{pmatrix}\cos t + \begin{pmatrix} 3 \\ 5 \\ -7 \\ 0 \end{pmatrix}\sin t\right], \ \mathbf{x}_2(t) = e^{-t}\left[\begin{pmatrix} -3 \\ -5 \\ 7 \\ 0 \end{pmatrix}\cos t + \begin{pmatrix} -11 \\ -5 \\ -1 \\ 10 \end{pmatrix}\sin t\right].$$

Next, we look at the eigenvalue $\lambda = -2$. For $\lambda = -2$,

$$A - \lambda I = \begin{pmatrix} -6 & -16 & -16 & -17 \\ -2 & -8 & -8 & -7 \\ -2 & 0 & 0 & -3 \\ 6 & 14 & 14 & 16 \end{pmatrix} \rightarrow \begin{pmatrix} 1 & 0 & 0 & 3/2 \\ 0 & 1 & 1 & 1/2 \\ 0 & 0 & 0 & 0 \\ 0 & 0 & 0 & 0 \end{pmatrix}.$$

Therefore,

$$\mathbf{v}_3 = \begin{pmatrix} 3 \\ 1 \\ 0 \\ -2 \end{pmatrix}, \quad \mathbf{v}_4 = \begin{pmatrix} 3 \\ 0 \\ 1 \\ -2 \end{pmatrix}$$

are two linearly independent solutions of $(A - \lambda I)\mathbf{v} = 0$. Then by Theorem 6.7.1, for $\lambda = -2$, $k = 3, 4$,

$$\mathbf{x}_k(t) = e^{-2t}\mathbf{v}_k$$

are two more linearly independent solutions of the system. Plugging in the values for $\mathbf{v}_3, \mathbf{v}_4$, we have

$$\mathbf{x}_3(t) = e^{-2t} \begin{pmatrix} 3 \\ 1 \\ 0 \\ -2 \end{pmatrix}, \quad \mathbf{x}_4(t) = e^{-2t} \begin{pmatrix} 3 \\ 0 \\ 1 \\ -2 \end{pmatrix}.$$

Therefore,

$$\mathbf{X}(t) = \begin{pmatrix} -11e^{-t}\cos t + 3e^{-t}\sin t & -3e^{-t}\cos t - 11e^{-t}\sin t & 3e^{-2t} & 3e^{-2t} \\ -5e^{-t}\cos t + 5e^{-t}\sin t & -5e^{-t}\cos t - 5e^{-t}\sin t & e^{-2t} & 0 \\ -e^{-t}\cos t - 7e^{-t}\sin t & 7e^{-t}\cos t - e^{-t}\sin t & 0 & e^{-2t} \\ 10e^{-t}\cos t & 10e^{-t}\sin t & -2e^{-2t} & -2e^{-2t} \end{pmatrix}$$

is a fundamental matrix for this system.

9. We find the eigenvalues first.

$$A - \lambda I = \begin{pmatrix} 1 - \lambda & -4 \\ 4 & -7 - \lambda \end{pmatrix}.$$

Therefore, $\det(A - \lambda I) = \lambda^2 + 6\lambda + 9 = (\lambda + 3)^2$. Thus $\lambda = -3$ is the only eigenvalue of A. For $\lambda = -3$,

$$(A - \lambda I)^2 = \begin{pmatrix} 0 & 0 \\ 0 & 0 \end{pmatrix}.$$

Therefore,

$$\mathbf{v}_1 = \begin{pmatrix} 1 \\ 0 \end{pmatrix}, \quad \mathbf{v}_2 = \begin{pmatrix} 0 \\ 1 \end{pmatrix}$$

are linearly independent solutions of $(A - \lambda I)^2 \mathbf{v} = 0$. Then, by Theorem 6.7.1, for $\lambda = -3$, $k = 1, 2$, $\mathbf{x}_k(t) = e^t[\mathbf{v}_k + t(A - \lambda I)\mathbf{v}_k]$ are linearly independent solutions of the system. Plugging in for $\mathbf{v}_k$ and $A - \lambda I$, we have

$$\mathbf{x}_1(t) = e^{-3t}\left[\begin{pmatrix} 1 \\ 0 \end{pmatrix} + t\begin{pmatrix} 4 \\ 4 \end{pmatrix}\right] \quad \text{and} \quad \mathbf{x}_2(t) = e^{-3t}\left[\begin{pmatrix} 0 \\ 1 \end{pmatrix} + t\begin{pmatrix} -4 \\ -4 \end{pmatrix}\right].$$

Therefore,

$$\mathbf{X}(t) = e^{-3t}\begin{pmatrix} 1 + 4t & -4t \\ 4t & 1 - 4t \end{pmatrix}$$

is a fundamental matrix; also, $\mathbf{X}(0) = I$, thus $e^{At} = \mathbf{X}(t)$. Therefore, the solution of this initial value problem is given by

$$\mathbf{x}(t) = e^{At}\mathbf{x}_0 = e^{-3t}\begin{pmatrix} 1 + 4t & -4t \\ 4t & 1 - 4t \end{pmatrix}\begin{pmatrix} 7 \\ 1 \end{pmatrix} = e^{-3t}\begin{pmatrix} 7 + 24t \\ 1 + 24t \end{pmatrix}.$$

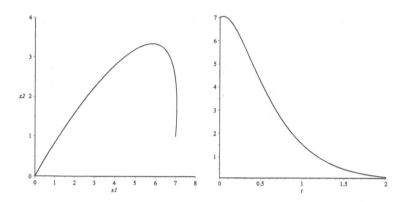

11.   The eigenvalues of A are $\lambda = 2$ which has multiplicity one and $\lambda = 1$ which has multiplicity two. First, for $\lambda = 2$,

$$\mathbf{v}_1 = \begin{pmatrix} 0 \\ -3 \\ 1 \end{pmatrix}$$

is a corresponding eigenvector, and, therefore,

$$\mathbf{x}_1(t) = e^{2t} \begin{pmatrix} 0 \\ -3 \\ 1 \end{pmatrix}$$

is one solution of the system. Next, for $\lambda = 1$,

$$A - \lambda I = \begin{pmatrix} 3 & 1 & 3 \\ 6 & 3 & 6 \\ -5 & -2 & -5 \end{pmatrix}, \qquad (A - \lambda I)^2 = \begin{pmatrix} 0 & 0 & 0 \\ 6 & 3 & 6 \\ -2 & -1 & -2 \end{pmatrix}.$$

Thus

$$\mathbf{v}_2 = \begin{pmatrix} 0 \\ -2 \\ 1 \end{pmatrix}, \qquad \mathbf{v}_3 = \begin{pmatrix} 1 \\ -2 \\ 0 \end{pmatrix}$$

are two linearly independent solutions of $(A - \lambda I)^2 \mathbf{v} = 0$. Therefore, by Theorem 6.7.1, for $\lambda = 1$, $k = 2, 3$, $\mathbf{x}_k(t) = e^t [\mathbf{v}_k + t(A - I)\mathbf{v}_k]$ are two linearly independent solutions of the system. Plugging in for $\mathbf{v}_k$ and $A - \lambda I$, we have the following two solutions,

$$\mathbf{x}_2(t) = e^t \left[\begin{pmatrix} 0 \\ -2 \\ 1 \end{pmatrix} + t \begin{pmatrix} 1 \\ 0 \\ -1 \end{pmatrix} \right] \quad \text{and} \quad \mathbf{x}_3(t) = e^t \left[\begin{pmatrix} 1 \\ -2 \\ 0 \end{pmatrix} + t \begin{pmatrix} 1 \\ 0 \\ -1 \end{pmatrix} \right].$$

Therefore,

$$\mathbf{X}(t) = \begin{pmatrix} 0 & te^t & e^t + te^t \\ -3e^{2t} & -2e^t & -2e^t \\ e^{2t} & e^t - te^t & -te^t \end{pmatrix}$$

is a fundamental matrix. Now

$$\mathbf{X}(0) = \begin{pmatrix} 0 & 0 & 1 \\ -3 & -2 & -2 \\ 1 & 1 & 0 \end{pmatrix}, \quad \text{so} \quad \mathbf{X}^{-1}(0) = \begin{pmatrix} -2 & -1 & -2 \\ 2 & 1 & 3 \\ 1 & 0 & 0 \end{pmatrix}.$$

Thus

$$e^{At} = \mathbf{X}(t)\mathbf{X}^{-1}(0) = \begin{pmatrix} 3te^t + e^t & te^t & 3te^t \\ 6e^{2t} - 6e^t & 3e^{2t} - 2e^t & 6e^{2t} - 6e^t \\ -2e^{2t} + 2e^t - 3te^t & -e^{2t} + e^t - te^t & -2e^{2t} + 3e^t - 3te^t \end{pmatrix}.$$

Therefore, the solution of this initial value problem is given by

$$\begin{aligned} \mathbf{x}(t) &= e^{At}\mathbf{x}_0 = \begin{pmatrix} 3te^t + e^t & te^t & 3te^t \\ 6e^{2t} - 6e^t & 3e^{2t} - 2e^t & 6e^{2t} - 6e^t \\ -2e^{2t} + 2e^t - 3te^t & -e^{2t} + e^t - te^t & -2e^{2t} + 3e^t - 3te^t \end{pmatrix} \begin{pmatrix} 1 \\ -2 \\ 5 \end{pmatrix} \\ &= \begin{pmatrix} 16te^t + e^t \\ 30e^{2t} - 32e^t \\ -10e^{2t} + 15e^t - 16te^t \end{pmatrix}. \end{aligned}$$

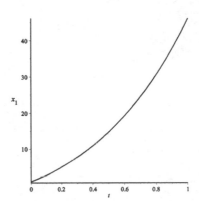

Chapter 7

Nonlinear Differential Equations and Stability

7.1 Autonomous Systems and Stability

1.(a) $-2y + xy = 0$ implies $y(-2 + x) = 0$ implies $x = 2$ or $y = 0$. Then, $x + 4xy = 0$ implies $x(1 + 4y) = 0$ implies $x = 0$ or $y = -1/4$. Therefore, the critical points are $(2, -1/4)$ and $(0, 0)$.

(b)

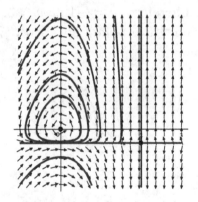

(c) The critical point $(0, 0)$ is a center, therefore, stable. The critical point $(2, -1/4)$ is a saddle point, therefore, unstable.

3.(a) The equation $2x - x^2 - xy = 0$ implies $x = 0$ or $x + y = 2$. The equation $3y - 2y^2 - 3xy = 0$ implies $y = 0$ or $3x + 2y = 3$. Solving these equations, we have the critical points $(0, 0)$, $(0, 3/2)$, $(2, 0)$, and $(-1, 3)$.

(b)

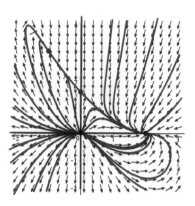

(c) The critical point $(0,0)$ is an unstable node. The critical point $(0, 3/2)$ is a saddle point, therefore, unstable. The critical point $(2, 0)$ is an asymptotically stable node. The critical point $(-1, 3)$ is an asymptotically stable node.

(d) For $(2, 0)$, the basin of attraction is the first quadrant plus the region in the fourth quadrant bounded by the trajectories heading away from $(0, 0)$ but looping back towards $(2, 0)$. For $(-1, 3)$, the basin of attraction is bounded to the right by the y−axis and to the left by those trajectories leaving $(0, 0)$ but looping back towards $(-1, 3)$.

5.(a) The equation $x(6 - x - y) = 0$ implies $x = 0$ or $x + y = 6$. If $x = 0$, the equation $-x + 7y - 2xy = 0$ implies $y = 0$. If $x + y = 6$, then the equation $-x + 7y - 2xy = 0$ can be reduced to $y^2 - 2y - 3 = 0$. Therefore, $y = -1$ or $y = 3$. Now if $y = -1$, then $x = 7$. If $y = 3$, then $x = 3$. Therefore, the critical points are $(0, 0)$, $(7, -1)$, and $(3, 3)$.

(b)

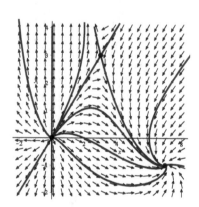

(c) The critical point $(0, 0)$ is an unstable node. The critical point $(3, 3)$ is a saddle point, therefore, unstable. The critical point $(7, -1)$ is an asymptotically stable spiral point.

(d) For $(7, -1)$, the basin of attraction is bounded to the left by the y−axis and above by the trajectories heading into the unstable critical point $(3, 3)$.

7.(a) The equation $(2 - y)(x - y) = 0$ implies $y = 2$ or $x = y$. The equation $(1 + x)(x + y) = 0$ implies $x = -1$ or $x = -y$. The solutions of those two equations are the critical points $(0, 0)$, $(-1, 2)$, $(-2, 2)$, and $(-1, -1)$.

(b)

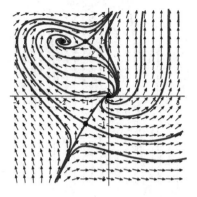

(c) The critical point $(0,0)$ is an unstable spiral point. The critical point $(-1,2)$ is a saddle point, therefore, unstable. The critical point $(-2,2)$ is an asymptotically stable spiral point. The critical point $(-1,-1)$ is a saddle point, therefore, unstable.

(d) For $(-2,2)$, the basin of attraction is a circular region lying mainly in the second quadrant, close to the x and y-axes, bounded by trajectories leaving the unstable spiral point $(0,0)$ and approaching the saddle $(-1,2)$.

9.(a) The equation $(3+x)(2y-x) = 0$ implies $x = -3$ or $x = 2y$. The equation $(2-x)(y-x) = 0$ implies $x = 2$ or $x = y$. The solutions of these two equations are the critical points $(0,0)$, $(2,1)$, and $(-3,-3)$.

(b)

(c) The critical point $(0,0)$ is an asymptotically stable spiral point. The critical point $(2,1)$ is a saddle point, therefore, unstable. The critical point $(-3,-3)$ is also a saddle point, therefore, unstable.

(d) The basin of attraction of $(0,0)$ is bounded by the trajectories approaching the saddle point $(2,1)$.

11.(a) The equation $-y = 0$ implies $y = 0$. Then if $y = 0$, the equation $-3y - x(x-1)(x-2) = 0$ reduces to $x = 0,1,2$. Therefore, the critical points are $(0,0)$, $(1,0)$, and $(2,0)$.

(b)

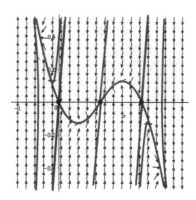

(c) The critical point $(0,0)$ is a saddle point, therefore, unstable. The critical point $(1,0)$ is an asymptotically stable node. The critical point $(2,0)$ is a saddle point, therefore, unstable.

(d) The basin of attraction for $(1,0)$ is bounded on the left by trajectories heading towards and then away from the unstable critical point $(0,0)$. The basin is bounded on the right by trajectories traveling towards and then away from the unstable critical point $(2,0)$.

13.(a) The equation $dx/dt = 0$ implies $x = 0$ or $x + 3y = 8$. The equation $dy/dt = 0$ implies $y = 0$, $x = 3$, or $x = -2$. Solving these two equations simultaneously, we see that the critical points are $(0,0)$, $(8,0)$, $(3,5/3)$, and $(-2,10/3)$.

(b)

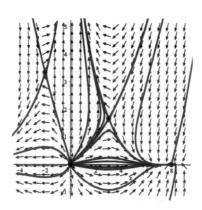

(c) The critical point $(0,0)$ is an unstable node. The critical point $(3,5/3)$ is a saddle point, therefore, unstable. The critical point $(8,0)$ is an asymptotically stable node. The critical point $(-2,10/3)$ is a saddle point, therefore, unstable.

(d) The basin of attraction for $(8,0)$ is the fourth quadrant plus the part of the first quadrant to the right of the two solutions approaching the saddle point $(3,5/3)$.

15.(a) The equation $dx/dt = 0$ implies $y = 1$, $x = 1$ or $x = -2$. The equation $dy/dt = 0$ implies $x = 0$ or $y = 2$. Solving these two equations simultaneously, we see that the critical points are $(0,1)$, $(1,2)$, and $(-2,2)$.

(b)

(c) The critical point $(0, 1)$ is a center, therefore, stable. The critical point $(1, 2)$ is a saddle point, therefore, unstable. The critical point $(-2, 2)$ is a saddle point, therefore, unstable.

17.(a) The equation $dx/dt = 0$ implies $y = 0$. The equation $dy/dt = 0$ implies $x = 0$ or $x = \pm\sqrt{6}$. Solving these two equations simultaneously, we see that the critical points are $(0, 0)$, $(\sqrt{6}, 0)$, and $(-\sqrt{6}, 0)$.

(b)

(c) The critical point $(0, 0)$ is a saddle point, therefore, unstable. The critical point $(\sqrt{6}, 0)$ is a center, therefore, stable. The critical point $(-\sqrt{6}, 0)$ is a center, therefore, stable.

19.(a) The trajectories are solutions of the differential equation

$$\frac{dy}{dx} = \frac{-\omega^2 \sin x}{y}.$$

Rewriting this equation as $\omega^2 \sin x\, dx + y\, dy = 0$, we see that this differential equation is exact with

$$\frac{\partial H}{\partial x} = \omega^2 \sin x \quad \text{and} \quad \frac{\partial H}{\partial y} = y.$$

Integrating the first equation, we find that $H(x, y) = -\omega^2 \cos x + f(y)$. Differentiating this equation with respect to y, we find that $H_y = f'(y) = y$ which implies that $f(y) = y^2/2 + C$. Therefore, the solutions of the differential equation are level curves of

$$H(x, y) = -\omega^2 \cos x + \frac{y^2}{2}.$$

Adding an arbitrary constant does not affect the trajectories. Therefore, the trajectories can be written as

$$\frac{1}{2}y^2 + \omega^2(1 - \cos x) = C,$$

where C is an arbitrary constant.

(b) Multiplying by mL^2 and reverting to the original physical variables, we obtain

$$\frac{1}{2}mL^2\left(\frac{d\theta}{dt}\right)^2 + mL^2\omega^2(1 - \cos\theta) = mL^2c.$$

Since $\omega^2 = g/L$, this equation can be written as

$$\frac{1}{2}mL^2\left(\frac{d\theta}{dt}\right)^2 + mgL(1 - \cos\theta) = E,$$

where $E = mL^2c$.

(c) The absolute velocity of the point mass is given by $v = Ld\theta/dt$. The kinetic energy of the mass is $T = mv^2/2$. Choosing the rest position as the datum, that is, the level of zero potential energy, the gravitational potential energy of the point mass is $V = mgL(1 - \cos\theta)$. It follows that the total energy, $T + V$, is constant along the trajectories.

21.(a) For initial velocity $v = 3$, the pendulum will swing back and forth about its equilibrium position. For initial velocity $v = 6$, the pendulum will not swing back and forth, but instead will continue to rotate about the origin indefinitely.

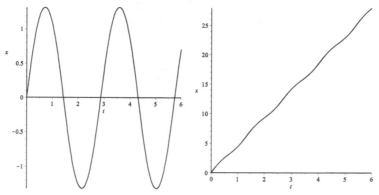

(b) By plotting the graphs of the solutions for different values of v, we conclude that the transition takes place at $v_c \approx 4.90$.

23. We compute:

$$\frac{d\Phi}{dt} = \frac{d\phi}{dt}(t - s) = F(\phi(t - s), \psi(t - s)) = F(\Phi, \Psi) = F(x, y)$$

and

$$\frac{d\Psi}{dt} = \frac{d\psi}{dt}(t - s) = G(\phi(t - s), \psi(t - s)) = F(\Phi, \Psi) = G(x, y).$$

Therefore, $\Phi(t), \Psi(t)$ is a solution for $\alpha + s < t < \beta + s$.

25. If we assume that a trajectory can reach a critical point (x_0, y_0) in a finite length of time, then we would have two trajectories passing through the same point. This contradicts the result in problem 24.

7.2   Almost Linear Systems

1.(a) The equation $dx/dt = 0$ implies $y = 2x$. The equation $dy/dt = 0$ implies $y = x^2$. Therefore, for these equations to both be satisfied, we need $2x = x^2$ which means $x = 0$ or $x = 2$. Thus the two critical points are $(0,0)$ and $(2,4)$.

(b) Here, we have $F(x,y) = -2x + y$ and $G(x,y) = x^2 - y$. Therefore, the Jacobian matrix for this system is

$$\mathbf{J}(x,y) = \begin{pmatrix} F_x & F_y \\ G_x & G_y \end{pmatrix} = \begin{pmatrix} -2 & 1 \\ 2x & -1 \end{pmatrix}.$$

Near the critical point $(0,0)$, the Jacobian matrix is

$$\mathbf{J}(0,0) = \begin{pmatrix} F_x(0,0) & F_y(0,0) \\ G_x(0,0) & G_y(0,0) \end{pmatrix} = \begin{pmatrix} -2 & 1 \\ 0 & -1 \end{pmatrix}$$

and the corresponding linear system near $(0,0)$ is

$$\frac{d}{dt} \begin{pmatrix} x \\ y \end{pmatrix} = \begin{pmatrix} -2 & 1 \\ 0 & -1 \end{pmatrix} \begin{pmatrix} x \\ y \end{pmatrix}.$$

Near the critical point $(2,4)$, the Jacobian matrix is

$$\mathbf{J}(2,4) = \begin{pmatrix} F_x(2,4) & F_y(2,4) \\ G_x(2,4) & G_y(2,4) \end{pmatrix} = \begin{pmatrix} -2 & 1 \\ 4 & -1 \end{pmatrix}$$

and the corresponding linear system near $(2,4)$ is

$$\frac{d}{dt} \begin{pmatrix} u \\ v \end{pmatrix} = \begin{pmatrix} -2 & 1 \\ 4 & -1 \end{pmatrix} \begin{pmatrix} u \\ v \end{pmatrix}$$

where $u = x - 2$ and $v = y - 4$.

(c) The eigenvalues of the linear system near $(0,0)$ are $\lambda = -1, -2$. From this, we can conclude that $(0,0)$ is an asymptotically stable node for the nonlinear system. The eigenvalues of the linear system near $(2,4)$ are $(-3 \pm \sqrt{17})/2$. Since one of these eigenvalues is positive and one is negative, the critical point $(2,4)$ is an unstable saddle point for the nonlinear system.

(d)

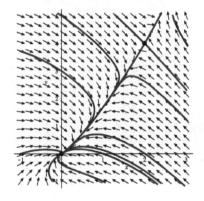

(e) The basin of attraction for the asymptotically stable critical point $(0,0)$ is bounded on the right by trajectories heading towards the critical point $(2,4)$.

3.(a) To find the critical points, we need to solve the equations $x = -y^2$ and $x = -2y$. In order for these two equations to be satisfied simultaneously, we need $y^2 = 2y$. Therefore, $y = 0$ or $y = 2$. Therefore, the two critical points are $(0,0)$ and $(-4,2)$.

(b) Here, we have $F(x,y) = x + y^2$ and $G(x,y) = x + 2y$. Therefore, the Jacobian matrix for this system is

$$\mathbf{J}(x,y) = \begin{pmatrix} F_x & F_y \\ G_x & G_y \end{pmatrix} = \begin{pmatrix} 1 & 2y \\ 1 & 2 \end{pmatrix}.$$

Near the critical point $(0,0)$, the Jacobian matrix is

$$\mathbf{J}(0,0) = \begin{pmatrix} F_x(0,0) & F_y(0,0) \\ G_x(0,0) & G_y(0,0) \end{pmatrix} = \begin{pmatrix} 1 & 0 \\ 1 & 2 \end{pmatrix}$$

and the corresponding linear system near $(0,0)$ is

$$\frac{d}{dt}\begin{pmatrix} x \\ y \end{pmatrix} = \begin{pmatrix} 1 & 0 \\ 1 & 2 \end{pmatrix}\begin{pmatrix} x \\ y \end{pmatrix}.$$

Near the critical point $(-4,2)$, the Jacobian matrix is

$$\mathbf{J}(-4,2) = \begin{pmatrix} F_x(-4,2) & F_y(-4,2) \\ G_x(-4,2) & G_y(-4,2) \end{pmatrix} = \begin{pmatrix} 1 & 4 \\ 1 & 2 \end{pmatrix}$$

and the corresponding linear system near $(-4,2)$ is

$$\frac{d}{dt}\begin{pmatrix} u \\ v \end{pmatrix} = \begin{pmatrix} 1 & 4 \\ 1 & 2 \end{pmatrix}\begin{pmatrix} u \\ v \end{pmatrix}$$

where $u = x + 4$ and $v = y - 2$.

(c) The eigenvalues of the linear system near $(0,0)$ are $\lambda = 1, 2$. From this, we can conclude that $(0,0)$ is an unstable node for the nonlinear system. The eigenvalues of the linear system near $(-4,2)$ are $\lambda = (3 \pm \sqrt{17})/2$. Since one of these eigenvalues is positive and one is negative, the critical point $(-4,2)$ is an unstable saddle point for the nonlinear system.

(d)

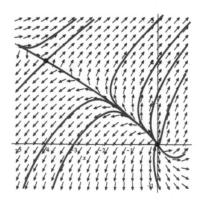

5.(a) To find the critical points, we need to solve the equations $(4 + x)(y - x) = 0$ and $(10 - x)(y + x) = 0$. Solving this system of equations, we see that the critical points are given by $(0, 0)$, $(10, 10)$, and $(-4, 4)$.

(b) Here, we have $F(x, y) = (4 + x)(y - x)$ and $G(x, y) = (10 - x)(y + x)$. Therefore, the Jacobian matrix for this system is

$$\mathbf{J}(x, y) = \begin{pmatrix} F_x & F_y \\ G_x & G_y \end{pmatrix} = \begin{pmatrix} -4 - 2x + y & 4 + x \\ 10 - y - 2x & 10 - x \end{pmatrix}.$$

Near the critical point $(0, 0)$, the Jacobian matrix is

$$\mathbf{J}(0, 0) = \begin{pmatrix} F_x(0, 0) & F_y(0, 0) \\ G_x(0, 0) & G_y(0, 0) \end{pmatrix} = \begin{pmatrix} -4 & 4 \\ 10 & 10 \end{pmatrix}$$

and the corresponding linear system near $(0, 0)$ is

$$\frac{d}{dt}\begin{pmatrix} x \\ y \end{pmatrix} = \begin{pmatrix} -4 & 4 \\ 10 & 10 \end{pmatrix}\begin{pmatrix} x \\ y \end{pmatrix}.$$

Near the critical point $(-4, 4)$, the Jacobian matrix is

$$\mathbf{J}(-4, 4) = \begin{pmatrix} F_x(-4, 4) & F_y(-4, 4) \\ G_x(-4, 4) & G_y(-4, 4) \end{pmatrix} = \begin{pmatrix} 8 & 0 \\ 14 & 14 \end{pmatrix}$$

and the corresponding linear system near $(-4, 4)$ is

$$\frac{d}{dt}\begin{pmatrix} u \\ v \end{pmatrix} = \begin{pmatrix} 8 & 0 \\ 14 & 14 \end{pmatrix}\begin{pmatrix} u \\ v \end{pmatrix}$$

where $u = x + 4$ and $v = y - 4$. Near the critical point $(10, 10)$, the Jacobian matrix is

$$\mathbf{J}(10, 10) = \begin{pmatrix} F_x(10, 10) & F_y(10, 10) \\ G_x(10, 10) & G_y(10, 10) \end{pmatrix} = \begin{pmatrix} -14 & 14 \\ -20 & 0 \end{pmatrix}$$

and the corresponding linear system near $(10, 10)$ is

$$\frac{d}{dt}\begin{pmatrix} u \\ v \end{pmatrix} = \begin{pmatrix} -14 & 14 \\ -20 & 0 \end{pmatrix}\begin{pmatrix} u \\ v \end{pmatrix}$$

where $u = x - 10$ and $v = y - 10$.

(c) The eigenvalues of the linear system near $(0, 0)$ are $\lambda = 3 \pm \sqrt{89}$. From this, we can conclude that $(0, 0)$ is an unstable saddle point for the nonlinear system. The eigenvalues of the linear system near $(-4, 4)$ are $\lambda = 8, 14$. From this, we can conclude that $(-4, 4)$ is an unstable node for the nonlinear system. The eigenvalues of the linear system near $(10, 10)$ are $\lambda = -7 \pm i\sqrt{231}$. From this, we can conclude that $(10, 10)$ is an asymptotically stable spiral point the nonlinear system.

(d)

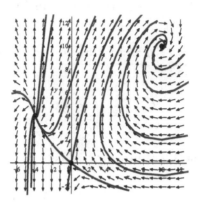

(e) The basin of attraction for the asymptotically stable point $(10, 10)$ is bounded below by trajectories heading in towards the origin and bounded on the left by trajectories heading away from $(-4, 4)$.

7.(a) To find the critical points, we need to solve the equations $1 - y = 0$ and $x^2 - y^2 = 0$. Solving this system of equations, we see that the critical points are given by $(-1, 1)$ and $(1, 1)$.

(b) Here, we have $F(x, y) = 1 - y$ and $G(x, y) = x^2 - y^2$. Therefore, the Jacobian matrix for this system is

$$\mathbf{J}(x, y) = \begin{pmatrix} F_x & F_y \\ G_x & G_y \end{pmatrix} = \begin{pmatrix} 0 & -1 \\ 2x & -2y \end{pmatrix}.$$

Near the critical point $(-1, 1)$, the Jacobian matrix is

$$\mathbf{J}(-1, 1) = \begin{pmatrix} 0 & -1 \\ -2 & -2 \end{pmatrix}.$$

Near the critical point $(1, 1)$, the Jacobian matrix is

$$\mathbf{J}(1, 1) = \begin{pmatrix} 0 & -1 \\ 2 & -2 \end{pmatrix}.$$

(c) The eigenvalues of the linear system near $(-1, 1)$ are $\lambda = -1 \pm \sqrt{3}$. From this, we can conclude that $(-1, 1)$ is a saddle point for the nonlinear system. The eigenvalues of the linear system near $(1, 1)$ are $\lambda = -1 \pm i$. From this, we can conclude that $(1, 1)$ is an asymptotically stable spiral point for the nonlinear system.

(d)

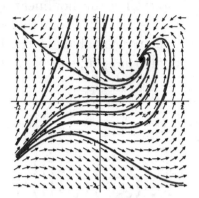

(e) The basin of attraction for the asymptotically stable point $(1,1)$ is bounded on the left by trajectories just to the left of the y-axis and below by trajectories heading away from the saddle point $(-1,1)$ and back towards $(1,1)$.

9.(a) To find the critical points, we need to solve the equations $-(x-y)(6-x-y) = 0$ and $x(4+y) = 0$. Solving these equations, we find that the critical points are $(0,0)$, $(-4,-4)$, $(0,6)$, and $(10,-4)$.

(b) Here, we have $F(x,y) = -(x-y)(6-x-y)$ and $G(x,y) = x(4+y)$. Therefore, the Jacobian matrix for this system is

$$\mathbf{J}(x,y) = \begin{pmatrix} F_x & F_y \\ G_x & G_y \end{pmatrix} = \begin{pmatrix} 2x-6 & 6-2y \\ 4+y & x \end{pmatrix}.$$

Near the critical point $(0,0)$, the Jacobian matrix is

$$\mathbf{J}(0,0) = \begin{pmatrix} -6 & 6 \\ 4 & 0 \end{pmatrix}.$$

Near the critical point $(0,6)$, the Jacobian matrix is

$$\mathbf{J}(0,6) = \begin{pmatrix} -6 & -6 \\ 10 & 0 \end{pmatrix}.$$

Near the critical point $(-4,-4)$, the Jacobian matrix is

$$\mathbf{J}(-4,-4) = \begin{pmatrix} -14 & 14 \\ 0 & -4 \end{pmatrix}.$$

Near the critical point $(10,-4)$, the Jacobian matrix is

$$\mathbf{J}(10,-4) = \begin{pmatrix} 14 & 14 \\ 0 & 10 \end{pmatrix}.$$

(c) The eigenvalues of the linear system near $(0,0)$ are $\lambda = -3 \pm \sqrt{33}$. From this, we can conclude that $(0,0)$ is an unstable saddle point for the nonlinear system. The eigenvalues of the linear system near $(0, 6)$ are $\lambda = -3 \pm i\sqrt{51}$. From this, we can conclude that $(0, 1)$ is an asymptotically stable spiral for the nonlinear system. The eigenvalues of the linear system near $(-4, -4)$ are $\lambda = -4, -14$. From this, we can conclude that $(-4, -4)$ is an asymptotically stable node. The eigenvalues of the linear system near $(10, -4)$ are $\lambda = 10, 14$. From this, we can conclude that $(10, -4)$ is an unstable node.

(d)

(e) Tha basin of attraction for $(0, 4)$ is bounded below by the two trajectories approaching $(0, 0)$. The basin of attraction for $(-4, -4)$ is bounded above by the same two trajectories.

11.(a) To find the critical points, we need to solve the equations $2x + y + xy^3 = 0$ and $x - 2y - xy = 0$. Substituting $y = x/(x+2)$ into the first equation results in

$$3x^4 + 13x^3 + 28x^2 + 20x = 0.$$

One root of this equation is $x = 0$. The only other real root is

$$x = \frac{1}{9} \left[\left(287 + 18\sqrt{2019} \right)^{1/3} - 83 \left(287 + 18\sqrt{2019} \right)^{-1/3} - 13 \right].$$

Therefore, the critical points are $(0, 0)$ and (approximately) $(-1.19345, -1.4797)$.

(b) Here, we have $F(x, y) = 2x + y + xy^3$ and $G(x, y) = x - 2y - xy$. Therefore, the Jacobian matrix for this system is

$$\mathbf{J}(x, y) = \begin{pmatrix} F_x & F_y \\ G_x & G_y \end{pmatrix} = \begin{pmatrix} 2 + y^3 & 1 + 3xy^2 \\ 1 - y & -2 - x \end{pmatrix}.$$

Near the critical point $(0, 0)$, the Jacobian matrix is

$$\mathbf{J}(0, 0) = \begin{pmatrix} 2 & 1 \\ 1 & -2 \end{pmatrix}.$$

Near the critical point $(-1.19345, -1.4797)$, the Jacobian matrix is

$$\mathbf{J}(-1.19345, -1.4797) = \begin{pmatrix} -1.2399 & -6.8393 \\ 2.4797 & -0.8065 \end{pmatrix}.$$

(c) The eigenvalues of the linear system near $(0,0)$ are $\lambda = \pm\sqrt{5}$. From this, we can conclude that $(0,0)$ is an unstable saddle point for the nonlinear system. The eigenvalues of the linear system near $(-1.19345, -1.4797)$ are $\lambda = -1.0232 \pm 4.1125i$. From this, we can conclude that $(-1.19345, -1.4797)$ is an asymptotically stable spiral point for the nonlinear system.

(d)

(e) The basin of attraction for the stable spiral point is bounded by the two trajectories approaching $(0,0)$.

13.(a) To find the critical points, we need to solve the equations $x - y^2 = 0$ and $y - x^2 = 0$. Substituting $y = x^2$ into the first equation, results in $x - x^4 = 0$ which has real roots $x = 0, 1$. Therefore, the critical points are $(0,0)$ and $(1,1)$.

(b) Here, we have $F(x, y) = x - y^2$ and $G(x, y) = y - x^2$. Therefore, the Jacobian matrix for this system is

$$\mathbf{J}(x,y) = \begin{pmatrix} F_x & F_y \\ G_x & G_y \end{pmatrix} = \begin{pmatrix} 1 & -2y \\ -2x & 1 \end{pmatrix}.$$

Near the critical point $(0,0)$, the Jacobian matrix is

$$\mathbf{J}(0,0) = \begin{pmatrix} 1 & 0 \\ 0 & 1 \end{pmatrix}.$$

Near the critical point $(1,1)$, the Jacobian matrix is

$$\mathbf{J}(1,1) = \begin{pmatrix} 1 & -2 \\ -2 & 1 \end{pmatrix}.$$

(c) There is a repeated eigenvalue $\lambda = 1$ for the linear system near $(0,0)$. Based on this information, we cannot make a conclusion about the nature of the critical point near $(0,0)$ for the nonlinear system, only that it is unstable and it is either a node or a spiral. The eigenvalues of the linear system near $(1,1)$ are $\lambda = 3, -1$. From this, we can conclude that the critical point $(1,1)$ is a saddle.

(d)

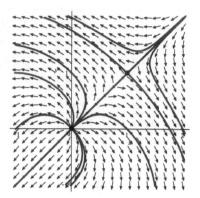

Upon looking at the phase portrait, we see that the critical point $(0,0)$ is an unstable node.

15.(a) To find the critical points, we need to solve the equations

$$-2x - y - x(x^2 + y^2) = 0$$
$$x - y + y(x^2 + y^2) = 0.$$

It is clear that $(0,0)$ is a critical point. Solving the first equation for y, we find that

$$y = \frac{-1 \pm \sqrt{1 - 8x^2 - 4x^4}}{2x}.$$

Substitution of these relations into the second equation results in two equations of the form $f_1(x) = 0$ and $f_2(x) = 0$. Plotting these functions, we note that only $f_1(x) = 0$ has real roots given by $x \approx \pm 0.33076$. It follows that the additional critical points are (approximately) $(-0.33076, 1.0924)$ and $(0.33076, -1.0924)$.

(b) Here, we have $F(x,y) = -2x - y - x(x^2 + y^2)$ and $G(x,y) = x - y + y(x^2 + y^2)$. Therefore, the Jacobian matrix for this system is

$$\mathbf{J}(x,y) = \begin{pmatrix} F_x & F_y \\ G_x & G_y \end{pmatrix} = \begin{pmatrix} -2 - 3x^2 - y^2 & -1 - 2xy \\ 1 + 2xy & -1 + x^2 + 3y^2 \end{pmatrix}.$$

Near the critical point $(0,0)$, the Jacobian matrix is

$$\mathbf{J}(0,0) = \begin{pmatrix} -2 & -1 \\ 1 & -1 \end{pmatrix}.$$

Near the critical points $(-0.33076, 1.0924)$ and $(0.33076, -1.0924)$, the Jacobian matrix is

$$\mathbf{J}(-0.33076, 1.0924) = \mathbf{J}(0.33076, -1.0924) = \begin{pmatrix} -3.5216 & -0.27735 \\ 0.27735 & 2.6895 \end{pmatrix}.$$

(c) The eigenvalues of the linear system near $(0,0)$ are $\lambda = (-3 \pm i\sqrt{3})/2$. Therefore, $(0,0)$ is an asymptotically stable spiral point for the nonlinear system. The eigenvalues of the linear system near $(-0.33076, 1.0924)$ and $(0.33076, -1.0924)$ are $-3.5092, 2.6771$, respectively. From this, we can conclude that these two critical points are saddles.

(d)

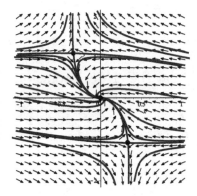

(e) The basin of attraction for $(0,0)$ is bounded above and below by the solutions approaching the two saddles, respectively.

17.(a) The critical points occur when either $y = -2$ or $y = 0.5x$ and either $x = 2$ or $y = -0.5x$. Solving these equations simultaneously, we have the critical points $(0,0)$, $(2,-2)$, $(2,1)$, and $(4,-2)$.

(b) Here, we have $F(x,y) = (2+y)(y-0.5x)$ and $G(x,y) = (2-x)(y+0.5x)$. Therefore, the Jacobian matrix for this system is

$$\mathbf{J}(x,y) = \begin{pmatrix} F_x & F_y \\ G_x & G_y \end{pmatrix} = \begin{pmatrix} -1 - 0.5y & 2 + 2y - 0.5x \\ -y + 1 - x & 2 - x \end{pmatrix}.$$

Near the critical point $(0,0)$, the Jacobian matrix is

$$\mathbf{J}(0,0) = \begin{pmatrix} F_x(0,0) & F_y(0,0) \\ G_x(0,0) & G_y(0,0) \end{pmatrix} = \begin{pmatrix} -1 & 2 \\ 1 & 2 \end{pmatrix}$$

and the corresponding linear system near $(0,0)$ is

$$\frac{d}{dt}\begin{pmatrix} x \\ y \end{pmatrix} = \begin{pmatrix} -1 & 2 \\ 1 & 2 \end{pmatrix}\begin{pmatrix} x \\ y \end{pmatrix}.$$

Near the critical point $(2,-2)$, the Jacobian matrix is

$$\mathbf{J}(2,-2) = \begin{pmatrix} F_x(2,-2) & F_y(2,-2) \\ G_x(2,-2) & G_y(2,-2) \end{pmatrix} = \begin{pmatrix} 0 & -3 \\ 1 & 0 \end{pmatrix}$$

and the corresponding linear system near $(2,-2)$ is

$$\frac{d}{dt}\begin{pmatrix} u \\ v \end{pmatrix} = \begin{pmatrix} 0 & -3 \\ 1 & 0 \end{pmatrix}\begin{pmatrix} u \\ v \end{pmatrix}$$

where $u = x - 2$ and $v = y + 2$. Near the critical point $(2,1)$, the Jacobian matrix is

$$\mathbf{J}(2,1) = \begin{pmatrix} F_x(2,1) & F_y(2,1) \\ G_x(2,1) & G_y(2,1) \end{pmatrix} = \begin{pmatrix} -3/2 & 3 \\ -2 & 0 \end{pmatrix}$$

and the corresponding linear system near $(2, 1)$ is

$$\frac{d}{dt}\begin{pmatrix} u \\ v \end{pmatrix} = \begin{pmatrix} -3/2 & 3 \\ -2 & 0 \end{pmatrix}\begin{pmatrix} u \\ v \end{pmatrix}$$

where $u = x - 2$ and $v = y - 1$. Near the critical point $(4, -2)$, the Jacobian matrix is

$$\mathbf{J}(4, -2) = \begin{pmatrix} F_x(4, -2) & F_y(4, -2) \\ G_x(4, -2) & G_y(4, -2) \end{pmatrix} = \begin{pmatrix} 0 & -4 \\ -1 & -2 \end{pmatrix}$$

and the corresponding linear system near $(4, -2)$ is

$$\frac{d}{dt}\begin{pmatrix} u \\ v \end{pmatrix} = \begin{pmatrix} 0 & -4 \\ -1 & -2 \end{pmatrix}\begin{pmatrix} u \\ v \end{pmatrix}$$

where $u = x - 4$ and $v = y + 2$.

(c) The eigenvalues of the linear system near $(0, 0)$ are $\lambda = (1 \pm \sqrt{17})/2$. From this, we can conclude that $(0, 0)$ is an unstable saddle point for the nonlinear system. The eigenvalues of the linear system near $(2, -2)$ are $\lambda = \pm\sqrt{3}i$. From this, we can only say that $(2, -2)$ is a center or spiral point for the nonlinear system, and we do cannot determine the stability. The eigenvalues of the linear system near $(2, 1)$ are $\lambda = (-3 \pm \sqrt{87}i)/4$. From this, we can conclude that $(2, 1)$ is an asymptotically stable spiral point. The eigenvalues of the linear system near $(4, -2)$ are $\lambda = -1 \pm \sqrt{5}$. From this, we can conclude that $(4, -2)$ is an unstable saddle point.

(d)

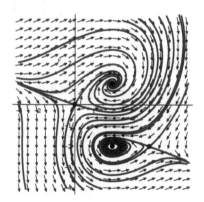

From the phase portrait, we can see that $(2, -2)$ is a stable center for the nonlinear system.

(e) The basin of attraction for the asymptotically stable point $(2, 1)$ is bounded to the left and below by trajectories heading towards the unstable saddle point $(0, 0)$. The basin of attraction is bounded to the right by trajectories heading around the left side of the center $(2, -2)$.

19.(a) The critical points occur when either $y = 1$ or $y = 2x$ and either $x = -2$ or $x = 2y$. Therefore, we see that the critical points are $(0, 0)$, $(2, 1)$, $(-2, 1)$, and $(-2, -4)$.

(b) Here, we have $F(x, y) = (1 - y)(2x - y)$ and $G(x, y) = (2 + x)(x - 2y)$. Therefore, the Jacobian matrix for this system is

$$\mathbf{J}(x, y) = \begin{pmatrix} F_x & F_y \\ G_x & G_y \end{pmatrix} = \begin{pmatrix} 2 - 2y & -2x - 1 + 2y \\ 2 + 2x - 2y & -4 - 2x \end{pmatrix}.$$

Near the critical point $(0, 0)$, the Jacobian matrix is

$$\mathbf{J}(0, 0) = \begin{pmatrix} F_x(0, 0) & F_y(0, 0) \\ G_x(0, 0) & G_y(0, 0) \end{pmatrix} = \begin{pmatrix} 2 & -1 \\ 2 & -4 \end{pmatrix}$$

and the corresponding linear system near $(0, 0)$ is

$$\frac{d}{dt} \begin{pmatrix} x \\ y \end{pmatrix} = \begin{pmatrix} 2 & -1 \\ 2 & -4 \end{pmatrix} \begin{pmatrix} x \\ y \end{pmatrix}.$$

Near the critical point $(2, 1)$, the Jacobian matrix is

$$\mathbf{J}(2, 1) = \begin{pmatrix} F_x(2, 1) & F_y(2, 1) \\ G_x(2, 1) & G_y(2, 1) \end{pmatrix} = \begin{pmatrix} 0 & -3 \\ 4 & -8 \end{pmatrix}$$

and the corresponding linear system near $(2, 1)$ is

$$\frac{d}{dt} \begin{pmatrix} u \\ v \end{pmatrix} = \begin{pmatrix} 0 & -3 \\ 4 & -8 \end{pmatrix} \begin{pmatrix} u \\ v \end{pmatrix}$$

where $u = x - 2$ and $v = y - 1$. Near the critical point $(-2, 1)$, the Jacobian matrix is

$$\mathbf{J}(-2, 1) = \begin{pmatrix} F_x(-2, 1) & F_y(-2, 1) \\ G_x(-2, 1) & G_y(-2, 1) \end{pmatrix} = \begin{pmatrix} 0 & 5 \\ -4 & 0 \end{pmatrix}$$

and the corresponding linear system near $(-2, 1)$ is

$$\frac{d}{dt} \begin{pmatrix} u \\ v \end{pmatrix} = \begin{pmatrix} 0 & 5 \\ -4 & 0 \end{pmatrix} \begin{pmatrix} u \\ v \end{pmatrix}$$

where $u = x - 2$ and $v = y - 1$. Near the critical point $(-2, -4)$, the Jacobian matrix is

$$\mathbf{J}(-2, -4) = \begin{pmatrix} F_x(-2, -4) & F_y(-2, -4) \\ G_x(-2, -4) & G_y(-2, -4) \end{pmatrix} = \begin{pmatrix} 10 & -5 \\ 6 & 0 \end{pmatrix}$$

and the corresponding linear system near $(-2, -4)$ is

$$\frac{d}{dt} \begin{pmatrix} u \\ v \end{pmatrix} = \begin{pmatrix} 10 & -5 \\ 6 & 0 \end{pmatrix} \begin{pmatrix} u \\ v \end{pmatrix}$$

where $u = x + 2$ and $v = y + 4$.

(c) The eigenvalues of the linear system near $(0, 0)$ are $\lambda = -1 \pm \sqrt{7}$. From this, we can conclude that $(0, 0)$ is an unstable saddle point for the nonlinear system. The eigenvalues of the linear system near $(2, 1)$ are $\lambda = -2, -6$. From this, we can conclude that $(2, 1)$ is an asymptotically stable node for the nonlinear system. The eigenvalues of the linear system near $(-2, 1)$ are $\lambda = \pm 2\sqrt{5}i$. From this, we can only conclude that $(-2, 1)$ is a either a center or a spiral point, and we cannot determine its stability. The eigenvalues of the linear system near $(-2, -4)$ are $\lambda = 5 \pm \sqrt{5}i$. From this, we can conclude that $(-2, -4)$ is an unstable spiral point.

(d)

From the phase portrait above, we can see that $(-2, 1)$ is a stable center for the nonlinear system.

(e) The basin of attraction for the asymptotically stable node $(2, 1)$ is bounded on the left by trajectories approaching the stable center $(-2, 1)$ and also by those trajectories heading to the left away from the unstable spiral point $(-2, -4)$.

21.(a) The critical points occur when $y = 0$ and $x + 2x^3 = 0$. From the second equation, we need $x = 0$ or $1 + 2x^2 = 0$. The only real solution of those equations is $x = 0$. Therefore, the only critical point is $(0, 0)$.

(b) The Jacobian matrix is given by

$$\mathbf{J}(x, y) = \begin{pmatrix} F_x & F_y \\ G_x & G_y \end{pmatrix} = \begin{pmatrix} 0 & 1 \\ 1 + 6x^2 & 0 \end{pmatrix}.$$

Therefore, the coefficient matrix of the linearized system near $(0, 0)$ is

$$\mathbf{J}(0, 0) = \begin{pmatrix} 0 & 1 \\ 1 & 0 \end{pmatrix}.$$

That is, $dx/dt = y$, $dy/dt = x$. Thus $dy/dx = x/y$. Separating variables, this equation can be written as $y\, dy = x\, dx$. Integrating this equation, we have $y^2 = x^2 + C$. Therefore, the trajectories consist of a family of hyperbolas.

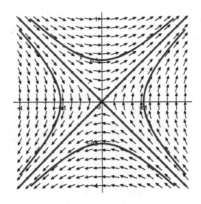

(c) For the nonlinear system, the equation for dy/dx is given by $dy/dx = (x + 2x^3)/y$. Separating variables, this equation can be written as $y\,dy = (x + 2x^3)\,dx$. Integrating this equation, we have $y^2 = x^2 + x^4 + C$. Therefore, the trajectories are level curves of $H(x, y) = x^2 + x^4 - y^2$. The trajectories that enter and leave the origin are no longer straight lines.

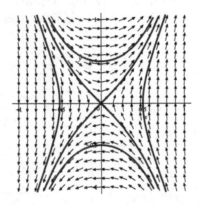

23.(a) The first graph represents a pendulum which will swing back and forth about the equilibrium position $\theta = 0$. The second graph represents a pendulum which will swing twice completely around the origin before swinging back and forth about the equilibrium position $\theta = 4\pi$.

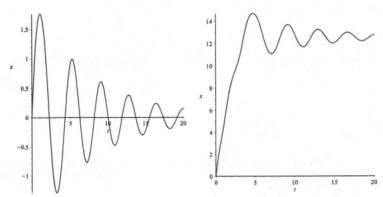

(b) Considering the graph of x versus t for various values of t, we see that the critical value of v occurs at $v_c \approx 3.98$.

(c) Below we show the graphs of x versus t for $\gamma = 1/2$ ($v = 3$, and $v = 6$, respectively).

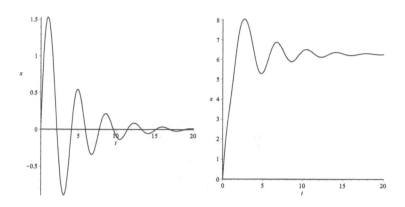

For $\gamma = 1/2$, the critical value of v will be $v_c \approx 4.51$. As γ increases, the critical value v_c increases.

25. The characteristic equation for the coefficient matrix is $\lambda^2 + 1 = 0$, which has roots $\lambda = \pm i$. Therefore, the critical point at the origin is a center. For the perturbed matrix, the characteristic equation is $\lambda^2 - 2\epsilon\lambda + 1 + \epsilon^2 = 0$. This equation has roots $\lambda = \epsilon \pm i$. Therefore, as long as $\epsilon \neq 0$, the critical point of the perturbed system is a spiral point. Its stability depends on the sign of ϵ.

27.(a) Let $y = x'$. Then $y' = x'' = -c(x)x' - g(x) = -c(x)y - g(x)$. Therefore, we can write the system as

$$\frac{dx}{dt} = y$$
$$\frac{dy}{dt} = -c(x)y - g(x).$$

(b) Clearly, we see that $(0,0)$ is a critical point of this system. The Jacobian matrix for the system is given by

$$\mathbf{J}(0,0) = \begin{pmatrix} 0 & 1 \\ -g'(0) & -c(0) \end{pmatrix}.$$

Therefore, we can rewrite the system as

$$\begin{pmatrix} x \\ y \end{pmatrix}' = \begin{pmatrix} 0 & 1 \\ -g'(0) & -c(0) \end{pmatrix} \begin{pmatrix} x \\ y \end{pmatrix} + \begin{pmatrix} 0 \\ g'(0)x + c(0)y - c(x)y - g(x) \end{pmatrix}.$$

We see that

$$\frac{g_2(x,y)}{r} = \frac{g'(0)x + c(0)y - c(x)y - g(x)}{r}$$
$$= \frac{g'(0)r\cos\theta + c(0)r\sin\theta - c(r\cos\theta)r\sin\theta - g(r\cos\theta)}{r}$$
$$= g'(0)\cos\theta + c(0)\sin\theta - c(r\cos\theta)\sin\theta - \frac{g(r\cos\theta)}{r}.$$

Then, since $c(r\cos\theta) \to c(0)$ as $r \to 0$ and $g(r\cos\theta)/r\cos\theta \to g'(0)$ as $r \to 0$, we conclude that $g_2(x,y)/r \to 0$ as $r \to 0$.

(c) The eigenvalues satisfy $\lambda^2 + c(0)\lambda + g'(0) = 0$. Therefore, the eigenvalues are given by

$$\lambda = \frac{-c(0) \pm \sqrt{c(0)^2 - 4g'(0)}}{2}.$$

Consider the case when $c(0) > 0$ and $g'(0) > 0$. If $c(0)^2 - 4g'(0) \geq 0$, then we have real, negative roots. If $c(0)^2 - 4g'(0) < 0$, then we have complex roots with negative real part. In the first of these cases, the origin is a stable node. In the second case, the origin is a stable spiral. In either case, the critical point $(0,0)$ is asymptotically stable. Next, consider the case when $c(0) < 0$ and $g'(0) < 0$. If $c(0)^2 - 4g'(0) > 0$, then we have real roots of opposite sign. If $c(0)^2 - 4g'(0) < 0$, then we have complex roots with a positive real part. In either of these cases, the critical point $(0,0)$ is unstable.

7.3 Competing Species

1.(a)

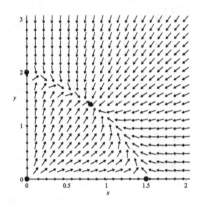

(b) The critical points are solutions of the system

$$x(1.5 - x - 0.5y) = 0$$
$$y(2 - y - 0.75x) = 0.$$

The four critical points are $(0,0)$, $(0,2)$, $(1.5,0)$, and $(0.8,1.4)$.

(c) The Jacobian matrix is

$$\mathbf{J}(x,y) = \begin{pmatrix} 3/2 - 2x - y/2 & -x/2 \\ -3y/4 & 2 - 3x/4 - 2y \end{pmatrix}.$$

At $(0,0)$,

$$\mathbf{J}(0,0) = \begin{pmatrix} 3/2 & 0 \\ 0 & 2 \end{pmatrix}.$$

The associated eigenvalues and eigenvectors are $\lambda_1 = 3/2$, $\mathbf{v}_1 = (1,0)^T$ and $\lambda_2 = 2$, $\mathbf{v}_2 = (0,1)^T$. The eigenvalues are positive. Therefore, the origin is an unstable node. At $(0,2)$,

$$\mathbf{J}(0,2) = \begin{pmatrix} 1/2 & 0 \\ -3/2 & -2 \end{pmatrix}.$$

The associated eigenvalues and eigenvectors are $\lambda_1 = 1/2$, $\mathbf{v}_1 = (1, -0.6)^T$ and $\lambda_2 = -2$, $\mathbf{v}_2 = (0, 1)^T$. The eigenvalues have opposite sign. Therefore, $(0, 2)$ is a saddle, which is unstable. At $(1.5, 0)$,

$$\mathbf{J}(1.5, 0) = \begin{pmatrix} -1.5 & -0.75 \\ 0 & 0.875 \end{pmatrix}.$$

The associated eigenvalues and eigenvectors are $\lambda_1 = -1.5$, $\mathbf{v}_1 = (1, 0)^T$ and $\lambda_2 = 0.875$, $\mathbf{v}_2 = (-0.31579, 1)^T$. The eigenvalues are opposite sign. Therefore, $(1.5, 0)$ is a saddle, which is unstable. At $(0.8, 1.4)$,

$$\mathbf{J}(0.8, 1.4) = \begin{pmatrix} -0.8 & -0.4 \\ -1.05 & -1.4 \end{pmatrix}.$$

The associated eigenvalues and eigenvectors are $\lambda_1 = (-11 + \sqrt{51})/10$, $\mathbf{v}_1 = (1, (3 - \sqrt{51})/4)^T$ and $\lambda_2 = (-11 - \sqrt{51})/10$, $\mathbf{v}_2 = (1, (3 + \sqrt{51})/4)^T$. The eigenvalues are negative. Therefore, $(0.8, 1.4)$ is a stable node, which is asymptotically stable.

(d,e)

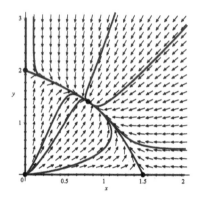

(f) Except for initial conditions lying on the coordinate axes, all trajectories converge to the stable node $(0.8, 1.4)$.

3.(a)

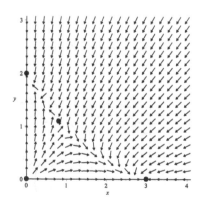

(b) The critical points are solutions of the system

$$x(1.5 - 0.5x - y) = 0$$
$$y(2 - y - 1.125x) = 0.$$

The four critical points are $(0,0)$, $(0,2)$, $(3,0)$, and $(4/5, 11/10)$.

(c) The Jacobian matrix is

$$\mathbf{J}(x, y) = \begin{pmatrix} 3/2 - x - y & -x \\ -1.125y & 2 - 2y - 1.125x \end{pmatrix}.$$

At $(0,0)$,

$$\mathbf{J}(0,0) = \begin{pmatrix} 3/2 & 0 \\ 0 & 2 \end{pmatrix}.$$

The associated eigenvalues and eigenvectors are $\lambda_1 = 3/2$, $\mathbf{v}_1 = (1,0)^T$ and $\lambda_2 = 2$, $\mathbf{v}_2 = (0,1)^T$. The eigenvalues are positive. Therefore, the origin is an unstable node. At $(0,2)$,

$$\mathbf{J}(0,2) = \begin{pmatrix} -1/2 & 0 \\ -9/4 & -2 \end{pmatrix}.$$

The associated eigenvalues and eigenvectors are $\lambda_1 = -1/2$, $\mathbf{v}_1 = (1, -3/2)^T$ and $\lambda_2 = -2$, $\mathbf{v}_2 = (0,1)^T$. The eigenvalues are both negative. Therefore, $(0,2)$ is a a stable node, which is asymptotically stable. At $(3,0)$,

$$\mathbf{J}(3,0) = \begin{pmatrix} -3/2 & -3 \\ 0 & -11/8 \end{pmatrix}.$$

The associated eigenvalues and eigenvectors are $\lambda_1 = -3/2$, $\mathbf{v}_1 = (1,0)^T$ and $\lambda_2 = -11/8$, $\mathbf{v}_2 = (-24, 1)^T$. The eigenvalues are both negative. Therefore, this critical point is a stable node, which is asymptotically stable. At $(4/5, 11/10)$,

$$\mathbf{J}(4/5, 11/10) = \begin{pmatrix} -2/5 & -4/5 \\ -99/80 & -11/10 \end{pmatrix}.$$

The associated eigenvalues and eigenvectors are $\lambda_1 = -3/4 + \sqrt{445}/20$, $\mathbf{v}_1 = (1, (7 - \sqrt{445})/16)^T$ and $\lambda_2 = -3/4 - \sqrt{445}/20$, $\mathbf{v}_2 = (0, (7 + \sqrt{445})/16)^T$. The eigenvalues are of opposite sign. Therefore, $(4/5, 11/10)$ is a saddle, which is unstable.

(d,e)

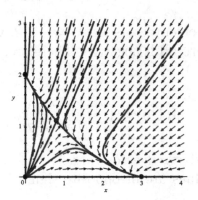

(f) Trajectories approaching the critical point $(4/5, 11/10)$ form a separatrix. Solutions on either side of the separatrix approach either $(3,0)$ or $(0,2)$.

5.(a)

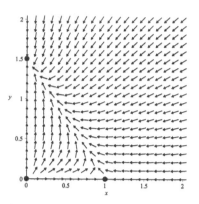

(b) The critical points are solutions of the system

$$x(1 - x - y) = 0$$
$$y(1.5 - y - x) = 0.$$

The three critical points are $(0, 0)$, $(0, 3/2)$, and $(1, 0)$.

(c) The Jacobian matrix is

$$\mathbf{J}(x, y) = \begin{pmatrix} 1 - 2x - y & -x \\ -y & 1.5 - 2y - x \end{pmatrix}.$$

At $(0, 0)$,

$$\mathbf{J}(0, 0) = \begin{pmatrix} 1 & 0 \\ 0 & 1.5 \end{pmatrix}.$$

The associated eigenvalues and eigenvectors are $\lambda_1 = 1$, $\mathbf{v}_1 = (1, 0)^T$ and $\lambda_2 = 1.5$, $\mathbf{v}_2 = (0, 1)^T$. The eigenvalues are positive. Therefore, the origin is an unstable node. At $(0, 3/2)$,

$$\mathbf{J}(0, 3/2) = \begin{pmatrix} -1/2 & 0 \\ -3/2 & -3/2 \end{pmatrix}.$$

The associated eigenvalues and eigenvectors are $\lambda_1 = -1/2$, $\mathbf{v}_1 = (1, -3/2)^T$ and $\lambda_2 = -3/2$, $\mathbf{v}_2 = (0, 1)^T$. The eigenvalues are both negative. Therefore, $(0, 3/2)$ is a stable node, which is asymptotically stable. At $(1, 0)$,

$$\mathbf{J}(1, 0) = \begin{pmatrix} -1 & -1 \\ 0 & 1/2 \end{pmatrix}.$$

The associated eigenvalues and eigenvectors are $\lambda_1 = -1$, $\mathbf{v}_1 = (1, 0)^T$ and $\lambda_2 = 1/2$, $\mathbf{v}_2 = (1, -3/2)^T$. The eigenvalues are of opposite sign. Therefore, this critical point is a saddle, which is unstable.

(d,e)

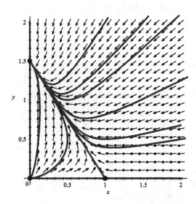

(f) All trajectories (not starting on the x-axis) converge to the stable node $(0, 1.5)$.

7. We see that

$$(\sigma_1 X + \sigma_2 Y)^2 - 4\sigma_1\sigma_2 XY = \sigma_1^2 X^2 + 2\sigma_1\sigma_2 XY + \sigma_2^2 Y^2 - 4\sigma_1\sigma_2 XY = (\sigma_1 X - \sigma_2 Y)^2.$$

The stated identity follows from the identity above. Since all parameters and variables are positive,

$$(\sigma_1 X + \sigma_2 Y)^2 - 4(\sigma_1\sigma_2 - \alpha_1\alpha_2)XY \geq 0.$$

Therefore, the eigenvalues can never be complex.

9.(a) The critical points are solutions of

$$\begin{aligned} x(1 - x - y) + \delta a &= 0 \\ y(0.75 - y - 0.5x) + \delta b &= 0. \end{aligned}$$

We assume the solutions can be written in the form

$$\begin{aligned} x &= x_0 + x_1\delta + x_2\delta^2 + \dots \\ y &= y_0 + y_1\delta + y_2\delta^2 + \dots \end{aligned}$$

Substituting these series expansions results in

$$\begin{aligned} x_0(1 - x_0 - y_0) + (x_1 - 2x_1x_0 - x_0y_1 - x_1y_0 + a)\delta + \dots &= 0 \\ y_0(0.75 - y_0 - 0.5x_0) + (0.75y_1 - 2y_0y_1 - x_1y_0/2 - x_0y_1/2 + b)\delta + \dots &= 0. \end{aligned}$$

(b) As $\delta \to 0$, the equations reduce to the original system of equations in (3). In that case, the critical points are $x_0 = y_0 = 0.5$.

(c) Setting the coefficients of the linear terms equal to zero, we find that

$$\begin{aligned} -y_1/2 - x_1/2 + a &= 0 \\ -x_1/4 - y_1/2 + b &= 0, \end{aligned}$$

which has solutions $x_1 = 4a - 4b$ and $y_1 = -2a + 4b$.

(d) If $b < a$, then there will be an increase in the level of species 1. On the other hand, at points where $b > a$, $x_1\delta < 0$. Similarly, if $2b > a$, then there will be an increase in the level of species 2. On the other hand, at points where $2b < a$, $y_1\delta < 0$. It follows that if $b < a < 2b$, the level of both species will increase. Otherwise, the level of one species will increase while the level of the other species decreases.

11.(a) Below we sketch some nullclines, corresponding to $\alpha = 16/3, 6$, and $20/3$.

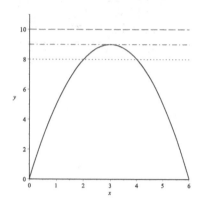

(b) The critical points are solutions of

$$-6x + y + x^2 = 0$$
$$3\alpha/2 - y = 0.$$

The solutions of these equations are $\left(3 \pm \sqrt{9 - 3\alpha/2}, 3\alpha/2\right)$ and exist for $\alpha \leq 6$.

(c) For $\alpha = 2$, the critical points are $(3 - \sqrt{6}, 3)$ and $(3 + \sqrt{6}, 3)$. The Jacobian matrix is

$$\mathbf{J}(x, y) = \begin{pmatrix} -6 + 2x & 1 \\ 0 & -1 \end{pmatrix}.$$

At $(3 - \sqrt{6}, 3)$,

$$\mathbf{J}(3 - \sqrt{6}, 3) = \begin{pmatrix} -2\sqrt{6} & 1 \\ 0 & -1 \end{pmatrix}.$$

The eigenvalues are $\lambda = -2\sqrt{6}, -1$. Since they are both negative, $(3 - \sqrt{6}, 3)$ is a stable node, which is asymptotically stable. At $(3 + \sqrt{6}, 3)$,

$$\mathbf{J}(3 + \sqrt{6}, 3) = \begin{pmatrix} 2\sqrt{6} & 1 \\ 0 & -1 \end{pmatrix}.$$

The eigenvalues are $\lambda = 2\sqrt{6}, -1$. Since they are of opposite sign $(3 + \sqrt{6}, 3)$ is a saddle, which is unstable.

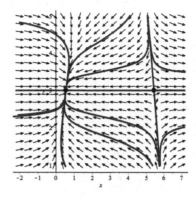

(d) The bifurcation value is $\alpha_0 = 6$. At this value α_0, the critical point is $(3, 9)$. The Jacobian matrix is

$$\mathbf{J}(3, 9) = \begin{pmatrix} 0 & 1 \\ 0 & -1 \end{pmatrix}.$$

The eigenvalues are $\lambda = 0, -1$.

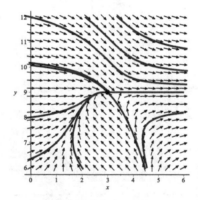

(e) Below we show the phase portrait for $\alpha = 8$.

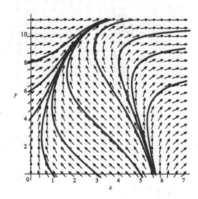

13.(a) Below we sketch some nullclines, corresponding to $\alpha = 0, 1$ and 2.

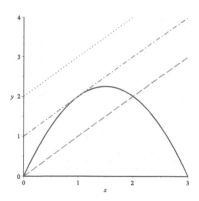

(b) The critical points are solutions of

$$-3x + y + x^2 = 0$$
$$-\alpha - x + y = 0.$$

The solutions of these equations are

$$x_0 = 1 - \sqrt{1-\alpha}, \qquad y_0 = \alpha + 1 - \sqrt{1-\alpha}$$

and

$$x_0 = 1 + \sqrt{1-\alpha}, \qquad y_0 = \alpha + 1 + \sqrt{1-\alpha}$$

and exist for $\alpha \leq 1$.

(c) For $\alpha = 2$, there are no critical points.

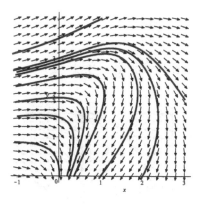

(d) The bifurcation value is $\alpha_0 = 1$. At this value α_0, the critical point is $(1, 2)$. The Jacobian matrix is

$$\mathbf{J}(1,2) = \begin{pmatrix} -1 & 1 \\ -1 & 1 \end{pmatrix}.$$

The eigenvalues are both $\lambda = 0$.

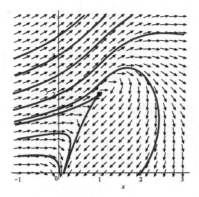

(e) Below we show the phase portrait for $\alpha = 3$.

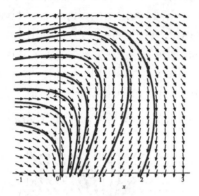

15.(a) The equation $x' = 0$ implies $x = -3$ or $x - y = 1$. The equation $y' = 0$ implies $y = 1$ or $x + \alpha y = -1$. Solving these equations, we see that the critical points are $(2, 1)$, $(-3, 1)$, $(-3, 2/\alpha)$, and $((-1 + \alpha)/(1 + \alpha), -2/(1 + \alpha))$.

(b) When $\alpha_0 = 2$, the second and third critical points listed above coincide. When $\alpha_0 = -3$, the first and fourth critical points listed above coincide, but here we are only considering $\alpha > 0$. Therefore, $\alpha_0 = 2$.

(c) Here, we have $F(x, y) = (3 + x)(1 - x + y)$ and $G(x, y) = (y - 1)(1 + x + \alpha y)$. Therefore, the Jacobian matrix for this system is

$$\mathbf{J}(x, y) = \begin{pmatrix} F_x & F_y \\ G_x & G_y \end{pmatrix} = \begin{pmatrix} -2 - 2x + y & 3 + x \\ y - 1 & 1 + x + 2\alpha y - \alpha \end{pmatrix}.$$

We will look at the linear systems near the second and third critical points above, namely $(-3, 1)$ and $(-3, 2/\alpha)$. Near the critical point $(-3, 1)$, the Jacobian matrix is

$$\mathbf{J}(-3, 1) = \begin{pmatrix} F_x(-3, 1) & F_y(-3, 1) \\ G_x(-3, 1) & G_y(-3, 1) \end{pmatrix} = \begin{pmatrix} 5 & 0 \\ 0 & -2 + \alpha \end{pmatrix}$$

and the corresponding linear system near $(-3, 1)$ is

$$\frac{d}{dt} \begin{pmatrix} u \\ v \end{pmatrix} = \begin{pmatrix} 5 & 0 \\ 0 & -2 + \alpha \end{pmatrix} \begin{pmatrix} u \\ v \end{pmatrix}$$

where $u = x + 3$ and $v = y - 1$. Near the critical point $(-3, 2/\alpha)$, the Jacobian matrix is

$$\mathbf{J}(-3, 2/\alpha) = \begin{pmatrix} F_x(-3, 2/\alpha) & F_y(-3, 2/\alpha) \\ G_x(-3, 2/\alpha) & G_y(-3, 2/\alpha) \end{pmatrix} = \begin{pmatrix} 4 + 2/\alpha & 0 \\ -1 + 2/\alpha & 2 - \alpha \end{pmatrix}$$

and the corresponding linear system near $(-3, 2/\alpha)$ is

$$\frac{d}{dt}\begin{pmatrix} u \\ v \end{pmatrix} = \begin{pmatrix} 4 + 2/\alpha & 0 \\ -1 + 2/\alpha & 2 - \alpha \end{pmatrix}\begin{pmatrix} u \\ v \end{pmatrix}$$

where $u = x + 3$ and $v = y - 2/\alpha$. The eigenvalues for the linearized system near $(-3, 1)$ are given by $\lambda = 5, -2 + \alpha$. The eigenvalues for the linearized system near $(-3, 2/\alpha)$ are given by $\lambda = 4 + 2/\alpha, 2 - \alpha$. In part (b), we determined that the bifurcation point was $\alpha_0 = 2$. Here, we see that if $\alpha > 2$, then $(-3, 1)$ will have two positive eigenvalues associated with it, and, therefore, be an unstable node, while $(-3, 2/\alpha)$ will have eigenvalues of opposite signs, and, therefore, by an unstable saddle point. If $\alpha < 2$, then $(-3, 1)$ will have eigenvalues of the opposite sign, and, therefore, be an unstable saddle point, while $(-3, 2/\alpha)$ will have two positive eigenvalues, and, therefore, be an unstable node.

(d) The phase portraits below are for $\alpha = 1, 2$ and 4, respectively.

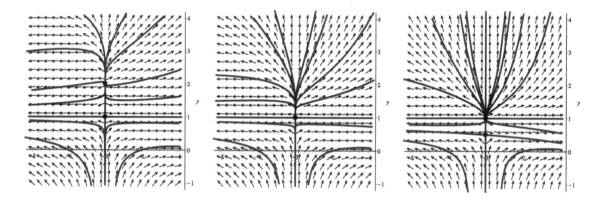

17.(a) The critical points need to satisfy the system of equations

$$\begin{aligned} x(4 - x - y) &= 0 \\ y(2 + 2\alpha - y - \alpha x) &= 0. \end{aligned}$$

The four critical points are $(0, 0)$, $(0, 2 + 2\alpha)$, $(4, 0)$, and $(2, 2)$.

(b) The Jacobian matrix is given by

$$\mathbf{J}(x, y) = \begin{pmatrix} 4 - 2x - y & -x \\ -\alpha y & 2 + 2\alpha - 2y - \alpha x \end{pmatrix}.$$

Therefore, at $(2, 2)$,

$$\mathbf{J}(2, 2) = \begin{pmatrix} -2 & -2 \\ -2\alpha & -2 \end{pmatrix}.$$

For $\alpha = 0.75$,

$$\mathbf{J}(2,2) = \begin{pmatrix} -2 & -2 \\ -3/2 & -2 \end{pmatrix}.$$

The eigenvalues of this matrix are $\lambda = -2 \pm \sqrt{3}$. Since both of these eigenvalues are negative, for $\alpha = 0.75$, $(2,2)$ is a stable node, which is asymptotically stable. For $\alpha = 1.25$,

$$\mathbf{J}(2,2) = \begin{pmatrix} -2 & -2 \\ -5/2 & -2 \end{pmatrix}.$$

The eigenvalues of this matrix are $\lambda = -2 \pm \sqrt{5}$. Since these eigenvalues are of opposite sign, for $\alpha = 1.25$, $(2,2)$ is a saddle point which is unstable. In general, the eigenvalues at $(2,2)$ are given by $-2 \pm 2\sqrt{\alpha}$. The nature of the critical point will change when $2\sqrt{\alpha} = 2$; that is, at $\alpha_0 = 1$. At this value of α, the number of negative eigenvalues changes from two to one.

(c) From the Jacobian matrix in part (b), we see that the approximate linear system is

$$\begin{pmatrix} u \\ v \end{pmatrix}' = \begin{pmatrix} -2 & -2 \\ -2\alpha & -2 \end{pmatrix} \begin{pmatrix} u \\ v \end{pmatrix}.$$

(d) As shown in part (b), the value $\alpha_0 = 1$.

(e) In the three phase portraits, we take $\alpha = 0.75, 1$ and 1.25, respectively.

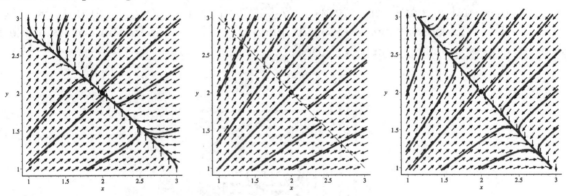

7.4 Predator-Prey Equations

1.(a)

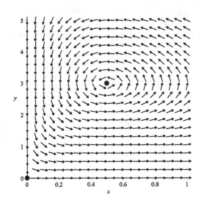

(b) The critical points are solutions of the system

$$x(1.5 - 0.5y) = 0$$
$$y(-0.5 + x) = 0.$$

The two critical points are $(0,0)$ and $(0.5, 3)$.

(c) The Jacobian matrix is

$$\mathbf{J}(x, y) = \begin{pmatrix} 3/2 - y/2 & -x/2 \\ y & -1/2 + x \end{pmatrix}.$$

At $(0,0)$,

$$\mathbf{J}(0,0) = \begin{pmatrix} 3/2 & 0 \\ 0 & -1/2 \end{pmatrix}.$$

The eigenvalues and eigenvectors are $\lambda_1 = 3/2$, $\mathbf{v}_1 = (1,0)^T$ and $\lambda_2 = -1/2$, $\mathbf{v}_2 = (0,1)^T$. The eigenvalues are of opposite sign. Therefore, $(0,0)$ is a saddle point, which is unstable. At $(0.5, 3)$,

$$\mathbf{J}(0.5, 3) = \begin{pmatrix} 0 & -1/4 \\ 3 & 0 \end{pmatrix}.$$

The eigenvalues and eigenvectors are $\lambda_1 = i\sqrt{3}/2$, $\mathbf{v}_1 = (1, -2i\sqrt{3})^T$ and $\lambda_2 = -i\sqrt{3}/2$, $\mathbf{v}_2 = (1, 2i\sqrt{3})^T$. The eigenvalues are purely imaginary. Using the method of Example 1, we get that $(0.5, 3)$ is a center, which is stable.

(d,e)

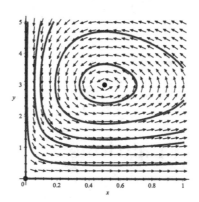

(f) Except for solutions along the coordinate axes, the other trajectories are closed curves about the critical point $(0.5, 3)$.

3.(a)

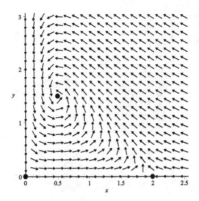

(b) The critical points are solutions of the system

$$x(1 - 0.5x - 0.5y) = 0$$
$$y(-0.25 + 0.5x) = 0.$$

The three critical points are $(0, 0)$, $(2, 0)$, and $(1/2, 3/2)$.

(c) The Jacobian matrix is

$$\mathbf{J}(x, y) = \begin{pmatrix} 1 - x - y/2 & -x/2 \\ y/2 & -1/4 + x/2 \end{pmatrix}.$$

At $(0, 0)$,

$$\mathbf{J}(0, 0) = \begin{pmatrix} 1 & 0 \\ 0 & -1/4 \end{pmatrix}.$$

The eigenvalues and eigenvectors are $\lambda_1 = 1$, $\mathbf{v}_1 = (1, 0)^T$ and $\lambda_2 = -1/4$, $\mathbf{v}_2 = (0, 1)^T$. The eigenvalues are of opposite sign. Therefore, $(0, 0)$ is a saddle point, which is unstable. At $(2, 0)$,

$$\mathbf{J}(2, 0) = \begin{pmatrix} -1 & -1 \\ 0 & 3/4 \end{pmatrix}.$$

The eigenvalues and eigenvectors are $\lambda_1 = -1$, $\mathbf{v}_1 = (1, 0)^T$ and $\lambda_2 = 3/4$, $\mathbf{v}_2 = (1, -7/4)^T$. The eigenvalues have opposite sign. Therefore, $(2, 0)$ is a saddle point, which is unstable. At $(1/2, 3/2)$,

$$\mathbf{J}(1/2, 3/2) = \begin{pmatrix} -1/4 & -1/4 \\ 3/4 & 0 \end{pmatrix}.$$

The eigenvalues and eigenvectors are $\lambda_1 = (-1 + i\sqrt{11})/8$, $\mathbf{v}_1 = ((-1 + i\sqrt{11})/6, 1)^T$ and $\lambda_2 = (-1 - i\sqrt{11})/8$, $\mathbf{v}_2 = ((-1 - i\sqrt{11})/6, 1)^T$. The eigenvalues have negative real part. Therefore, $(1/2, 3/2)$ is a stable spiral, which is asymptotically stable.

(d,e)

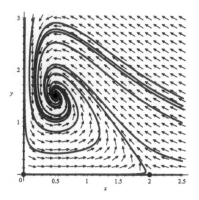

(f) Except for solutions along the coordinate axes, the other trajectories spiral towards the critical point $(1/2, 3/2)$.

5.(a)

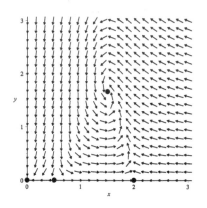

(b) The critical points are solutions of the system

$$
\begin{aligned}
x(-1 + 2.5x - 0.3y - x^2) &= 0 \\
y(-1.5 + x) &= 0.
\end{aligned}
$$

The four critical points are $(0, 0)$, $(1/2, 0)$, $(2, 0)$, and $(3/2, 5/3)$.

(c) The Jacobian matrix is

$$
\mathbf{J}(x, y) = \begin{pmatrix} -1 + 5x - 3x^2 - 3y/10 & -3x/10 \\ y & -3/2 + x \end{pmatrix}.
$$

At $(0, 0)$,

$$
\mathbf{J}(0, 0) = \begin{pmatrix} -1 & 0 \\ 0 & -3/2 \end{pmatrix}.
$$

The eigenvalues and eigenvectors are $\lambda_1 = -1$, $\mathbf{v}_1 = (1, 0)^T$ and $\lambda_2 = -3/2$, $\mathbf{v}_2 = (0, 1)^T$. The eigenvalues are both negative. Therefore, $(0, 0)$ is a stable node, which is asymptotically stable. At $(1/2, 0)$,

$$
\mathbf{J}(1/2, 0) = \begin{pmatrix} 3/4 & -3/20 \\ 0 & -1 \end{pmatrix}.
$$

The eigenvalues and eigenvectors are $\lambda_1 = 3/4$, $\mathbf{v}_1 = (1,0)^T$ and $\lambda_2 = -1$, $\mathbf{v}_2 = (3,35)^T$. The eigenvalues have opposite sign. Therefore, $(1/2,0)$ is a saddle point, which is unstable. At $(2,0)$,

$$\mathbf{J}(2,0) = \begin{pmatrix} -3 & -3/5 \\ 0 & 1/2 \end{pmatrix}.$$

The eigenvalues and eigenvectors are $\lambda_1 = -3$, $\mathbf{v}_1 = (1,0)^T$ and $\lambda_2 = 1/2$, $\mathbf{v}_2 = (6,-35)^T$. The eigenvalues have opposite sign. Therefore, $(2,0)$ is a saddle, which is unstable. At $(3/2, 5/3)$,

$$\mathbf{J}(3/2, 5/3) = \begin{pmatrix} -3/4 & -9/20 \\ 5/3 & 0 \end{pmatrix}.$$

The eigenvalues and eigenvectors are $\lambda_1 = (-3 + i\sqrt{39})/8$, $\mathbf{v}_1 = ((-9 + i3\sqrt{39})/40, 1)^T$ and $\lambda_2 = (-3 - i\sqrt{39})/8$, $\mathbf{v}_2 = ((-9 - i3\sqrt{39})/40, 1)^T$. The eigenvalues have negative real part. Therefore, $(3/2, 5/3)$ is a stable spiral, which is asymptotically stable.

(d,e)

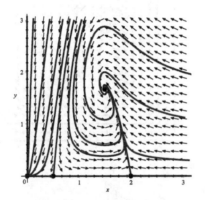

(f) The solution curve which is heading towards $(1/2, 0)$ is a separatrix. Trajectories on either side of that solution curve converge to the node $(0,0)$ or the stable spiral at $(3/2, 5/3)$.

7.(a) Looking at the coefficient of the trigonometric functions in equations (24), we see that the ratio will be given by

$$\frac{(c\gamma)}{(a/\alpha)\sqrt{c/a}} = \frac{\alpha\sqrt{c}}{\gamma\sqrt{a}}.$$

(b) For system (2), $a = 1$, $\alpha = 0.5$, $c = 0.75$ and $\gamma = 0.25$. Therefore, the ratio is $0.5\sqrt{0.75}/0.25 = \sqrt{3}$.

(c) The amplitude for the prey function in Figure 7.4.3 is approximately 2.5 and the amplitude for the predator function in Figure 7.4.3 is approximately 1.5. Using these approximations, we say that the ratio is approximately 5/3 which is close to $\sqrt{3}$.

(d) We show graphs of the prey and predators for initial conditions $x(0) = 1$, $y(0) = 1/2$. In this case, we emphasize that the ratio of the amplitudes is approximately $5.3/3 \approx 1.77$. In particular, we note that the ratio of the amplitudes slowly increases as the initial condition (x_0, y_0) moves away from the equilibrium point $(3, 2)$.

9.(a)

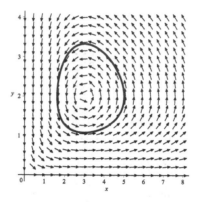

The period of oscillation is determined by observing when the trajectory becomes a closed curve. In this case, $T \approx 6.45$.

(b) For $a = 3$, $T \approx 3.69$. For $a = 1/3$, $T \approx 11.45$.

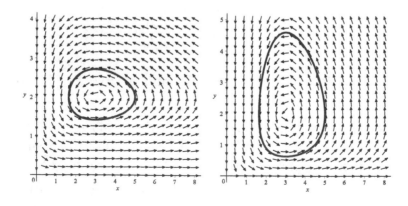

(c) For $b = 3$, $T \approx 3.82$. For $b = 1/3$, $T \approx 11.06$.

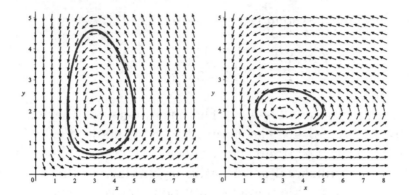

(d) If one parameter is fixed, the period varies inversely with the other parameter.

11.(a) Looking at the equation for $x' = 0$, we need $x = 0$ or $\sigma x + 0.5y = 0$ Looking at the equation for $y' = 0$, we need $y = 0$ or $x = 3$. Therefore, the critical points are given by $(0, 0)$, $(1/\sigma, 0)$, and $(3, 2 - 6\sigma)$. As σ increases from zero, the critical point $(1/\sigma, 0)$ approaches the origin, and the critical point $(3, 2 - 6\sigma)$ will eventually leave the first quadrant and enter the fourth quadrant.

(b) Here, we have $F(x, y) = x(1 - \sigma x - 0.5y)$ and $G(x, y) = y(-0.75 + 0.25x)$. Therefore, the Jacobian matrix for this system is

$$\mathbf{J}(x, y) = \begin{pmatrix} F_x & F_y \\ G_x & G_y \end{pmatrix} = \begin{pmatrix} 1 - 2\sigma x - 0.5y & -0.5x \\ 0.25y & -0.75 + 0.25x \end{pmatrix}.$$

We will look at the linear systems near the critical points above. Near the critical point $(0, 0)$, the Jacobian matrix is

$$\mathbf{J}(0, 0) = \begin{pmatrix} F_x(0, 0) & F_y(0, 0) \\ G_x(0, 0) & G_y(0, 0) \end{pmatrix} = \begin{pmatrix} 1 & 0 \\ 0 & -3/4 \end{pmatrix}$$

and the corresponding linear system near $(0, 0)$ is

$$\frac{d}{dt} \begin{pmatrix} x \\ y \end{pmatrix} = \begin{pmatrix} 1 & 0 \\ 0 & -3/4 \end{pmatrix} \begin{pmatrix} x \\ y \end{pmatrix}.$$

Near the critical point $(1/\sigma, 0)$, the Jacobian matrix is

$$\mathbf{J}(1/\sigma, 0) = \begin{pmatrix} F_x(1/\sigma, 0) & F_y(1/\sigma, 0) \\ G_x(1/\sigma, 0) & G_y(1/\sigma, 0) \end{pmatrix} = \begin{pmatrix} -1 & -1/2\sigma \\ 0 & (1 - 3\sigma)/4\sigma \end{pmatrix}$$

and the corresponding linear system near $(1/\sigma, 0)$ is

$$\frac{d}{dt} \begin{pmatrix} u \\ v \end{pmatrix} = \begin{pmatrix} -1 & -1/2\sigma \\ 0 & (1 - 3\sigma)/4\sigma \end{pmatrix} \begin{pmatrix} u \\ v \end{pmatrix}$$

where $u = x - 1/\sigma$ and $v = y$. Near the critical point $(3, 2 - 6\sigma)$, the Jacobian matrix is

$$\mathbf{J}(3, 2 - 6\sigma) = \begin{pmatrix} F_x(3, 2 - 6\sigma) & F_y(3, 2 - 6\sigma) \\ G_x(3, 2 - 6\sigma) & G_y(3, 2 - 6\sigma) \end{pmatrix} = \begin{pmatrix} -3\sigma & -3/2 \\ 1/2 - 3\sigma/2 & 0 \end{pmatrix}$$

and the corresponding linear system near $(3, 2 - 6\sigma)$ is

$$\frac{d}{dt} \begin{pmatrix} u \\ v \end{pmatrix} = \begin{pmatrix} -3\sigma & -3/2 \\ 1/2 - 3\sigma/2 & 0 \end{pmatrix} \begin{pmatrix} u \\ v \end{pmatrix}$$

where $u = x - 3$ and $v = y - 2 + 6\sigma$. The eigenvalues for the linearized system near $(0, 0)$ are given by $\lambda = 1, -3/4$. Therefore, $(0, 0)$ is a saddle point. The eigenvalues for the linearized system near $(1/\sigma, 0)$ are given by $\lambda = -1, (1 - 3\sigma)/(4\sigma)$. For $\sigma < 1/3$, there will be one positive eigenvalue and one negative eigenvalue. In this case, $(1/\sigma, 0)$ will be a saddle point. For $\sigma > 1/3$, both eigenvalues will be negative, in which case $(1/\sigma, 0)$ will be an asymptotically stable node. The eigenvalues for the linearized system near $(3, 2 - 6\sigma)$ are $\lambda = (-3\sigma \pm \sqrt{9\sigma^2 + 9\sigma - 3})/2$. Solving the polynomial equation $9\sigma^2 + 9\sigma - 3 = 0$, we see that the eigenvalues will have non-zero imaginary part if $0 < \sigma < (\sqrt{21} - 3)/6$. In this case, since the real part, -3σ will be negative, the point $(3, 2 - 6\sigma)$ will be an asymptotically stable spiral point. If $\sigma > (\sqrt{21} - 3)/6$, then the eigenvalues will both be real. We just need to determine whether they will have the same sign or opposite signs. Solving the equation $-3\sigma + \sqrt{9\sigma^2 + 9\sigma - 3} = 0$, we see that the cut-off is $\sigma = 1/3$. In particular, we conclude that if $(\sqrt{21} - 3)/6 < \sigma < 1/3$, then this critical point will have two real-valued eigenvalues which are negative, in which case this critical point will be an asymptotically stable node. If $\sigma > 1/3$, however, the eigenvalues will be real-valued, but with opposite signs, in which case $(3, 2 - 6\sigma)$ will be a saddle point. We see that the critical point $(3, 2 - 6\sigma)$ is the critical point in the first quadrant if $0 < \sigma < 1/3$. From the analysis above, we see that the nature of the critical point changes at $\sigma_1 = (\sqrt{21} - 3)/6$. In particular, at this value of σ, the critical point switches from an asymptotically stable spiral point to an asymptotically stable node.

(c) The two phase portraits below are shown $\sigma = 0.1$ and $\sigma = 0.3$, respectively.

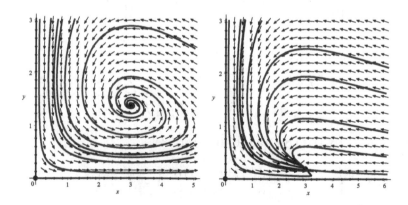

(d) As σ increases, the spiral behavior disappears. For smaller values of σ, the number of prey will decrease, causing a decrease in the number of predators, but then triggering an increase in the number of prey and eventually an increase in the number of predators. This cycle will continue to repeat as the system approaches the equilibrium point. As the value for σ increases, the cycling behavior between the predators and prey goes away.

13.(a) Solving the equations $x' = 0$ and $y' = 0$ simultaneously, we arrive at the critical points $(0, 0)$, $(5, 0)$, and $(2, 2.4)$.

(b) Here, we have $F(x, y) = x(1 - 0.2x - \dfrac{2y}{x+6})$ and $G(x, y) = y(-0.25 + \dfrac{x}{x+6})$. Therefore, the Jacobian matrix for this system is

$$\mathbf{J}(x, y) = \begin{pmatrix} F_x & F_y \\ G_x & G_y \end{pmatrix} = \begin{pmatrix} 1 - 0.4x - 12y/(x+6)^2 & -2x/(x+6) \\ 6y/(x+6)^2 & -1/4 + x/(x+6) \end{pmatrix}.$$

We will look at the linear systems near the critical points above. Near the critical point $(0, 0)$, the Jacobian matrix is

$$\mathbf{J}(0, 0) = \begin{pmatrix} F_x(0, 0) & F_y(0, 0) \\ G_x(0, 0) & G_y(0, 0) \end{pmatrix} = \begin{pmatrix} 1 & 0 \\ 0 & -1/4 \end{pmatrix}$$

and the corresponding linear system near $(0, 0)$ is

$$\frac{d}{dt} \begin{pmatrix} x \\ y \end{pmatrix} = \begin{pmatrix} 1 & 0 \\ 0 & -1/4 \end{pmatrix} \begin{pmatrix} x \\ y \end{pmatrix}.$$

Near the critical point $(5, 0)$, the Jacobian matrix is

$$\mathbf{J}(5, 0) = \begin{pmatrix} F_x(5, 0) & F_y(5, 0) \\ G_x(5, 0) & G_y(5, 0) \end{pmatrix} = \begin{pmatrix} -1 & -10/11 \\ 0 & 9/44 \end{pmatrix}$$

and the corresponding linear system near $(5, 0)$ is

$$\frac{d}{dt} \begin{pmatrix} u \\ v \end{pmatrix} = \begin{pmatrix} -1 & -10/11 \\ 0 & 9/44 \end{pmatrix} \begin{pmatrix} u \\ v \end{pmatrix}$$

where $u = x - 5$ and $v = y$. Near the critical point $(2, 2.4)$, the Jacobian matrix is

$$\mathbf{J}(2, 2.4) = \begin{pmatrix} F_x(2, 2.4) & F_y(2, 2.4) \\ G_x(2, 2.4) & G_y(2, 2.4) \end{pmatrix} = \begin{pmatrix} -1/4 & -1/2 \\ 9/40 & 0 \end{pmatrix}$$

and the corresponding linear system near $(2, 2.4)$ is

$$\frac{d}{dt} \begin{pmatrix} u \\ v \end{pmatrix} = \begin{pmatrix} -1/4 & -1/2 \\ 9/40 & 0 \end{pmatrix} \begin{pmatrix} u \\ v \end{pmatrix}$$

where $u = x - 2$ and $v = y - 2.4$. The eigenvalues for the linearized system near $(0, 0)$ are $\lambda = 1, -1/4$. Therefore, $(0, 0)$ is a saddle point. The eigenvalues for the linearized system near $(5, 0)$ are $\lambda = -1, 9/44$. Therefore, $(5, 0)$ is a saddle point. The eigenvalues for the linearized system near $(2, 2.4)$ are $\lambda = (-5 \pm i\sqrt{155})/40$. Since the real part of these eigenvalues is negative, $(2, 2.4)$ is an asymptotically stable spiral point.

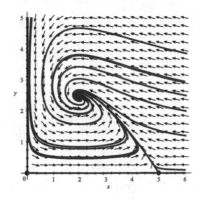

15.(a) Solving the equations $x' = 0$ and $y' = 0$, we see that the equilibrium solution (in which we have a non-zero number of predators and prey) is given by $((c + E_2)/\gamma, (a\gamma - E_1\gamma - \sigma c - \sigma E_2)/(\alpha\gamma))$. Therefore, if $E_1 > 0$, but $E_2 = 0$, then the number of prey stays the same, while the number of predators decreases.

(b) If $E_1 = 0$, but $E_2 > 0$, then the number of prey increases, while the number of predators decreases.

(c) If both $E_1 > 0$ and $E_2 > 0$, then the number of prey increases, while the number of predators decreases.

7.5   Periodic Solutions and Limit Cycles

1. The equilibrium solutions of the differential equation are $r = 0$ and $r = 1$. We notice that for $0 < r < 1$, $dr/dt > 0$, while for $r > 1$, $dr/dt < 0$. Therefore, $r = 0$ is an unstable critical point, while $r = 1$ is an asymptotically stable critical point. A limit cycle is given by $r = 1$, $\theta = t + t_0$, which is asymptotically stable.

3. The equilibrium solutions are given by $r = 0$, $r = 2$ and $r = 5$. We notice that $dr/dt > 0$ for $0 < r < 2$ and $r > 5$, while $dr/dt < 0$ for $2 < r < 5$. Therefore, $r = 0$ is an unstable critical point, $r = 2$ is an asymptotically stable critical point, and $r = 5$ is an unstable critical point. A limit cycle is given by $r = 2$, $\theta = t + t_0$, which is asymptotically stable. Another limit cycle is given by $r = 5$, $\theta = t + t_0$. This limit cycle is unstable.

5. The equilibrium solutions of the differential equation are given by $r = n/3$, $n = 0, 1, 2, \ldots$ Based on the sign of dr/dt in the neighborhood of each critical value, we see that the equilibrium solutions corresponding to $r = 2k/3$ for $k = 1, 2, \ldots$ are unstable periodic solutions with $\theta = t + t_0$. The equilibrium solutions $r = (2k + 1)/3$ for $k = 0, 1, 2, \ldots$ correspond to stable limit cycles with $\theta = t + t_0$. The solution $r = 0$ corresponds to an unstable critical point.

7. We compute:

$$y\frac{dx}{dt} - x\frac{dy}{dt} = r\sin\theta\left[\frac{dr}{dt}\cos\theta - r\sin\theta\frac{d\theta}{dt}\right] - r\cos\theta\left[\frac{dr}{dt}\sin\theta + r\cos\theta\frac{d\theta}{dt}\right]$$

$$= -r^2\sin^2\theta\frac{d\theta}{dt} - r^2\cos^2\theta\frac{d\theta}{dt} = -r^2\frac{d\theta}{dt}.$$

9. Using the fact that

$$r\frac{dr}{dt} = x\frac{dx}{dt} + y\frac{dy}{dt},$$

we have

$$r\frac{dr}{dt} = x(-y) + \frac{x^2}{\sqrt{x^2 + y^2}}(x^2 + y^2 - 5) + xy + \frac{y^2}{\sqrt{x^2 + y^2}}(x^2 + y^2 - 5) = r(r^2 - 5).$$

Therefore, $dr/dt = r^2 - 5$. Thus we have one critical point at $r = \sqrt{5}$. We see that $dr/dt > 0$ if $r > \sqrt{5}$ and $dr/dt < 0$ if $r < \sqrt{5}$. Therefore, $r = \sqrt{5}$ is an unstable periodic solution. To find the direction of motion on the closed trajectories, we use the fact that

$$-r^2 \frac{d\theta}{dt} = y\frac{dx}{dt} - x\frac{dy}{dt}.$$

Therefore, here we have

$$-r^2 \frac{d\theta}{dt} = -y^2 - x^2 = -r^2.$$

Thus $d\theta/dt = 1$, which implies $\theta = t + t_0$.

11. Given that $F(x,y) = 4x + y + 3x^3 - y^2$ and $G(x,y) = -x + 5y + x^2y + y^3/3$, $F_x + G_y = 9 + 10x^2 + y^2$ is positive for all (x,y). Therefore, by Theorem 7.5.2, the system cannot have a nontrivial periodic solution.

13. We parametrize the curve C by t. Therefore, we can rewrite the line integral as

$$\int_C [F(x,y)\, dy - G(x,y)\, dx] = \int_t^{t+T} [F(\phi(t), \psi(t))\psi'(t) - G(\phi(t), \psi(t))\phi'(t)]\, dt$$

$$= \int_t^{t+T} [\phi'(t)\psi'(t) - \psi'(t)\phi'(t)]\, dt = 0.$$

Then, using Green's Theorem, we must have

$$\iint_R [F_x(x,y) + G_y(x,y)]\, dA = 0.$$

If this integral is zero, there must be at least one point in A for which $F_x(x,y) + G_y(x,y) = 0$. Therefore, if we have a nontrivial periodic solution, we must have $F_x + G_y = 0$ somewhere in A.

15.(a) Letting $x = u$ and $y = u'$, we obtain the system

$$\frac{dx}{dt} = y$$

$$\frac{dy}{dt} = -x + \mu\left(1 - \frac{1}{3}y^2\right)y.$$

(b) To find the critical points, we need to solve the system

$$y = 0$$

$$-x + \mu\left(1 - \frac{1}{3}y^2\right)y = 0.$$

We see that the only solution of this system is $(0,0)$. Therefore, the only critical point is $(0,0)$. We notice that this system is almost linear. Therefore, we look at the Jacobian matrix. We see that

$$\mathbf{J}(x,y) = \begin{pmatrix} 0 & 1 \\ -1 & \mu - \mu y^2 \end{pmatrix}.$$

Therefore,

$$\mathbf{J}(0,0) = \begin{pmatrix} 0 & 1 \\ -1 & \mu \end{pmatrix}.$$

The eigenvalues are $\lambda = (\mu \pm \sqrt{\mu^2 - 4})/2$. If $\mu = 0$, the equation reduces to the differential equation for the simple harmonic oscillator. In that case, the eigenvalues are purely imaginary and $(0,0)$ is a center, which is stable. If $0 < \mu < 2$, the eigenvalues have non-zero imaginary part with positive real part. In that case, the critical point $(0,0)$ is an unstable spiral. If $\mu > 0$, the eigenvalues are real and both positive. In that case, the origin is an unstable node.

(c) We will consider initial conditions $x(0) = 2$, $y(0) = 0$. For $\mu = 1.0$, $A \approx 2.16$ and $T \approx 6.65$:

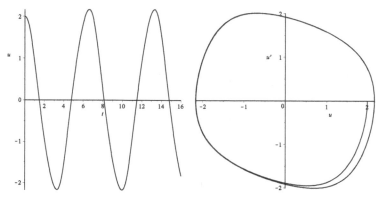

(d) For $\mu = 0.2$, $A \approx 1.99$ and $T \approx 6.31$:

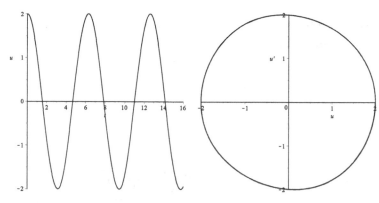

For $\mu = 0.5$, $A \approx 2.03$ and $T \approx 6.39$:

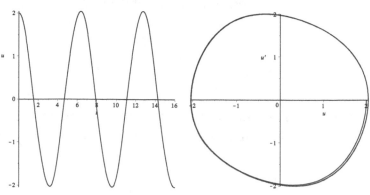

For $\mu = 2.0$, $A \approx 2.60$ and $T \approx 7.65$:

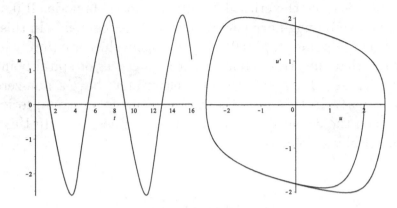

For $\mu = 5.0$, $A \approx 4.36$ and $T \approx 11.60$:

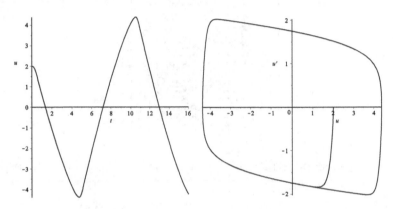

(e)

μ	A	T
0.2	1.99	6.31
0.5	2.03	6.39
1.0	2.16	6.65
2.0	2.60	7.65
5.0	4.36	11.60

17.(a) At a critical point, we need $x' = y = 0$. Then plugging $y = 0$ into the equation for $y' = 0$, we have $-x = 0$. Therefore, $(0,0)$ is the only critical point. To determine its type and stability, we need to look at the Jacobian matrix,

$$\mathbf{J}(x,y) = \begin{pmatrix} F_x & F_y \\ G_x & G_y \end{pmatrix} = \begin{pmatrix} 0 & 1 \\ -1 - 2\mu xy & \mu - \mu x^2 \end{pmatrix}.$$

Therefore, the Jacobian matrix near $(0,0)$ is

$$\mathbf{J}(0,0) = \begin{pmatrix} 0 & 1 \\ -1 & \mu \end{pmatrix}.$$

The eigenvalues of this matrix are $\lambda = (\mu \pm \sqrt{\mu^2 - 4})/2$. If $\mu \geq 2$, then the eigenvalues are real and positive. In this case, the critical point is an unstable node. If $0 < \mu < 2$, then the eigenvalues have a non-zero imaginary part with a positive real part. In this case, the critical point is an unstable spiral point. If $\mu = 0$, then the eigenvalues are purely imaginary. In this case, we can only say that the critical point is either a center or spiral point, and we cannot determine the stability. If $-2 < \mu < 0$, then the eigenvalues have a non-zero imaginary part with a negative real part. In this case, the critical point is an asymptotically stable spiral point. If $\mu \leq -2$, then the eigenvalues are both real and negative. In this case, the critical point is an asymptotically stable node.

(b)

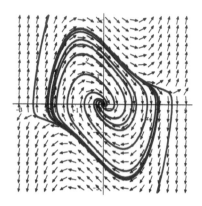

(c) Below we show the phase portraits in the case $\mu = -2, -3$, and -4, respectively. Observe that the y-axis is getting longer.

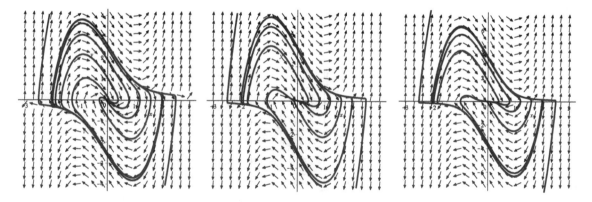

(d) Below we show the phase portraits in the cases $\mu = -0.5, -0.25, 0.25$, and 0.5, respectively.

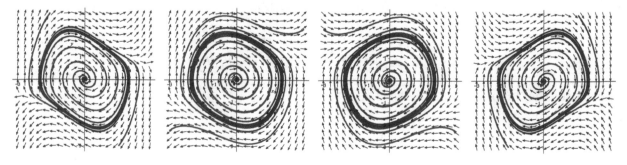

19.(a) The critical points are $(0,0)$, $(5a, 0)$, and $(2, 4a - 1.6)$.

(b) To determine the type and stability of the critical point in the interior of the first quadrant, we calculate the Jacobian matrix

$$\mathbf{J}(x, y) = \begin{pmatrix} F_x & F_y \\ G_x & G_y \end{pmatrix} = \begin{pmatrix} a - 0.4x - 12y/(x+6)^2 & -2x/(x+6) \\ 6y/(x+6)^2 & -0.25 + x/(x+6) \end{pmatrix}.$$

Therefore, the Jacobian matrix near $(2, 4a - 1.6)$ is

$$\mathbf{J}(2, 4a - 1.6) = \begin{pmatrix} a/4 - 1/2 & -1/2 \\ 3a/8 - 3/20 & 0 \end{pmatrix}.$$

The eigenvalues of this matrix are given by $\lambda = -0.25 + 0.125a \pm 0.025\sqrt{220 - 400a + 25a^2}$. We see that the term inside the square root will be negative for $8 - 2\sqrt{345}/5 < a < 8 + 2\sqrt{345}/5$. Therefore, for a in this range, the critical point will be a spiral point. We see that at $a = 2$, the real part of the eigenvalues will be zero. Therefore, at $a_0 = 2$, the critical point switches from an asymptotically stable spiral point to an unstable spiral point.

(c) For $a = 2.1$:

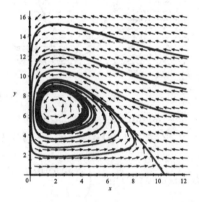

The limit cycle gets bigger as a increases.

21.(a) The critical points are solutions of

$$3\left(x + y - \frac{1}{3}x^3 - k\right) = 0$$
$$-\frac{1}{3}(x + 0.8y - 0.7) = 0.$$

Multiplying the second equation by 11.25 and adding to the first equation, we have

$$x^3 + 0.75x - 2.625 + 3k = 0.$$

Letting $f(x) = x^3 + 0.75x - 2.625 + 3k$, we see that $f'(x) = 3x^2 + 0.75 > 0$ for all x. Therefore, there is only one value of x which will satisfy the above system. For that value of x, we can calculate y and will find one critical point.

(b) Using the equation in part (a) for x and setting $k = 0$, we see that the $x-$coordinate of the critical point is $x = 1.19941$. Substituting that value for x into the second equation, we conclude that $y = -0.62426$. The Jacobian matrix is given by

$$\mathbf{J}(x, y) = \begin{pmatrix} 3 - 3x^2 & 3 \\ -1/3 & -4/15 \end{pmatrix}.$$

Therefore, at $(1.19941, -0.62426)$,

$$\mathbf{J}(1.19941, -0.62426) = \begin{pmatrix} -1.316 & 3 \\ -1/3 & -4/15 \end{pmatrix}.$$

The eigenvalues of this matrix are $\lambda = -0.791 \pm 0.851i$. Therefore, the critical point $(1.19941, -0.62426)$ is an asymptotically stable spiral point.

Now using the equation in part (a) and setting $k = 0.5$, we see that the $x-$coordinate of the critical point is $x = 0.80485$. Substituting that value for x into the second equation, we conclude that $y = -0.13106$. Therefore, at $(0.80485, -0.13106)$,

$$\mathbf{J}(0.80485, -0.13106) = \begin{pmatrix} 1.05665 & 3 \\ -1/3 & -4/15 \end{pmatrix}.$$

The eigenvalues of this matrix are $\lambda = 0.395 \pm 0.7498i$. Therefore, the critical point $(0.80485, -0.13106)$ is an unstable spiral point.

(c) By numerical calculation, we see that $k_0 \approx 0.3465$. For this value of k_0, we calculate the critical point as in part (b). In particular, we find that the critical point is $(0.9545, -0.31813)$.

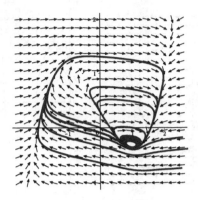

(d) In what follows, we consider initial conditions $x(0) = 0.5, y(0) = 0.5$. For $k = 0.4$, $T \approx 11.23$:

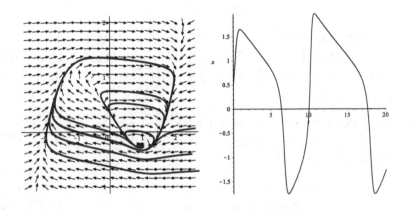

For $k = 0.5$, $T \approx 10.37$:

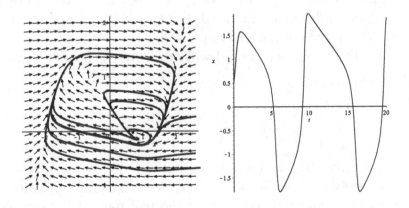

For $k = 0.6$, $T \approx 9.93$:

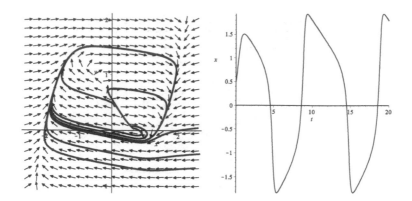

(e) By numerical analysis, we see that $k_1 \approx 1.4035$.

7.6   Chaos and Strange Attractors: The Lorenz Equations

1.(a) Since the eigenvalues must be solutions of $-(8/3+\lambda)(\lambda^2+11\lambda-10(r-1)) = 0$, we see that the eigenvalues are given by $\lambda = -8/3$ and the roots of $\lambda^2 + 11\lambda - 10(r-1) = 0$. The roots of the quadratic equation are $\lambda = (-11 \pm \sqrt{81+40r})/2$. Therefore, the eigenvalues are as stated in equation (7).

(b) The corresponding eigenvectors are given by $\mathbf{v}_1 = (0,0,1)^T$ for $\lambda_1 = -8/3$, $\mathbf{v}_2 = ((-9+\sqrt{81+40r})/2r, 1, 0)^T$ for $\lambda_2 = (-11+\sqrt{81+40r})/2$ and $\mathbf{v}_3 = ((-9-\sqrt{81+40r})/2r, 1, 0)^T$ for $\lambda_3 = (-11-\sqrt{81+40r})/2$

(c) For $r = 28$, using the formulas for the eigenvalues and eigenvectors from part (b), we see that $\lambda_1 = -8/3$ with eigenvector $\mathbf{v}_1 = (0,0,1)^T$, $\lambda_2 \approx 11.8277$ with eigenvector $\mathbf{v}_2 \approx (20, 43.6554, 0)^T$, and $\lambda_3 \approx -22.8277$ with eigenvector $\mathbf{v}_3 \approx (20, -25.6554, 0)^T$.

3.(a) For $r = 28$, in problem 2, we saw that the real part of the complex roots was positive. By numerical investigation, we see that the real part changes sign at $r \approx 24.737$.

(b) Suppose a cubic polynomial $x^3 + Ax^2 + Bx + C$ has one real zero and two pure imaginary zeros. Then the polynomial can be factored as $(x - \lambda_1)(x^2 + \lambda_2)$ where $\lambda_2 > 0$. Thus $x^3+Ax^2+Bx+C = (x-\lambda_1)(x^2+\lambda_2)$, which implies that $A = -\lambda_1, B = \lambda_2, C = -\lambda_1\lambda_2 = AB$. Therefore, if $AB \neq C$, the cubic polynomial will not have the specified type of roots.

(c) First, we rewrite the equation as

$$x^3 + \frac{41}{3}\lambda^2 + \frac{8(r+10)}{3} + \frac{160(r-1)}{3} = 0.$$

Using the result from part (b), we need to find when $AB = C$, where $A = 41/3$, $B = 8(r+10)/3$ and $C = 160(r-1)/3$. Setting $AB = C$, we have the equation $328(r+10)/9 = 160(r-1)/3$. Solving this equation, we see that the real part of the complex roots changes sign when $r = 470/19$.

5. First, we plot x versus t corresponding to $r = 28$ with an initial condition of $(5, 5, 5)$:

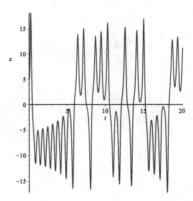

Next, we plot x versus t corresponding to $r = 28$ with an initial condition of $(5.01, 5, 5)$:

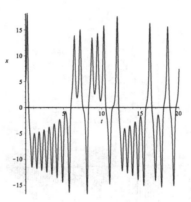

7.(a) We plot x versus t corresponding to $r = 21$ with an initial condition of $(3, 8, 0)$, $(5, 5, 5)$, and $(5, 5, 10)$, respectively:

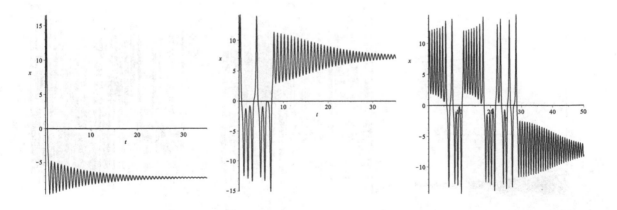

(b) We plot x versus t corresponding to $r = 22$ with an initial condition of $(3, 8, 0)$, $(5, 5, 5)$, and $(5, 5, 10)$, respectively:

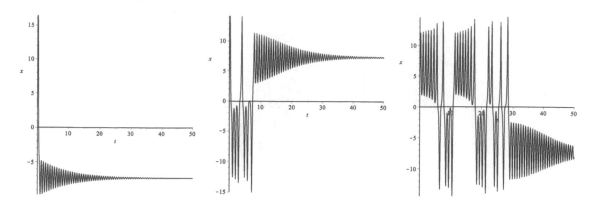

We plot x versus t corresponding to $r = 23$ with an initial condition of $(3, 8, 0)$, $(5, 5, 5)$, and $(5, 5, 10)$, respectively:

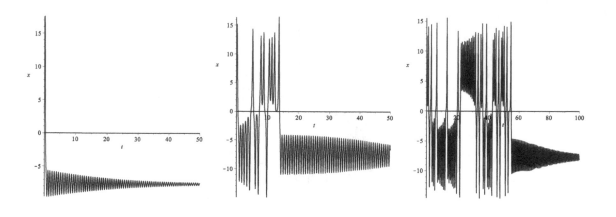

We plot x versus t corresponding to $r = 24$ with an initial condition of $(3, 8, 0)$, $(5, 5, 5)$, and $(5, 5, 10)$, respectively:

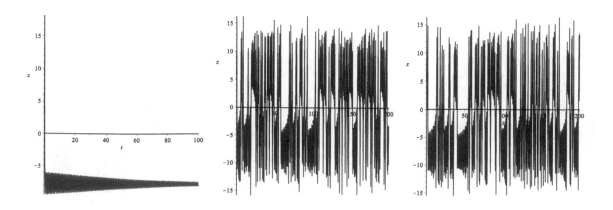

(c) We plot x versus t corresponding to $r = 24.2$ with an initial condition of $(3, 8, 0)$, $(5, 5, 5)$, and $(5, 5, 10)$, respectively:

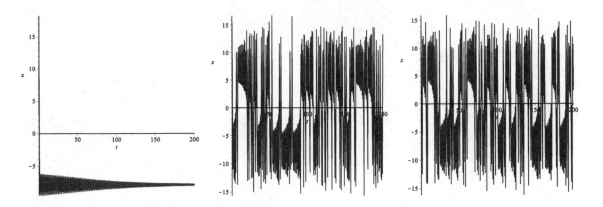

9.(a) For $r = 100.2, 100.4$ and 100.6, with initial condition $(5, 5, 5)$:

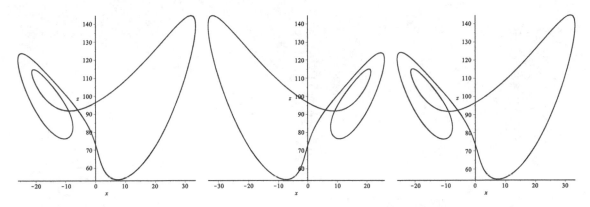

(b) For $r = 100.782, 100.785$ and 100.788, with initial condition $(5, 5, 5)$:

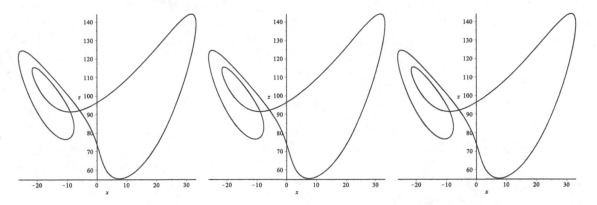

Chapter 8

Numerical Methods

8.1 Numerical Approximations: Euler's Method

1. The Euler formula is given by $y_{n+1} = y_n + h(3 + t_n - y_n) = (1 - h)y_n + h(3 + t_n)$.

(a) 1.2, 1.39, 1.571, 1.7439

(b) 1.1975, 1.38549, 1.56491, 1.73658

(c) 1.19631, 1.38335, 1.56200, 1.73308

(d) The differential equation is linear with solution $\phi(t) = 2 + t - e^{-t}$.

   1.19516, 1.38127, 1.55918, 1.72968

3. The Euler formula is given by $y_{n+1} = y_n + h(0.5 - t_n + 2y_n) = (1 + 2h)y_n + h(0.5 - t_n)$.

(a) 1.25, 1.54, 1.878, 2.2736

(b) 1.26, 1.5641, 1.92156, 2.34359

(c) 1.26551, 1.57746, 1.94586, 2.38287

(d) The differential equation is linear with solution $\phi(t) = 0.5t + e^{2t}$.

   1.2714, 1.59182, 1.97212, 2.42554

5.

The solutions converge for $y \geq 0$. Solutions are undefined for $y < 0$.

7.

All solutions converge.

9.

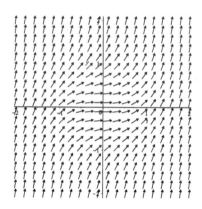

All solutions diverge.

11. The Euler formula is $y_{n+1} = y_n - 3h\sqrt{y_n} + 5h$, with $y_0 = 2$.

(a) 2.30800, 2.49006, 2.60023, 2.66773, 2.70939, 2.73521

(b) 2.30167, 2.48263, 2.59352, 2.66227, 2.70519, 2.73209

(c) 2.29864, 2.47903, 2.59024, 2.65958, 2.70310, 2.73053

(d) 2.29686, 2.47691, 2.58830, 2.65798, 2.70185, 2.72959

13. The Euler formula is $y_{n+1} = y_n + h(4 - t_n y_n)/(1 + y_n^2)$, with $(t_0, y_0) = (0, -2)$.

(a) $-1.48849, -0.412339, 1.04687, 1.43176, 1.54438, 1.51971$

(b) $-1.46909, -0.287883, 1.05351, 1.42003, 1.53000, 1.50549$

(c) $-1.45865, -0.217545, 1.05715, 1.41486, 1.52334, 1.49879$

(d) $-1.45212, -0.173376, 1.05941, 1.41197, 1.51949, 1.49490$

15. The Euler formula is $y_{n+1} = y_n + h3t_n^2/(3y_n^2 - 4)$, with $(t_0, y_0) = (1, 0)$.

(a) $-0.166134, -0.410872, -0.804660, 4.15867$

(b) $-0.174652, -0.434238, -0.889140, -3.09810$

(c) Since the line tangent to the solution is parallel to the y-axis when $y \approx \pm 1.155$, Euler's formula can be off by quite a bit. As the slope tends to ∞, using that slope as an approximation to the change in the function can cause a large error in the approximation.

17. The Euler formula is $y_{n+1} = y_n + h(y_n^2 + 2t_n y_n)/(3 + t_n^2)$ with $(t_0, y_0) = (1, 2)$. A reasonable estimate for y at $t = 2.5$ is between 18 and 19. No reliable estimate is possible at $t = 3$ from the specified data.

19.(a)

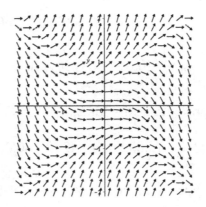

(b) The Euler formula is $y_{n+1} = y_n + h(y_n^2 - t_n^2)$. For $y_0 < 0.67$, the solutions seem to converge, while the solutions seem to diverge if $y_0 > 0.68$. Therefore, we conclude that $0.67 < \alpha_0 < 0.68$.

8.2 Accuracy of Numerical Methods

1. The Euler formula is $y_{n+1} = y_n + h(3 + t_n - y_n)$ in which $t_n = t_0 + nh$. Since $t_0 = 0$, we have $y_{n+1} = y_n(1 - h) + 3h + nh^2$.

(a) For $h = 0.05$, the Euler approximations for y_n at $n = 2, 4, 6, 8$ are given by

$$1.1975, \quad 1.38549, \quad 1.56491, \quad 1.73658.$$

(b) For $h = 0.025$, the Euler approximations for y_n at $n = 4, 8, 12, 16$ are given by

$$1.19631, \quad 1.38335, \quad 1.56200, \quad 1.73308.$$

3. The Euler formula is $y_{n+1} = y_n + h(2y_n - 3t_n)$ in which $t_n = t_0 + nh$. Since $t_0 = 0$, we have $y_{n+1} = y_n(1 + 2h) - 3nh^2$.

(a) For $h = 0.05$, the Euler approximations for y_n at $n = 2, 4, 6, 8$ are given by

$$1.2025, \quad 1.41603, \quad 1.64289, \quad 1.88590.$$

(b) For $h = 0.025$, the Euler approximations for y_n at $n = 4, 8, 12, 16$ are given by

$$1.20388, \quad 1.41936, \quad 1.64896, \quad 1.89572.$$

5. The Euler formula is $y_{n+1} = y_n + h(y_n^2 + 2t_ny_n)/(3 + t_n^2)$ in which $t_n = t_0 + nh$. Since $t_0 = 0$, we have $y_{n+1} = y_n + h(y_n^2 + 2nhy_n)/(3 + n^2h^2)$.

(a) For $h = 0.05$, the Euler approximations for y_n at $n = 2, 4, 6, 8$ are given by

$$0.509239, \quad 0.522187, \quad 0.539023, \quad 0.559936.$$

(b) For $h = 0.025$, the Euler approximations for y_n at $n = 4, 8, 12, 16$ are given by

$$0.509701, \quad 0.523155, \quad 0.540550, \quad 0.562089.$$

7. The Euler formula is $y_{n+1} = y_n + h(0.5 - t_n + 2y_n)$ in which $t_n = t_0 + nh$. Since $t_0 = 0$, we have $y_{n+1} = y_n + h(0.5 - nh + 2y_n)$.

(a) For $h = 0.025$, the Euler approximations for y_n at $n = 20, 40, 60, 80$ are given by

$$2.90330, \quad 7.53999, \quad 19.4292, \quad 50.5614.$$

(b) For $h = 0.0125$, the Euler approximations for y_n at $n = 40, 80, 120, 160$ are given by

$$2.93506, \quad 7.70957, \quad 20.1081, \quad 52.9779.$$

9. The Euler formula is $y_{n+1} = y_n + h\sqrt{t_n + y_n}$ in which $t_n = t_0 + nh$. Since $t_0 = 0$, we have $y_{n+1} = y_n + h\sqrt{nh + y_n}$.

(a) For $h = 0.025$, the Euler approximations for y_n at $n = 20, 40, 60, 80$ are given by

$$3.95713, \quad 5.09853, \quad 6.41548, \quad 7.90174.$$

(b) For $h = 0.0125$, the Euler approximations for y_n at $n = 40, 80, 120, 160$ are given by

$$3.95965, \quad 5.10371, \quad 6.42343, \quad 7.91255.$$

11. The Euler formula is $y_{n+1} = y_n + h(4 - t_ny_n)/(1 + y_n^2)$ in which $t_n = t_0 + nh$. Since $t_0 = 0$, we have $y_{n+1} = y_n + h(4 - nhy_n)/(1 + y_n^2)$.

(a) For $h = 0.025$, the Euler approximations for y_n at $n = 20, 40, 60, 80$ are given by

$$-1.45865, \quad -0.217545, \quad 1.05715, \quad 1.41487.$$

(b) For $h = 0.0125$, the Euler approximations for y_n at $n = 40, 80, 120, 160$ are given by

$$-1.45322, \quad -0.180813, \quad 1.05903, \quad 1.41244.$$

13. The Euler formula is $y_{n+1} = y_n + h(1 - t_n + 4y_n)$, in which $t_n = t_0 + nh$. Since $t_0 = 0$, we can write $y_{n+1} = y_n + h - nh^2 + 4hy_n$, with $y_0 = 1$. With $h = 0.01$, a total of 200 iterations is necessary to reach $\bar{t} = 2$. With $h = 0.001$, a total of 2000 iterations is necessary.

15. We know that $e_{n+1} = (1/2)\phi''(\bar{t}_n)h^2$ where $t_n < \bar{t}_n < t_{n+1}$. Here $\phi'(t) = 2\phi(t) - 1$. Therefore, $\phi''(t) = 2\phi'(t) = 2(2\phi(t) - 1) = 4\phi(t) - 2$, thus $e_{n+1} = (2\phi(\bar{t}_n) - 1)h^2$, and then $|e_{n+1}| \le |2M + 1|h^2$, where $M = \max_{0 \le t \le 1}|\phi(t)|$. The exact solution of this linear equation is $y(t) = 1/2 + 1/2e^{2t}$. Then, using the fact that the local truncation error is given by $e_{n+1} = (1/2)\phi(\bar{t}_n)h^2$ and $\phi(t) = 1/2 + 1/2e^{2t}$, we can conclude that $e_{n+1} = e^{2\bar{t}_n}h^2$. Therefore, $|e_1| \le e^{0.2}(0.1)^2 \approx 0.012$. Similarly, $|e_4| \le e^{0.8}(0.1)^2 \approx 0.022$.

17. We know that $e_{n+1} = (1/2)\phi''(\bar{t}_n)h^2$ where $t_n < \bar{t}_n < t_{n+1}$. Here $\phi'(t) = t^2 + (\phi(t))^2$. Therefore, $\phi''(t) = 2t + 2\phi(t)\phi'(t) = 2t + 2t^2\phi(t) + 2(\phi(t))^3$, thus $e_{n+1} = (\bar{t}_n + \bar{t}_n^2\phi(\bar{t}_n) + (\phi(\bar{t}_n))^3)h^2$, and then $|e_{n+1}| \le |t_{n+1} + t_{n+1}^2 M_{n+1} + M_{n+1}^3|h^2$, where $M_{n+1} = \max_{t_n \le t \le t_{n_1}}|\phi(t)|$.

19. We know that $e_{n+1} = (1/2)\phi''(\bar{t}_n)h^2$ where $t_n < \bar{t}_n < t_{n+1}$. Here $\phi'(t) = \sqrt{t + \phi(t)}$. Therefore,

$$\phi''(t) = \frac{1 + \phi'(t)}{2\sqrt{t + \phi(t)}} = \frac{1}{2\sqrt{t + \phi(t)}} + \frac{1}{2},$$

thus

$$e_{n+1} = \frac{1}{4}\left[1 + \frac{1}{\sqrt{\bar{t}_n + \phi(\bar{t}_n)}}\right]h^2.$$

21.(a) The solution is given by $\phi(t) = \dfrac{1}{5\pi}\sin(5\pi t) + 1$.

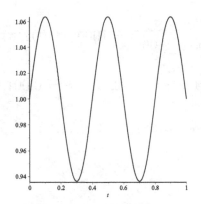

(b) Approximate values at $t = 0.2, 0.4, 0.6$ are given by $1.2, 1.0, 1.2$, respectively.

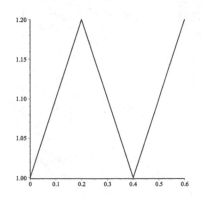

(c) Approximate values at $t = 0.2, 0.4, 0.6$ are given by $1.1, 1.0, 1.1$, respectively.

(d) Since $\phi''(t) = -5\pi \sin(5\pi t)$, the local truncation error for the Euler method is given by

$$e_{n+1} = -\frac{5\pi h^2}{2} \sin(5\pi \bar{t}_n).$$

In order to guarantee that $|e_{n+1}| < 0.05$, we need

$$\frac{5\pi h^2}{2} < 0.05.$$

Solving this inequality, we conclude that we would need $h < 1/\sqrt{50\pi} \approx 0.08$.

23.(a)

$$1000 \cdot \begin{vmatrix} 6.0 & 18 \\ 2.0 & 6.0 \end{vmatrix} = 1000 \cdot 0 = 0.$$

(b)

$$1000 \cdot \begin{vmatrix} 6.01 & 18.0 \\ 2.00 & 6.00 \end{vmatrix} = 1000(0.06) = 60.$$

(c)

$$1000 \cdot \begin{vmatrix} 6.010 & 18.04 \\ 2.004 & 6.000 \end{vmatrix} = 1000(-0.09216) = -92.16.$$

25.(a) The maximum errors occur at $t = 2$. For $h = 0.001, 0.01, 0.025, 0.05$, they are given by

$$56.0393, \quad 510.8722, \quad 1107.4123, \quad 1794.5339.$$

(b)

(c) Yes.

(d) Using a curve-fitting routine, the slope of the least squares line is ≈ 0.909.

8.3   Improved Euler and Runge-Kutta Methods

1.(a) The improved Euler formula is

$$y_{n+1} = y_n + h \left(3 + \frac{1}{2}t_n + \frac{1}{2}t_{n+1} - y_n \right) - \frac{h^2}{2}(3 + t_n - y_n).$$

Since $t_n = t_0 + nh$ and $t_0 = 0$, this formula can be simplified to

$$y_{n+1} = y_n + h(3 - y_n) + \frac{h^2}{2}(y_n - 2 + 2n) - \frac{nh^3}{2}$$

with $y_0 = 1$. With $h = 0.05$, the approximate values of the solution at $t = 0.1, 0.2, 0.3, 0.4$ are

$$1.19512, \quad 1.38120, \quad 1.55909, \quad 1.72956.$$

(b) Using $h = 0.025$, the approximate values of the solution at $t = 0.1, 0.2, 0.3, 0.4$ are

$$1.19515, \quad 1.38125, \quad 1.55916, \quad 1.72965.$$

(c) Using $h = 0.0125$, the approximate values of the solution at $t = 0.1, 0.2, 0.3, 0.4$ are

$$1.19516, \quad 1.38126, \quad 1.55918, \quad 1.72967.$$

(d) Using the Runge-Kutta method with $h = 0.1$, the approximate values of the solution at $t = 0.1, 0.2, 0.3, 0.4$ are

$$1.19516, \quad 1.38127, \quad 1.55918, \quad 1.72968.$$

(e) Using the Runge-Kutta method with $h = 0.05$, the approximate values of the solution at $t = 0.1, 0.2, 0.3, 0.4$ are

$$1.19516, \quad 1.38127, \quad 1.55918, \quad 1.72968.$$

3.(a) The improved Euler formula is

$$y_{n+1} = y_n + \frac{h}{2}(4y_n - 3t_n - 3t_{n+1}) + h^2(2y_n - 3t_n).$$

Since $t_n = t_0 + nh$ and $t_0 = 0$, this formula can be simplified to

$$y_{n+1} = y_n + 2hy_n + \frac{h^2}{2}(4y_n - 3 - 6n) - 3nh^3$$

with $y_0 = 1$. With $h = 0.05$, the approximate values of the solution at $t = 0.1, 0.2, 0.3, 0.4$ are

$$1.20526, \quad 1.42273, \quad 1.65511, \quad 1.90570.$$

(b) Using $h = 0.025$, the approximate values of the solution at $t = 0.1, 0.2, 0.3, 0.4$ are

$$1.20533, \quad 1.42290, \quad 1.65542, \quad 1.90621.$$

(c) Using $h = 0.0125$, the approximate values of the solution at $t = 0.1, 0.2, 0.3, 0.4$ are

$$1.20534, \quad 1.42294, \quad 1.65550, \quad 1.90634.$$

(d) Using the Runge-Kutta method with $h = 0.1$, the approximate values of the solution at $t = 0.1, 0.2, 0.3, 0.4$ are

$$1.20535, \quad 1.42295, \quad 1.65553, \quad 1.90638.$$

(e) Using the Runge-Kutta method with $h = 0.05$, the approximate values of the solution at $t = 0.1, 0.2, 0.3, 0.4$ are

$$1.20535, \quad 1.42296, \quad 1.65553, \quad 1.90638.$$

5.(a) The improved Euler formula is

$$y_{n+1} = y_n + \frac{h}{2}\left(\frac{y_n^2 + 2t_n y_n}{3 + t_n^2}\right) + \frac{h}{2}\left(\frac{K_n^2 + 2t_{n+1}K_n}{3 + t_{n+1}^2}\right)$$

where $K_n = y_n + h(y_n^2 + 2t_n y_n)/(3 + t_n^2)$. Since $t_n = t_0 + nh$ and $t_0 = 0$, this formula can be simplified to

$$y_{n+1} = y_n + \frac{h}{2}\left(\frac{y_n^2 + 2nhy_n}{3 + n^2h^2}\right) + \frac{h}{2}\left(\frac{K_n^2 + 2(n+1)hK_n}{3 + (n+1)^2h^2}\right)$$

with $y_0 = 0.5$. With $h = 0.05$, the approximate values of the solution at $t = 0.1, 0.2, 0.3, 0.4$ are

$$0.510164, \quad 0.524126, \quad 0.542083, \quad 0.564251.$$

(b) Using $h = 0.025$, the approximate values of the solution at $t = 0.1, 0.2, 0.3, 0.4$ are

$$0.510168, \quad 0.524135, \quad 0.542100, \quad 0.564277.$$

(c) Using $h = 0.0125$, the approximate values of the solution at $t = 0.1, 0.2, 0.3, 0.4$ are

$$0.510169, \quad 0.524137, \quad 0.542104, \quad 0.564284.$$

(d) Using the Runge-Kutta method with $h = 0.1$, the approximate values of the solution at $t = 0.1, 0.2, 0.3, 0.4$ are

$$0.510170, \quad 0.524138, \quad 0.542105, \quad 0.564286.$$

(e) Using the Runge-Kutta method with $h = 0.05$, the approximate values of the solution at $t = 0.1, 0.2, 0.3, 0.4$ are

$$0.520169, \quad 0.524138, \quad 0.542105, \quad 0.564286.$$

7.(a) The improved Euler formula is

$$y_{n+1} = y_n + \frac{h}{2}(0.5 - t_n + 2y_n) + \frac{h}{2}(0.5 - t_{n+1} + 2(y_n + h(0.5 - t_n + 2y_n))).$$

Since $t_n = t_0 + nh$ and $t_0 = 0$, this formula can be simplified to

$$y_{n+1} = y_n + h(2y_n + 0.5) + h^2(2y_n - n) - nh^3$$

with $y_0 = 1$. With $h = 0.025$, the approximate values of the solution at $t = 0.5, 1.0, 1.5, 2.0$ are

$$2.96719, \quad 7.88313, \quad 20.8114, \quad 55.5106.$$

(b) Using $h = 0.0125$, the approximate values of the solution at $t = 0.5, 1.0, 1.5, 2.0$ are

$$2.96800, \quad 7.88755, \quad 20.8294, \quad 55.5758.$$

(c) Using the Runge-Kutta method with $h = 0.1$, the approximate values of the solution at $t = 0.5, 1.0, 1.5, 2.0$ are

$$2.96825, \quad 7.88889, \quad 20.8349, \quad 55.5957.$$

(d) Using the Runge-Kutta method with $h = 0.05$, the approximate values of the solution at $t = 0.5, 1.0, 1.5, 2.0$ are

$$2.96828, \quad 7.88904, \quad 20.8355, \quad 55.5980.$$

9.(a) The improved Euler formula is

$$y_{n+1} = y_n + \frac{h}{2}\sqrt{t_n + y_n} + \frac{h}{2}\sqrt{t_{n+1} + K_n}$$

where $K_n = y_n + h\sqrt{t_n + y_n}$. Since $t_n = t_0 + nh$ and $t_0 = 0$, this formula can be simplified to

$$y_{n+1} = y_n + \frac{h}{2}\sqrt{nh + y_n} + \frac{h}{2}\sqrt{(n+1)h + K_n}$$

with $y_0 = 3$. With $h = 0.025$, the approximate values of the solution at $t = 0.5, 1.0, 1.5, 2.0$ are

$$3.96217, \quad 5.10887, \quad 6.43134, \quad 7.92332.$$

(b) Using $h = 0.0125$, the approximate values of the solution at $t = 0.5, 1.0, 1.5, 2.0$ are

$$3.96218, \quad 5.10889, \quad 6.43138, \quad 7.92337.$$

(c) Using the Runge-Kutta method with $h = 0.1$, the approximate values of the solution at $t = 0.5, 1.0, 1.5, 2.0$ are

$$3.96219, \quad 5.10890, \quad 6.43139, \quad 7.92338.$$

(d) Using the Runge-Kutta method with $h = 0.05$, the approximate values of the solution at $t = 0.5, 1.0, 1.5, 2.0$ are

$$3.96219, \quad 5.10890, \quad 6.43139, \quad 7.92338.$$

11.(a) The improved Euler formula is

$$y_{n+1} = y_n + \frac{h}{2}\left(\frac{4 - t_n y_n}{1 + y_n^2}\right) + \frac{h}{2}\left(\frac{4 - t_{n+1}K_n}{1 + K_n^2}\right)$$

where $K_n = y_n + h(4 - t_n y_n)/(1 + y_n^2)$. Since $t_n = t_0 + nh$ and $t_0 = 0$, this formula can be simplified to

$$y_{n+1} = y_n + \frac{h}{2}\left(\frac{4 - nhy_n}{1 + y_n^2}\right) + \frac{h}{2}\left(\frac{4 - h(n+1)K_n}{1 + K_n^2}\right)$$

with $y_0 = -2$. With $h = 0.025$, the approximate values of the solution at $t = 0.5, 1.0, 1.5, 2.0$ are

$$-1.44768, \quad -0.144478, \quad 1.06004, \quad 1.40960.$$

(b) Using $h = 0.0125$, the approximate values of the solution at $t = 0.5, 1.0, 1.5, 2.0$ are

$$-1.44765, \quad -0.143690, \quad 1.06072, \quad 1.40999.$$

(c) Using the Runge-Kutta method with $h = 0.1$, the approximate values of the solution at $t = 0.5, 1.0, 1.5, 2.0$ are

$$-1.44764, \quad -0.143543, \quad 1.06089, \quad 1.41008.$$

(d) Using the Runge-Kutta method with $h = 0.05$, the approximate values of the solution at $t = 0.5, 1.0, 1.5, 2.0$ are

$$-1.44764, \quad -0.143427, \quad 1.06095, \quad 1.41011.$$

13. The improved Euler method is

$$y_{n+1} = y_n + \frac{h}{2}(1 - t_n + 4y_n) + \frac{h}{2}[1 - (t_n + h) + 4K_n]$$

where $K_n = y_n + h(1 - t_n + 4y_n)$. Since $t_n = nh + t_0$ and $t_0 = 0$, this equation can be simplified to

$$y_{n+1} = y_n + \frac{h}{2}(1 - nh + 4y_n) + \frac{h}{2}[1 - h(n+1) + 4K_n]$$

with $y_0 = 1$.

15.(a)

(b) The following are the approximate values of the solution at $t = 0.8, 0.9, 0.95$ using the Runge-Kutta method with $h = 0.01$:

$$5.848616, 14.304785, 50.436365.$$

17.(a) First we notice that

$$\phi'(t_n)h - \frac{f(t_n, y_n)h}{2} = \phi'(t_n)h - \frac{y'_n h}{2} = \phi'(t_n)h - \frac{\phi'(t_n)h}{2} = \frac{\phi'(t_n)h}{2}.$$

Using this fact, it follows that $\phi(t_{n+1}) - y_{n+1}$ satisfies the given equation.

(b) First, using the Taylor approximation, we see that

$$f[t_n + h, y_n + hf(t_n, y_n)] - f(t_n, y_n) =$$
$$= f_t(t_n, y_n)h + f_y(t_n, y_n)hf(t_n, y_n) + \frac{1}{2!}(h^2 f_{tt} + 2hk f_{ty} + k^2 f_{yy})\Big|_{x=\xi, y=\eta}$$

Next, we see that

$$\phi''(t_n)h = f_t(t_n, \phi(t_n))h + f_y(t_n, \phi(t_n))\phi'(t_n)h = f_t(t_n, y_n)h + f_y(t_n, y_n)hf(t_n, y_n).$$

Therefore, we conclude that

$$\frac{1}{2!}\left[\phi''(t_n)h - \{f[t_n + h, y_n + hf(t_n, y_n)] - f(t_n, y_n)\}\right]h =$$
$$= h\left(\frac{1}{2!}(h^2 f_{tt} + 2hk f_{ty} + k^2 f_{yy})\Big|_{x=\xi, y=\eta}\right)$$

is proportional to h^3.

(c) If f is linear in t and y, then $f_{tt} = f_{ty} = f_{yy} = 0$. Therefore, the terms from part (b) above are all zero.

19. The exact solution of the initial value problem is $\phi(t) = (1 + e^{2t})/2$. Based on the result from problem 17(c), the local truncation error for a linear differential equation is

$$e_{n+1} = \frac{1}{6}\phi'''(\bar{t}_n)h^3.$$

Here $\phi' = e^{2t}$, $\phi'' = 2e^{2t}$, $\phi''' = 4e^{2t}$. Therefore,

$$e_{n+1} = \frac{2}{3}e^{2\bar{t}_n}h^3.$$

Further, on the interval $0 \le t \le 1$,

$$|e_{n+1}| \le \frac{2}{3}e^2 h^3 = 4.92604h^3.$$

Letting $h = 0.1$,

$$|e_1| \le \frac{2}{3}e^{2(0.1)}(0.1)^3 = 0.000814269.$$

Using the improved Euler method, with $h = 0.1$, we have $y_1 \approx 1.11000$. The exact value of the solution is $\phi(0.1) = 1.1107014$.

21. The Euler formula is

$$y_{n+1} = y_n + h(0.5 - t_n + 2y_n).$$

Since $t_0 = 0$, $y_0 = 1$ and $h = 0.1$, we have

$$y_1 = 1 + 0.1(0.5 - 0 + 2) = 1.25.$$

For $t_0 = 0$, the improved Euler formula is

$$y_{n+1} = y_n + h(2y_n + 0.5) + h^2(2y_n - n) - nh^3.$$

Therefore, for $y_0 = 1$ and $h = 0.1$,

$$y_1 = 1 + 0.1(2 + 0.5) + (0.1)^2(2 - 0) - 0(0.1)^3 = 1.27.$$

Therefore, the estimated error of the Euler method is $e_{n+1}^{\text{ext}} = 1.27 - 1.25 = 0.02$. If we want the error of the Euler method to be less than 0.0025, we need to multiply the original step size of 0.1 by the factor $\sqrt{0.0025/0.02} \approx 0.36$. Therefore, the required step size is estimated to be $h \approx (0.1)(0.36) = 0.036$.

23. For $t_0 = 0$, the Euler formula is

$$y_{n+1} = y_n + h\sqrt{nh + y_n}.$$

Therefore, for $y_0 = 3$ and $h = 0.1$, we have

$$y_1 = 3 + 0.1\sqrt{0 + 3} = 3.173205.$$

For $t_0 = 0$, the improved Euler formula is

$$y_{n+1} = y_n + \frac{h}{2}\sqrt{nh + y_n} + \frac{h}{2}\sqrt{(n+1)h + K_n}$$

where $K_n = y_n + h\sqrt{t_n + y_n}$. Therefore, for $y_0 = 3$ and $h = 0.1$,

$$y_1 = 3 + 0.05\sqrt{0 + 3} + 0.05\sqrt{0.1 + (3 + 0.1\sqrt{0 + 3})} = 3.177063.$$

Therefore, the estimated error of the Euler method is $e_{n+1}^{\text{ext}} = 3.177063 - 3.173205 = 0.003858$. If we want the error of the Euler method to be less than 0.0025, we need to multiply the original step size of 0.1 by the factor $\sqrt{0.0025/0.003858} \approx 0.805$. Therefore, the required step size is estimated to be $h \approx (0.1)(0.226) = 0.0805$.

8.4   Numerical Methods for Systems of First Order Equations

1. In vector notation, the initial value problem can be written as

$$\frac{d}{dt}\begin{pmatrix} x \\ y \end{pmatrix} = \begin{pmatrix} x + y + t \\ 4x - 2y \end{pmatrix}$$

with initial condition

$$\mathbf{x}(0) = \begin{pmatrix} 1 \\ 0 \end{pmatrix}.$$

(a) The Euler formula is

$$\begin{pmatrix} x_{n+1} \\ y_{n+1} \end{pmatrix} = \begin{pmatrix} x_n \\ y_n \end{pmatrix} + h \begin{pmatrix} x_n + y_n + t_n \\ 4x_n - 2y_n \end{pmatrix}.$$

With $h = 0.1$, the approximate values at $t_n = 0.2, 0.4, 0.6, 0.8, 1.0$ are given by

$$\begin{pmatrix} x_n \\ y_n \end{pmatrix} = \begin{pmatrix} 1.26 \\ 0.76 \end{pmatrix}, \begin{pmatrix} 1.7714 \\ 1.4824 \end{pmatrix}, \begin{pmatrix} 2.58991 \\ 2.3703 \end{pmatrix}, \begin{pmatrix} 3.82374 \\ 3.60413 \end{pmatrix}, \begin{pmatrix} 5.64246 \\ 5.38885 \end{pmatrix},$$

respectively.

(b) The Runge-Kutta method uses the following intermediate calculations:

$$\mathbf{k}_{n1} = (x_n + y_n + t_n, 4x_n - 2y_n)^T$$

$$\mathbf{k}_{n2} = \left[x_n + \frac{h}{2}k_{n1}^1 + y_n + \frac{h}{2}k_{n1}^2 + t_n + \frac{h}{2}, 4\left(x_n + \frac{h}{2}k_{n1}^1\right) - 2\left(y_n + \frac{h}{2}k_{n1}^2\right) \right]^T$$

$$\mathbf{k}_{n3} = \left[x_n + \frac{h}{2}k_{n2}^1 + y_n + \frac{h}{2}k_{n2}^2 + t_n + \frac{h}{2}, 4\left(x_n + \frac{h}{2}k_{n2}^1\right) - 2\left(y_n + \frac{h}{2}k_{n2}^2\right) \right]^T$$

$$\mathbf{k}_{n4} = [x_n + hk_{n3}^1 + y_n + hk_{n3}^2 + t_n + h, 4(x_n + hk_{n3}^1) - 2(y_n + hk_{n3}^2)]^T.$$

With $h = 0.2$, the approximate values at $t_n = 0.2, 0.4, 0.6, 0.8, 1.0$ are given by

$$\begin{pmatrix} x_n \\ y_n \end{pmatrix} = \begin{pmatrix} 1.32493 \\ 0.758933 \end{pmatrix}, \begin{pmatrix} 1.93679 \\ 1.57919 \end{pmatrix}, \begin{pmatrix} 2.93414 \\ 2.66099 \end{pmatrix}, \begin{pmatrix} 4.48318 \\ 4.22639 \end{pmatrix}, \begin{pmatrix} 6.84236 \\ 6.56452 \end{pmatrix},$$

respectively.

(c) With $h = 0.1$, the approximate values at $t_n = 0.2, 0.4, 0.6, 0.8, 1.0$ are given by

$$\begin{pmatrix} x_n \\ y_n \end{pmatrix} = \begin{pmatrix} 1.32489 \\ 0.759516 \end{pmatrix}, \begin{pmatrix} 1.9369 \\ 1.57999 \end{pmatrix}, \begin{pmatrix} 2.93459 \\ 2.66201 \end{pmatrix}, \begin{pmatrix} 4.48422 \\ 4.22784 \end{pmatrix}, \begin{pmatrix} 6.8444 \\ 6.56684 \end{pmatrix},$$

respectively.

3. In vector notation, the initial value problem can be written as

$$\frac{d}{dt}\begin{pmatrix} x \\ y \end{pmatrix} = \begin{pmatrix} -tx - y - 1 \\ x \end{pmatrix}$$

with initial condition

$$\mathbf{x}(0) = \begin{pmatrix} 1 \\ 1 \end{pmatrix}.$$

(a) The Euler formula is

$$\begin{pmatrix} x_{n+1} \\ y_{n+1} \end{pmatrix} = \begin{pmatrix} x_n \\ y_n \end{pmatrix} + h \begin{pmatrix} -t_n x_n - y_n - 1 \\ x_n \end{pmatrix}.$$

With $h = 0.1$, the approximate values at $t_n = 0.2, 0.4, 0.6, 0.8, 1.0$ are given by

$$\begin{pmatrix} x_n \\ y_n \end{pmatrix} = \begin{pmatrix} 0.582 \\ 1.18 \end{pmatrix}, \begin{pmatrix} 0.117969 \\ 1.27344 \end{pmatrix}, \begin{pmatrix} -0.336912 \\ 1.27382 \end{pmatrix}, \begin{pmatrix} -0.730007 \\ 1.18572 \end{pmatrix}, \begin{pmatrix} -1.02134 \\ 1.02371 \end{pmatrix},$$

respectively.

(b) The Runge-Kutta method uses the following intermediate calculations:

$$\mathbf{k}_{n1} = (-t_n x_n - y_n - 1, x_n)^T$$

$$\mathbf{k}_{n2} = \left[-\left(t_n + \frac{h}{2} \right)\left(x_n + \frac{h}{2}k_{n1}^1 \right) - \left(y_n + \frac{h}{2}k_{n1}^2 \right) - 1, x_n + \frac{h}{2}k_{n1}^1 \right]^T$$

$$\mathbf{k}_{n3} = \left[-\left(t_n + \frac{h}{2} \right)\left(x_n + \frac{h}{2}k_{n2}^1 \right) - \left(y_n + \frac{h}{2}k_{n2}^2 \right) - 1, x_n + \frac{h}{2}k_{n2}^1 \right]^T$$

$$\mathbf{k}_{n4} = \left[-\left(t_n + h \right)\left(x_n + hk_{n3}^1 \right) - \left(y_n + hk_{n3}^2 \right) - 1, x_n + hk_{n3}^1 \right]^T.$$

With $h = 0.2$, the approximate values at $t_n = 0.2, 0.4, 0.6, 0.8, 1.0$ are given by

$$\begin{pmatrix} x_n \\ y_n \end{pmatrix} = \begin{pmatrix} 0.568451 \\ 1.15775 \end{pmatrix}, \begin{pmatrix} 0.109776 \\ 1.22556 \end{pmatrix}, \begin{pmatrix} -0.32208 \\ 1.20347 \end{pmatrix}, \begin{pmatrix} -0.681296 \\ 1.10162 \end{pmatrix}, \begin{pmatrix} -0.937852 \\ 0.937852 \end{pmatrix},$$

respectively.

(c) With $h = 0.1$, the approximate values at $t_n = 0.2, 0.4, 0.6, 0.8, 1.0$ are given by

$$\begin{pmatrix} x_n \\ y_n \end{pmatrix} = \begin{pmatrix} 0.56845 \\ 1.15775 \end{pmatrix}, \begin{pmatrix} 0.109773 \\ 1.22557 \end{pmatrix}, \begin{pmatrix} -0.322081 \\ 1.20347 \end{pmatrix}, \begin{pmatrix} -0.681291 \\ 1.10161 \end{pmatrix}, \begin{pmatrix} -0.937841 \\ 0.93784 \end{pmatrix},$$

respectively.

5. In vector notation, the initial value problem can be written as

$$\frac{d}{dt} \begin{pmatrix} x \\ y \end{pmatrix} = \begin{pmatrix} x(1 - 0.5x - 0.5y) \\ y(-0.25 + 0.5x) \end{pmatrix}$$

with initial condition

$$\mathbf{x}(0) = \begin{pmatrix} 4 \\ 1 \end{pmatrix}.$$

(a) The Euler formula is

$$\begin{pmatrix} x_{n+1} \\ y_{n+1} \end{pmatrix} = \begin{pmatrix} x_n \\ y_n \end{pmatrix} + h \begin{pmatrix} x_n(1 - 0.5x_n - 0.5y_n) \\ y_n(-0.25 + 0.5x_n) \end{pmatrix}.$$

With $h = 0.1$, the approximate values at $t_n = 0.2, 0.4, 0.6, 0.8, 1.0$ are given by

$$\begin{pmatrix} x_n \\ y_n \end{pmatrix} = \begin{pmatrix} 2.96225 \\ 1.34538 \end{pmatrix}, \begin{pmatrix} 2.34119 \\ 1.67121 \end{pmatrix}, \begin{pmatrix} 1.90236 \\ 1.97158 \end{pmatrix}, \begin{pmatrix} 1.56602 \\ 2.23895 \end{pmatrix}, \begin{pmatrix} 1.29768 \\ 2.46732 \end{pmatrix},$$

respectively.

(b) Let $f(t, x, y) = x(1 - 0.5x - 0.5y)$, $g(t, x, y) = y(-0.25 + 0.5x)$. The Runge-Kutta method uses the following intermediate calculations:

$$\mathbf{k}_{n1} = [f(t_n, x_n, y_n), g(t_n, x_n, y_n)]^T$$

$$\mathbf{k}_{n2} = \left[f\left(t_n + \frac{h}{2}, x_n + \frac{h}{2}k_{n1}^1, y_n + \frac{h}{2}k_{n1}^2 \right), g\left(t_n + \frac{h}{2}, x_n + \frac{h}{2}k_{n1}^1, y_n + \frac{h}{2}k_{n1}^2 \right) \right]^T$$

$$\mathbf{k}_{n3} = \left[f\left(t_n + \frac{h}{2}, x_n + \frac{h}{2}k_{n2}^1, y_n + \frac{h}{2}k_{n2}^2 \right), g\left(t_n + \frac{h}{2}, x_n + \frac{h}{2}k_{n2}^1, y_n + \frac{h}{2}k_{n2}^2 \right) \right]^T$$

$$\mathbf{k}_{n4} = \left[f\left(t_n + h, x_n + hk_{n3}^1, y_n + hk_{n3}^2 \right), g\left(t_n + h, x_n + hk_{n3}^1, y_n + hk_{n3}^2 \right) \right]^T.$$

With $h = 0.2$, the approximate values at $t_n = 0.2, 0.4, 0.6, 0.8, 1.0$ are given by

$$\begin{pmatrix} x_n \\ y_n \end{pmatrix} = \begin{pmatrix} 3.06339 \\ 1.34858 \end{pmatrix}, \begin{pmatrix} 2.44497 \\ 1.68638 \end{pmatrix}, \begin{pmatrix} 1.9911 \\ 2.00036 \end{pmatrix}, \begin{pmatrix} 1.63818 \\ 2.27981 \end{pmatrix}, \begin{pmatrix} 1.3555 \\ 2.5175 \end{pmatrix},$$

respectively.

(c) With $h = 0.1$, the approximate values at $t_n = 0.2, 0.4, 0.6, 0.8, 1.0$ are given by

$$\begin{pmatrix} x_n \\ y_n \end{pmatrix} = \begin{pmatrix} 3.06314 \\ 1.34899 \end{pmatrix}, \begin{pmatrix} 2.44465 \\ 1.68699 \end{pmatrix}, \begin{pmatrix} 1.99075 \\ 2.00107 \end{pmatrix}, \begin{pmatrix} 1.63781 \\ 2.28057 \end{pmatrix}, \begin{pmatrix} 1.35514 \\ 2.51827 \end{pmatrix},$$

respectively.

7. Let $f(t, x, y) = -x + 4y$, $g(t, x, y) = x - y$. The Runge-Kutta method uses the following intermediate calculations:

$$\mathbf{k}_{n1} = [f(t_n, x_n, y_n), g(t_n, x_n, y_n)]^T$$

$$\mathbf{k}_{n2} = \left[f\left(t_n + \frac{h}{2}, x_n + \frac{h}{2}k_{n1}^1, y_n + \frac{h}{2}k_{n1}^2 \right), g\left(t_n + \frac{h}{2}, x_n + \frac{h}{2}k_{n1}^1, y_n + \frac{h}{2}k_{n1}^2 \right) \right]^T$$

$$\mathbf{k}_{n3} = \left[f\left(t_n + \frac{h}{2}, x_n + \frac{h}{2}k_{n2}^1, y_n + \frac{h}{2}k_{n2}^2 \right), g\left(t_n + \frac{h}{2}, x_n + \frac{h}{2}k_{n2}^1, y_n + \frac{h}{2}k_{n2}^2 \right) \right]^T$$

$$\mathbf{k}_{n4} = \left[f\left(t_n + h, x_n + hk_{n3}^1, y_n + hk_{n3}^2 \right), g\left(t_n + h, x_n + hk_{n3}^1, y_n + hk_{n3}^2 \right) \right]^T.$$

With $h = 0.2$, the approximate values of the solution at $t = 1$ are given by

$$\begin{pmatrix} x_n \\ y_n \end{pmatrix} = \begin{pmatrix} 1.43421 \\ 0.64202 \end{pmatrix}.$$

With $h = 0.1$, the approximate values of the solution at $t = 1$ are given by

$$\begin{pmatrix} x_n \\ y_n \end{pmatrix} = \begin{pmatrix} 1.43384 \\ 0.64222 \end{pmatrix}.$$

With $h = 0.05$ and $h = 0.025$ the approximate values of the solution (approximating up to the first 6 digits) at $t = 1$ are given by

$$\begin{pmatrix} x_n \\ y_n \end{pmatrix} = \begin{pmatrix} 1.43382 \\ 0.64223 \end{pmatrix}.$$

The exact solution is given by

$$\phi(1) = \frac{e + 3e^{-3}}{2} \approx 1.43382$$

$$\psi(1) = \frac{e - 3e^{-3}}{4} \approx 0.64223.$$

Printed in the USA
K080925SCI022118   01S29053000000002907